A Meeting *of* Land *and* Sea

A Meeting *of* Land *and* Sea

Nature *and the* Future *of* MARTHA'S VINEYARD

DAVID R. FOSTER

Illustrations by Brian R. Hall *Photography by* David R. Foster

Yale UNIVERSITY PRESS

New Haven and London

Published with assistance from the George G. and Doris B. Daniels Wildlife Trust and from the Edey Foundation.

Frontispiece: Beech uprooted by the 1944 hurricane on the Chilmark Town Forest currently supporting stems that were produced from branches along its trunk.
p. vi: Outwash deposits of alternating sands and gravels at Wasque, Chappaquiddick.
p. viii: Nor'easter in March 2014 driving waves onto North Neck on Chappaquiddick.
p. xii: Open young woods with a continuous tall layer of huckleberry.
p.1: A lighthouse view east across Aquinnah.
p. 39: Another perspective: prehistory shaped by nature. In Albert Bierstadt's (ca. 1858) rendition of Gosnold's encampment on Cuttyhunk, the landscape appears wild and fully forested by ancient trees, supporting diverse wildlife, and exhibiting no obvious native influence. Courtesy of the New Bedford Whaling Museum.
p. 105: Mill sluiceway and pond that powered the paint mill in Chilmark.
p. 213: Gay Head as tropical storm Irene passed the Island in August 2011.

This volume is a contribution of the Long Term Ecological Research (LTER) program funded by the National Science Foundation. The largest and longest-lived ecological network in the United States, the twenty-five LTER sites encompass diverse ecosystems from Alaska and Antarctica to islands in the Caribbean and the Pacific. The Harvard Forest has been an LTER site since 1988.

LTER

Yale University Press books may be purchased in quantity for educational, business, or promotional use. For information, please e-mail sales.press@yale.edu (U.S. office) or sales@yaleup.co.uk (U.K. office).

Designed by Richard Hendel
Set in Utopia and TheSans type by
Tseng Information Systems, Inc.

Printed in China

Library of Congress Control Number:
2016937232
ISBN 978-0-300-21417-8 (cloth : alk. paper)

A catalogue record for this book is available from the British Library.

This paper meets the requirements of ANSI/NISO Z39.48-1992 (Permanence of Paper).

10 9 8 7 6 5 4 3 2 1

To Anne and Pete Foster, who enabled it all to happen

CONTENTS

ix Preface
xiii Acknowledgments

I. THE LAY OF THE LAND
3 Landscape History and Ecology
17 The Setting and Stage

II. SHAPING THE ISLAND: PREHISTORY TO A.D. 1600
41 Geology and Sea Level Rise
59 Climate and Vegetation Dynamics
79 People on the Coast

III. SHAPING THE LAND: HISTORY THROUGH THE TWENTY-FIRST CENTURY
107 Winds That Roil the Island
127 Waves That Waste the Coast
147 1850: The Landscape That Became Today
174 Making the Modern Landscape
195 Conserving the Vineyard

IV. SHAPING THE PRESENT AND FUTURE
215 Exploring the Natural and Cultural Landscape
238 Wildlife
260 Planning the Island
281 A Conservation Future
303 Perspectives in Island Time

313 Bibliography
323 Index

PREFACE

Relentlessly, and unavoidably, Martha's Vineyard is disappearing. Every year, on average, the spectacular south-shore beaches on this magnificent island roll inland some three to five feet. As they do, the sea consumes fields, forest, coastal ponds, and houses, if not moved. This loss of land represents one of the fastest rates of erosion on the eastern seaboard. It totals more than a quarter of a mile since British explorer Bartholomew Gosnold first explored these waters and gave the Island and neighboring Cape Cod their names in 1602.* And it makes the Vineyard an extreme example of the challenges of coping with a changing climate, its rising seas, and its increasingly intense storms.

But the Island stands threatened in other ways. This landscape, which has supported people for more than ten thousand years, is being consumed by one of the most rapid paces of development in the northeast. The Vineyard's greatest assets—its natural beauty and quality of life—are being compromised by the desire of so many people to experience them. With the loss of land and nature, there are threats to many ways of life and connections to the land. Over the years, farms have become fewer, first lost to distant competition, then engulfed by forest over the past two centuries, and now replaced by a trophy house or six.

Though imperiled, this distinctive place displays a remarkable readiness to confront the forces of nature and humanity. The first call for Island-wide planning dates to 1871; today the six towns work individually, and collectively through an Island-wide commission, to integrate planning and regulation. In New England, a region distinguished by global leadership in conservation, the Vineyard stands out, having conserved some 40 percent of its land through the action of local and national groups and an Island-wide Land Bank. Learning from the history of active use of the land, there is also a growing resurgence of hands on the land. Agriculture is once

* In keeping with the custom introduced by editor Henry Beetle Hough at the *Vineyard Gazette* newspaper, Martha's Vineyard is distinguished here from all other islands by the use of a capital "I."

again expanding. Conservation lands are being opened to Island farmers. Old woodlots cut for centuries are being eyed for traditional and new wood products. And a culture of conserving and using the land sustainably is being incorporated into school programs motivated by a stewardship ethic that seeks to retain the Vineyard's striking beauty and biological diversity. Though in constant battle with many forces of change, the Island yields lessons worth reflection, appreciation, and replication elsewhere in New England and beyond.

This book explores this tension of loss, resilience, and strategy while filling a great hole in the understanding of this distinctive part of New England. Despite its many rich histories there is no story of the Vineyard landscape. No volume places the land center stage to address the simple question: What processes shaped this Island, made it what it is today, and will determine its future? Commencing with the deep history of ice and Island formation and coming through the days of native and colonial people to modern environmental struggles, this book explores the past to inform the present and guide the future.

When we pursue this goal we uncover much that is new. We also recognize that much of what we think we know about this Vineyard landscape is based on unfounded anecdote or short-term observation, rather than grounded in the depths of landscape history. Examples abound.

Exploring prehistory, we find no support for the prevailing storyline, on the Vineyard or New England broadly, of Native American farmers cultivating the mythical three sisters—corn, beans, and squash—and shaping the land with their hands and fire. In sharp contrast, the archaeological record paints a story of a people thriving for millennia entirely on the natural resources of land, waters, and sea. This notably peaceful and healthy lifestyle conveyed great flexibility in the face of environmental challenges and change, but left few traces on the land. This conclusion is accompanied by new insights into the nature of the landscape when Europeans arrived and the directions that it might head in the future. Once thickly forested and then cleared by colonists, Martha's Vineyard has been reverting to increasingly impressive woods for the past century and a half.

This historical perspective draws our attention to the critical importance of the mid-nineteenth-century height of New England agriculture. This brief period, when farmers and tens of thousands of sheep and other grazing animals intensively shaped the Vineyard, continues to control many characteristics of the modern landscape. Recognizing this legacy of past land use in shaping the land today, we come to an unexpected conclusion. Farming—the diversified crop and pasture-based life of the colonial past still pursued by a handful of farmers working lands from Chappaquiddick to Aquinnah—can be an instrumental tool in modern conservation. Indeed, when we look back to draw lessons from the land's history we discover that by melding farms, woodlots, and wilder nature we can support a diversity of life along with healthy lives for ourselves.

As we seek to learn from the land and its past, we become intrigued by previously untapped sources of information. One invaluable resource is a map from 1850 that until recently lay unrecognized on the shelves of the Martha's Vineyard Museum. Extolled by the *Vineyard Gazette* for its remarkable accuracy in the year of its release, the engraved chart captures every building, field, woodlot, road, shoreline, shoal, and body of water in unrivaled resolution. With it to guide us, we can seamlessly relate the details in our own landscape to their origins two centuries ago. Like the land, the map has its own fascinating tale. Produced by a federal agency conceived by Thomas Jefferson, the survey was guided across the Vineyard by Henry Whiting—the West Tisbury patriarch of a family that continues to farm, paint, conserve, and document the natural history of the Island today.

One motivation for our historical perspective is to assess the types and magnitudes of natural events that might befall the Island in the future. Here, we are richly rewarded. Given its coastal location, great storms loom in the Vineyard's past and many memories, with the hurricane of 1938 often providing a frame of reference.

Yet, when we scour deeply, even this tremendous storm pales by comparison. The Island's signature event—with winds that flattened forests and surging seas that reconfigured shorelines—was the Great Colonial Hurricane of 1635. As we begin to understand the force of this historic storm by employing a variety of ingenious as well as serendipitous avenues of investigation, we are compelled to readjust our expectations for future events.

From these and other tales emerges an appreciation for melding history, ecology, and conservation as we seek to interpret and manage the Vineyard, or any other landscape. Every spot on earth has a history of intertwined forces of human and natural processes that have shaped it over time. By exploring these forces and processes we can better understand the distinctive features of the land today and best support these qualities and anticipate their future.

But, why focus on the Vineyard?

The most obvious reason is that the Island is iconic. In a coastal region of globally distinctive landscapes—Cape Cod, Nantucket, Block Island, and the Elizabeth Islands—the Vineyard stands out for its extraordinary variation rooted in natural and cultural history. These qualities remain today due to an unrivaled ethic of conservation whose own history of success in the face of relentless pressure offers much to learn. Island-wide planning and conservation have been powerful agents of action on the Vineyard. Their origins trace to the confluence of many factors—an ages-old ethic, Ted Kennedy's failed legislation, a consultant's study that was rejected and then suppressed by local politicians, the determination of key individuals, the formation of Island-based organizations and agencies, and the enduring desire by residents and visitors alike to sustain the qualities of the place.

Recognition of these distinctive qualities of the land and people bring us to the most important reason for a focus on the Vineyard. The Island deserves this treatment. With its notoriety from scandal and presidential visits and its crush of rich and famous, the Vineyard may appear from afar as the epitome of an elite resort. It is true that there are yacht and country clubs, galas, exclusive beaches, private jets, and mega-mansions. But these lie at the periphery of the place. And this is not their story, though they are certainly part of the tale.

The land that this volume explores is that of Vineyarders who live and work to make the Island a thriving and solidly real place, day by day, summer after summer, and through the generally peaceful, though occasionally challenging stretches of delightfully varying winters. They are new and old, multigenerational and recently arrived, farmers, educators, tradespeople, civil servants, professionals, and, at some level and in many ways, almost all conservationists. Some are native, some are washed ashore, others are seasonally transient, and quite a surprising number are descendants of one of many racial, ethnic, or cultural waves that swept up on the Island over time. These people have inspired and guided me and lent a few of their stories. It is their words and the tales of many who have gone before them that I hear as I walk the land and seek to uncover its past and present. It is for them and the future of this land that this volume is intended.

ACKNOWLEDGMENTS

Assembling this volume has brought many pleasures. Foremost is the joy of discovery in walking, riding, and kayaking Island lands and waters and encountering new sights, uncovering leads in answering old questions, and appreciating the thoughtful care of the land I passed by many owners. There is no end of special places on the Vineyard—each one more intriguing than the last—and I owe much to the many people who have generously, and just as frequently unknowingly, opened the Island to me.

The strange rush of excitement that comes with intellectual discovery is a second joy, and one that appeared as often in musty basements, archives, and repositories of history as under open air. While many institutions have provided key resources, the most welcoming and illuminating experiences have come at the *Vineyard Gazette* with its eclectic archive and embracing staff, the Martha's Vineyard Museum, the libraries in Vineyard Haven, Edgartown, and West Tisbury (with their archaic microfiche machines and carefully protected Island collections), the Polly Hill Arboretum, the archives at Harvard University, and the Leventhal Map Center of the Boston Public Library.

Reworking text, photos, and materials for years in the quiet before early morning and then sharing the resulting insights, frustrations, and joys with family has been a key to progress. I thank Marianne, Christian, and Ava for sharing that space and many tales.

Most critically, in the quest to know a living and lived landscape, there is no resource more fundamental or enjoyable than the people who call the place home. A few local guides have helped forge my major connections and understanding. Through 5 a.m. emails, countless Black Dog woofers, and spontaneous as well as calculated introductions to people and places, Fritz Knight—banker, developer, and Oak Bluffs denizen—has revealed Vineyard qualities and personalities that are fundamental but easily overlooked. Through our walks and exchanges of their great knowledge of natural and cultural history, Steve Spongberg and John Flender have greatly enriched my perspectives. For an encyclopedic knowledge of people, institutions, and views on Island life there can be no greater or generous source than Tom Dunlop. For our many ongoing exchanges I am extremely grateful. Finally, Tim Boland renewed my interest in the Vineyard in 2008 with his simple request that I examine those dead oaks in West Tisbury. Tim, Tom Clark, Karin Stanley, and the Polly Hill Arboretum hosted many visits that led to our own permanent base and this project.

Many others have shared their deep knowledge readily, and often regularly: Edo and Bob Potter, John Varkonda, Dianna Doucette, Nancy Weaver, Brendan O'Neill, Adam Moore, Wes Ward, Bob Woodruff, Jon Previent, Tom Chase, Kate Scott, Rebecca Gilbert, Allan Keith, Robin Hyde, Duke Smith, Luanne Johnson, Jim Athearn, Chris Kennedy, Matt Pelikan, Steve Bernier, Dave Dandridge, Sarah McKay, Allen Healy, Caitlin Jones, Prudy Burt, Tweed Roosevelt, Chris Seidel, Hilary Wall, Bow Van Riper, and Matt Poole.

For access to property and insights to land that they know well I owe many thanks to Burt Fischer, Gerry Jeffers, Lisina Hoch, Charlie Pachico, Bob and Anne Ganz, Dick Leahy, Allen Norton, Wes Nagy, Arlene Conroy, Nik and Pauline Scott, Joan Smith, Matthew and Martha Stackpole, Matthew Dix, Rick Schifter, Eric Peters, Tom Flynn, Tucker Hubbell, John Scherlis, Suzann Bellancampi, Peter Knight, Dick Burt, Martha Cohan, Peter Goodale, Soo Whiting, Mitch Posin and Clarissa Allen, James Lengyel, Julie Russell, Chuck and Martha Schmidt, the Kloss family, Bil Dillon, Lily Walter, the Dunkls, Sam Stevenson, Jay Lagerman, Eric Glasgow, Claudia Miller, Mark London, and Tess and Kib Bramhall.

The research underlying the volume was funded by the A. W. Mellon Foundation, the National Science Foundation's Long-Term Ecological Research program and Ecology program, Highstead Foundation, Harvard University, and the Beech Tree Trust on Naushon Island. I owe great thanks to the Edey Foundation, Gus Daniels Wildlife Trust, and Elisabeth Dudley for supporting the production of the book.

Many people have contributed many hours reviewing small to entire chunks of the book as it developed. These include Nancy Weaver, Steve Spongberg, Steve Hurley, Jim Athearn, Ed Faison, John McDonald, Dianna Doucette, John Flender, Fritz Knight, Mark London, Jonathan Thompson, Thomas Millette, Adam Moore, Brendan O'Neill, Dick Burt, Jim Richardson, Prudy Burt, Kristin Fauteux, Jim Tuck, Emery Boose, and James Lengyel. My greatest thanks go to Steve Long, writer, editor, friend, and conservationist, who worked closely with me in improving every aspect of the text. Jean Thomson Black, Senior Executive Editor at Yale University Press, has offered invaluable insight and support, aided by Samantha Ostrowski and others in New Haven.

Over the years many others have opened our eyes to different parts of this coastal landscape: Wendy Breiby, Tina Conroy, Bea Phear, Mait Edey, Ann Fielder, Pam and Clark Goff, Paul Goldstein, Will Hersey, Dick Johnson, Erin Largay, Jim Lentowski, Jane Varkonda, Steve Masterson, George Mathiesen, Bob Mill, Chris Neill, Bill Patterson, Lloyd Raleigh, Megan Sargent, George David, Tim Simmons, Tom Rawinski, Sundy Smith, Linsey Lee, Maria McFarland, Peter Dunwiddie, Ernie Steinauer, Wes Tiffney, JoAnn Taylor, Rob Cimino, Margaret Curtin, Russ Hopping, Jocelyn Forbush, Barbara Erickson, John Vasconcellos, Becky Brown, Bob Childs, Happy Spongberg, Karen Ogden, Penny Uhlendorf, Kelly McClintock, Peter Van Tassel, Peter Fletcher, Keith Wilda, Max McCreery, Jon Atwood, Ed Jerome, Ginny Dautreuil, Paul Cavanaugh, David Stainbrook, Brian Hawthorne, Bill Rivers, Henry Woolsey, Mark Diabaise, and Dick Jennings.

I thank Ted Stanley for piloting and patience with my efforts to obtain aerial photographs of the coastal landscape and to Michael Kidder for the use of his plane.

At the Harvard Forest, three people have been invaluable: Glenn Motzkin for sharing early explorations of the coast and the State Forest, Brian Hall for endless help with historical research, field studies, and the creation of a novel set of maps and figures, and Jenny Hobson for support with tasks large and small, and insightful assistance with the photographs. I have also been greatly aided by Wyatt Oswald, Elaine Doughty, Bryan Shuman, Emery Boose, Aaron

Ellison, Edythe Ellin, Laurie Chiasson, Linda Hampson, Jonathan Thompson, John Wisnewski, Roland Meunier, Julie Pallant, Kathy Lambert, Audrey Barker Plotkin, Neil Pederson, Jim Tang, Debby Kaspari, Tim Parshall, Art Allen, Posy Busby, Erik Jorgensen, Ava Foster, Rob Eberhardt, Dana McDonald, Jon Harrod, Erin Largay, Natalie Drake, Susan Clayden, Sylvia Barry Musielewicz, Lindsey Day, Sara Trube, Danica Doroski, and Michael Lawrence.

My understanding and access to the archaeological landscape has been greatly expanded by Elizabeth Chilton, Dianna Doucette, Deena Duranleau, Mitch Mulholland, Dick Burt, Jim Richardson, Jim Tuck, Katie Kirakosian, Evan Taylor, and Brian Jones. My growing appreciation for the landscape at Naushon Island was made possible through the hospitality of and illuminating excursions with Gon, Holly, and Ren Leon, Paul Elias, Bruce Bagley, Jodie Frothingham, Narain Schroeder, Brandon McElroy, and David Gregg.

Finally, many organizations have provided information, access, and inspiration: Polly Hill Arboretum, Sheriff's Meadow Foundation, Massachusetts Department of Conservation and Recreation, The Trustees of Reservations, The Nature Conservancy, Martha's Vineyard Land Bank Commission, Martha's Vineyard Commission, Vineyard Conservation Society, Massachusetts Audubon Society, National Park Service, Chicama Vineyards, Farm Neck Golf Course, Red Gate Farm, Morning Glory Farm, Mermaid Farm, and the Island Grown Initiative.

Let us continue these explorations.

The Lay of the Land

Landscape History and Ecology

The more you know about the past, the better prepared you are for the future.
— Theodore Roosevelt

(*opposite*)
Lucy Vincent Beach from the Wequobsque cliffs, December 2012.

The storm had barely subsided before word began to spread. That, plus the quick posting of riveting photographs on newspaper websites, drew the first onlookers to the Chilmark coast. Hurricane Sandy had ravaged Lucy Vincent Beach. Much of what remained of that beloved sandy expanse now lay inland a few dozen yards. The surge-driven waves had stripped rock, vegetation, walkways, and fences from the revered town beach, depositing some in distant Chilmark Pond, burying much with deep sand, and simply disappearing the rest. At the far end of the beach, the solitary and majestic cliff of sand and gravel that distinguishes this compelling landscape from all others had collapsed further. The most magnificent beach in the Vineyard town envied for its spectacular beaches appeared devastated. "Wiped out," according to one comment on the *Vineyard Gazette*'s website. The event was tragic and yet quite natural, inevitable, and glorious.

Cries arose for "saving" or "restoring" the beach. In response, a few sage locals emerged to remind the distraught that the storm-strewn aftermath was simply the latest chapter in a story that has played out for decades on this stretch of coast named in honor of its previous owner, former town librarian Lucinda Poole Mosher Vincent. One could take an even broader and more thought-provoking view. We are witnessing the disintegration of a solitary hill that has touched the lives of people for thousands of years and has been shaped by physical forces even longer. This is a poignant chapter in the larger story of natural and human change on the island of Martha's Vineyard that we should all appreciate.

Indeed, the eroding hill and beach at Lucy Vincent can tell many stories. The wooden shack at the narrow entrance to the rough dirt parking lot tells the modern narrative of an exclusive beach that is restricted to Chilmark residents in summer months and then opened to the rest of us once the seasonal demand and value subsides. The old fence posts, scrubby vegetation, and dark uppermost soil layer on the exposed cliff face tell a narrative of historical times when the slopes and crest of the hill were grazed by sheep and the land improved by its farmer owners. The prehistory

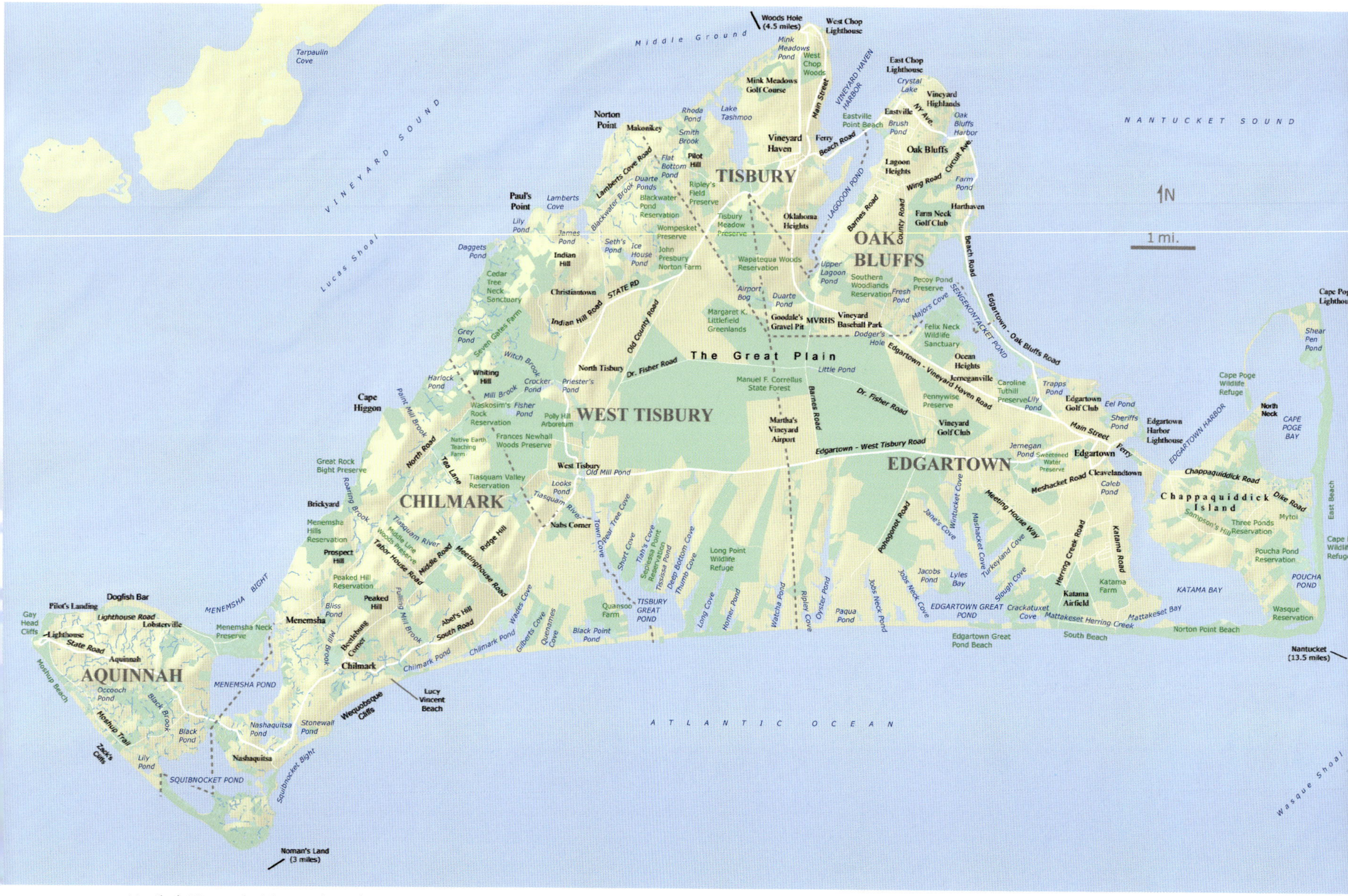

Martha's Vineyard with town boundaries, major roads, and many places referred to in the text. Protected areas conserved from future development are shaded in green.

of the site was revealed in the 1980s, followed by a surge of media attention. Two women strolling the shore encountered the skull of an ancient native that had been dislodged from the upper reaches of the cliff by the latest bout of erosion.

That discovery triggered a series of painstaking explorations by archaeologists from Harvard and the University of Massachusetts. They revealed that the great solitary hill had long been sacred to the local Wampanoag people, who focused on it for ceremonies, food preparation, and burials. The significance of this distinctive landform in the lives of ancient people was exposed through further explorations that hurried forward as the cliff continued to collapse into the sea. More than twenty burials and countless artifacts were uncovered. Transient glimpses of bone, shell middens, and stains continue to appear high on the cliff face and then disappear with the next episode of slumping earth. Wampanoag descendants visit the site regularly and discreetly, reverently caring for their ancestors and the precious materials left behind. The dated range of artifacts—pottery, human and animal bones, shells, stone utensils of all types, and personal items of adornment—extends back some ten thousand years. Well before the days of beach stickers, the Vincent family, or colonial farmers—indeed, back many millennia when the hill lay distant from the sea—this landform drew people to it and was shaped by them.

Inside the hill lie even earlier chapters of this narrative preceding human engagement. Within the cliff's massive elevation, a confusion of layered sands, gravel, and contorted blocks of clay and conglomerate tells the geological tale of its formation. Ice a mile thick pushed out from arctic lands to scour, engulf, and shove forward rocky debris that was disgorged and stranded at the ice's southernmost extent. This action formed the Vineyard, Cape, and other coastal lands. Glacially bulldozed debris built the nearby towering ridges and cliffs, whose names—Wequobsque, Squibnocket, and Noman's—imply tales of subsequent human connections and whose heights yield bird's-eye views. Examine a hunk of the glacial debris and you may discover a mystifying

Eroded shore at Lucy Vincent Beach following the passage of Hurricane Sandy in 2012.

collection of ancient animal and plant fossils derived from truly deep geological time when these fine materials were deposited in shallow seas to the north. There they lay for millions of years until scoured and pushed forward by the ice and then, in a geological second, began to erode and dissolve back into the rising sea that undermines the site today.

A focus on the encroaching sea helps tie this narrative back to the newspaper headlines and shift our attention to the present and future. It also raises the issue of our role in understanding and working with nature. Quite gradually, and yet unequivocally, Lucy Vincent Beach, along with the entire Island of Martha's Vineyard, is disappearing. For some twenty thousand years, the ocean has risen and the land left by the ice has shrunk as glaciers melted, the world's oceans grew, and storm waves eroded the beaches and

Grass- and shrub-covered dunes, wetlands, and ponds on Red Gate Farm, Aquinnah.

cliffs and swept them along the coast and out to sea. Today, this process is accelerating as temperatures rise rapidly, more intense storms arrive, and the sea creeps upward, driven by the melting of ice thousands of miles away. The shoreline retreats in fits and starts. With an intense storm like Sandy, the beaches and cliffs jump back and then settle down to an imperceptible pace of retreat. With the next storm they will retreat another notch. It is a ratcheting progression of change. From this perspective there is clearly no saving Lucy Vincent Beach.

In many ways the story and the dilemma of this singular spot define the scope and challenge of this volume. My guiding interest is to put a landscape that we live in and cherish in a broad and historical context so that we can understand it today, recognize the forces and hands that have shaped it in the past, and then conceive what it may be and how we might help guide it thoughtfully into the future. As the landscape at Lucy Vincent illustrates so nicely, pursuit of this goal certainly involves an understanding of geology, ecology, and natural processes. But it also requires a deep appreciation for people, their history, and the novel processes they introduce that drive changes in nature. Wherever we look on the Vineyard (and indeed anywhere on earth), whether at a gnarly forest, sandplain grassland, or eroding cliff face, we can only truly interpret the present setting when we consider the range of processes and dynamics that have shaped it through time. In turn, as we set our goals for conservation and management of these landscapes, it is critical that we are informed by a perspective of how they became what they are and how this informs their future condition.

Of course, as Lucy Vincent Beach highlights so well, those perspectives and decisions change over time. A few thousand years ago, we would likely have joined quiet ceremonies on an open knoll in a sea of forest, stared out toward the ocean lying a mile to the south, and revered the scene of natural beauty. If questioned a century and a half ago when the Island was in its agricultural and maritime heyday, we may have cherished the summit's closely cropped pasture and encouraged its owners to manage their herds well and improve their wool yield with new breeds. Three decades ago we would have cheered local efforts to preserve the beach for town use and appreciated it as a highly vaunted perk of residency and privilege. Today, we skirt the crumbling remains after passing signs warning of falling rock and the threat of a $25,000 fine for the removal of archaeological materials. And, as we look to the future, our advice is to allow nature to continue the job that it has pursued forever, heedless of all of our designs. All the while we seek to preserve the beauty and magic of the place in photographs, paintings, words, and memories.

The central goal of this book is to interpret the Island of Martha's Vineyard ecologically with an eye toward contributing to its appreciation and conservation. As we begin that quest, it is useful to introduce the conceptual understanding of nature that guides this

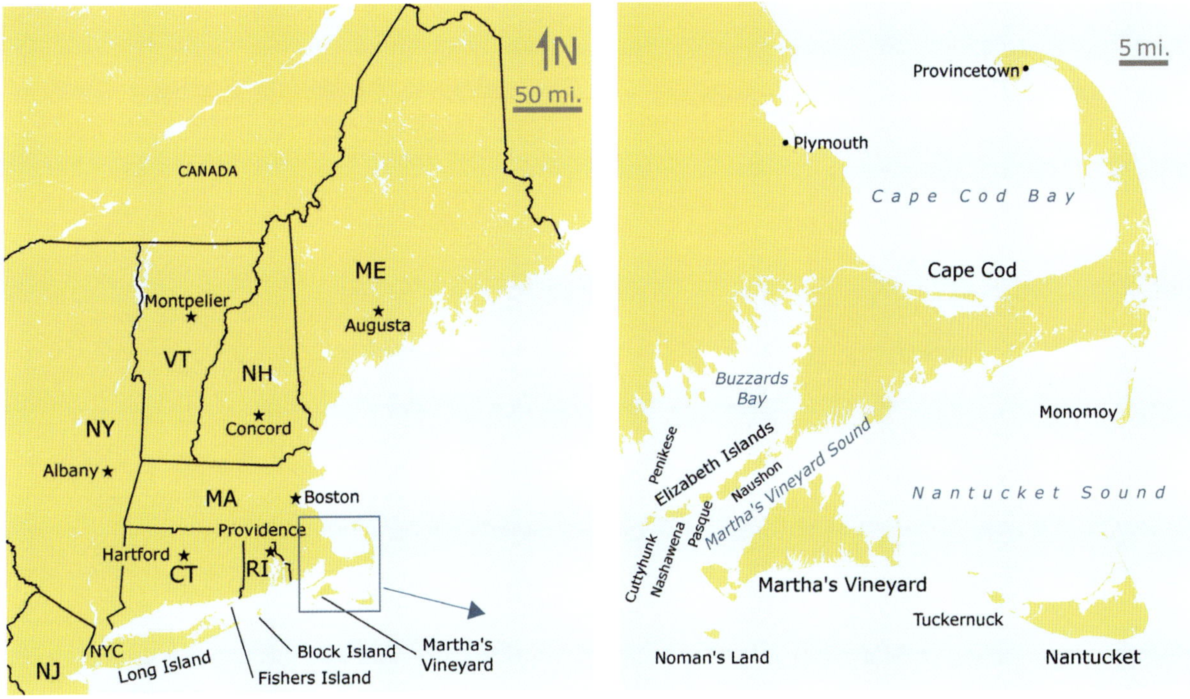

New England, the southeastern coast of Massachusetts, and major waters and land areas surrounding the island of Martha's Vineyard.

effort. These ideas inform the way that our Harvard Forest group operates in all of our research, they shape the organization of this volume, and they guide the interpretations that are made about the past as well as the insights that are offered concerning the future. They also can be applied to any landscape anywhere.

The first principle is the most important. *The ecological study of landscapes is an inherently historical as well as scientific undertaking.* To understand a place, it is necessary to pursue a combination of physical, biological, and social sciences, all firmly rooted in the depths of time. We need the many sciences because of the wide range of processes that shape the earth. But, we must also immerse ourselves in time, because the past conditions what we have today and will influence the future. History yields great insights into the workings of nature.

When thinking about the ways in which the past shapes the present, two concepts are also useful—*legacy,* a product of the past, and *inertia,* the tendency for a process initiated in the past to continue. Legacies are physical structures and imprints in the land generated by forces or people from the past and oftentimes absent today. While this concept applies intuitively to geological features—for example, the landmass of Martha's Vineyard is a legacy of the ice age—it also pertains to features arising from human activity. The cavernous hole of Goodale's gravel pit in the center of the Great Plain is a growing legacy of many efforts to build the Island's manmade infrastructure; the lone chimney, foundations, and scattered equipment at the mouth of Roaring Brook on the Vineyard's west coast are the legacy of a once thriving brickyard and the Island's modest nineteenth-century industrial history; and the stonewalls that line the roads and wander through forests today are remnants of fences that once restrained sheep, cattle, and hogs

Goodale's Pit on the Great Plain covers seventy-five acres and provides most of the Vineyard's sand and gravel, October 2011.

and bounded ownerships. Forests are filled with many legacies that we take for granted and yet yield fascinating stories of earlier days: stumps cut by a previous landowner; the uproot mounds, pits, and decaying wood generated by the 1938 or 1944 hurricanes; or the Indian shell middens and homogenized soil plowed by colonial farmers atop the knoll at Lucy Vincent.

Many living legacies remain from past environments. Among those most striking are magical forests of gnarly trees with far-reaching branches that grace coastal sites like Quansoo Farm and the Menemsha Hills Reservation. These trees were shaped under wide-open wind-swept conditions that left these landscapes decades ago, as testified by the younger straight, tall trees that now engulf them. Another example is what the Brits call "stools," the fairy rings of sprouting trees scattered across the Great Plain. These

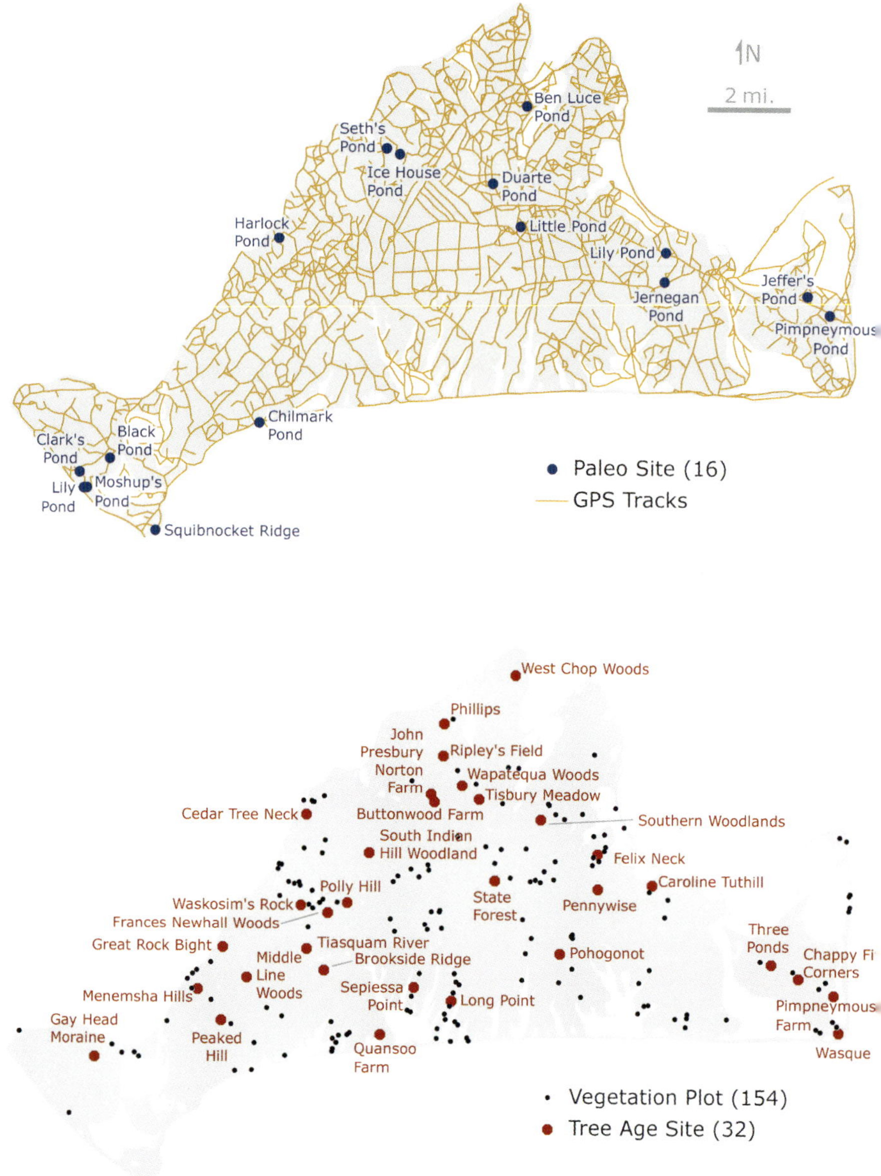

Sites studied and explored for this volume. *Above:* Sites cored for paleoecological studies and GPS tracks of areas investigated by foot, bike, and kayak. *Below:* Field sites for studies of the vegetation and tree ages.

Oaks on the Woods Preserve killed by caterpillar outbreaks and a lengthy drought, October 2014. The next generation of oaks and beech crowds the understory.

Hand-drilled and split boulder in a hundred-year-old forest on the Woods Preserve, West Tisbury.

striking circles of white, post, and black oak formed in response to many centuries of cutting and burning.

Every legacy is transient. All will eventually disappear, but each endures for a varying length of time. Leaves and acorns dropping to the ground in the fall persist over winter before succumbing to decay or an animal's hunger. Windthrown trees can be detected for decades as their woody carcasses gradually melt into the soil. The soil mounds and pits formed by the uprooting of a forest giant may remain as visible microtopography for centuries. And the drill holes in partially mined boulders in the West Tisbury hills and Goodale's cavernous gravel pit should persist until the Island's last days.

Legacies are more than simple historical curiosities—picturesque stone walls, old cellar holes and dams, flag-formed trees, or sharp breaks in the woods between pines and oak. These relics exert a strong influence over the way the landscape operates today. They also serve as a mindful warning to us that each of our actions today generates its own enduring legacies for the future. The stumps that we leave or the ornamental spruces we plant in the woods; the paths we carve through marshes and dunes; the clearings and fields we cut in our forests; and the residences we plunk down into nature will populate the land for a long time to come.

Inertia is the process analog to legacy the product—a tendency for a motion to continue or a process to play out long after the initial action occurs and may even be forgotten. Pollution provides an unfortunate example. The solvents spewed into the ground from gas spills and buried fuel tanks and the nitrogen released from lawn fertilizers and septic systems leach into the soil and percolate into groundwater, where the chemicals spread slowly and inexorably out toward drinking supplies, coastal ponds, and estuaries. Decades after the source may be terminated or the site remediated, the spread continues and its consequences continue to play out. While we may reduce our nitrogen or greenhouse gas outputs today, our coastal ponds and climate will display the consequences of our past actions for decades or longer.

The same is true of innumerable processes initiated by human

Open-grown oak on Peaked Hill with part of the Harvard Forest crew. *Top (l–r):* Eric Jorgensen, Ava Foster; *bottom (l–r):* Dave Orwig, and David Foster.

and natural activities. When a forest is blown or cut down, new trees establish rapidly and a new stand emerges in just a few decades. But that forest will continue to develop and display the evidence of the hurricane or logging for centuries to come. Recognition of inertia therefore has a profound influence on how we interpret the world. While we tend to investigate the present in terms of conditions today, much of what we see around us was determined in the past. As we wonder how nature is responding to the current environment, the land is everywhere engaged in a multifaceted state of recovering from everything that humans and nature inflicted on it in the past. The present is ever dynamic, for it comprises all the different trajectories of biological, physical, and chemical reactions that shaped the world minutes, a year, decades, or millennia ago.

Old sheep fencing partially engulfed by oaks off Kings Highway, Chilmark.

Remnant chimneys and lupines at Mohu, the old estate adjacent Lamberts Cove.

There is one preeminent inertial force that controls many of the ecological qualities of the Vineyard and New England landscape: recovery from colonial deforestation and agriculture. Though rooted in previous centuries, this process shapes most of what we see on the Island today and much of what will transpire in the future. For millennia following the ice age, the Vineyard was heavily forested and shaped by gradual shifts in climate and the light impress of a small native population. Then, for the first three centuries following European settlement, the land was cut, cleared, tamed, and structured by the daily activities of thousands of people (and tens of thousands of animals) who derived their living largely from the earth and surrounding sea. For three centuries, farming and wood chopping ruled the land. Whenever a farmer slowed in his work, woody nature crept back in. That process of nature reasserting itself commenced on a grand scale in the last half of the nineteenth century when our relationship with the land changed and agriculture and many Islanders moved west.

As much of New England's population uprooted to join newly arriving immigrants in bustling cities and factories of the Industrial Revolution or in claiming a stake in western lands, farming in New England declined. Scythes, axes, and mowers were laid down, fields were abandoned, sheep and cattle herds shrank, and weeds, shrubs, and trees marched across the landscape.

The evidence for this explosion of nature abounds across the Island, and the process unfolds incessantly today. Saplings relentlessly push out from forest margins; dense tangles of catbrier, grape, other vines, and shrubs fill backyard woods; and trees progressively obscure the views from homes and roadways. In the past century the Vineyard has become heavily forested with increasingly majestic stands of oak, pine, beech, and beetlebung, the Vineyard name for black gum. A process initiated long before any of us were born now patterns our landscape. This inertial force of forest regrowth will continue far into the future. Our forests and

Pitch pines spreading across the former Kupersmith development in the Southern Woodlands of Oak Bluffs, July 2012.

landscape are two to three hundred years away from anything like a stable condition. And, that quasi-stable state will only be realized if we or climate change do nothing further. In the meanwhile, the old cranberry bogs, pastures, hayfields, woodlots, mill ponds, and ditched wetlands that distinguished the nineteenth-century landscape remain, progressively obscured by this new and wilder nature.

For the foreseeable future, the enormous forces unleashed by the landscape's recovery from centuries of intensive sheep grazing and forest cutting will outweigh those anticipated from any other force—greater even than global climate change and rivaled only by the new waves of destruction of nature by human hands. To evaluate this statement, just think of the changes that the past centuries have brought. A century and a half ago, more than half of the Vineyard was bare of trees; dotted with boulders, sheep, and cattle; and only interrupted by an occasional house and fencelines that trailed off beyond distant bare hills. The remainder was heavily cut woods that kept the population warm and fueled operations

Cultural artifacts add color to a beech grove behind Arrowhead Farm, West Tisbury.

A "stool" of white oak sprouts formed through repeated cutting or burning of trees on the Great Plain.

like the brickyard. Today, distinctive, sometimes peculiar, and often magnificent forests characterize nearly every part of the Island from Aquinnah to West Chop and over to Chappaquiddick. For future centuries these Vineyard forests will grow and mature. Every year we collectively expend phenomenal energy and resources just to battle this immense inertial force around the edges—mowing lawns and grazing fields, bush-hogging shrub lands and scrub oak barrens, brushing out trails, harvesting woodlots, trimming roadside trees, and scorching back the growth with prescribed fires. Any place where that human activity ceases, the shrubs, vines, trees, and forest spread incessantly to engulf the land.

Thus, when we consider the forces that are shaping our land today, we should not simply think of the environment or activities that surround us. We also need to recognize a set of more subtle but relentless processes—soil formation, tree and forest growth, and the arrival, expansion, and declines of wildlife and plant populations—rooted in the past. And as we look to the future, both to anticipate as well as to manage the land, we need to recognize that our new activities—our mowing, grazing, cutting, introduction of new

Root cellar on the Woods Preserve, part of the massive stoneworks built by Dr. Daniel Fisher from boulders quarried on the property.

species, or simply stepping back to watch and admire—are layered onto processes initiated by prior generations and ancient events.

From the history at Lucy Vincent and these considerations of legacy and inertia, a final concept emerges that will help us to understand and conserve the Vineyard. *Humans are very much a part of the natural world.* Across the Island, New England, and most of the globe a combination of cultural *and* natural forces has shaped and condition the land. Indeed, one of the great joys of exploring the Vineyard is the inextricable way in which these twin forces come together in the landscape. As I walk the Island discovering and deciphering new scenes, I encounter captivating stories of human and natural history at every turn. The immense beech and oak in the old forests of Seven Gates are interspersed with remnants of drilled and cut boulders and ancient gravel pits; the aged pine woods on West and East Chops are dissected by equally ancient paths; shores from Lamberts Cove to Cape Poge display the sight of waves breaking on remnant stones of old jetties, wharfs, and military installations; and stonewall and ditch and ridge remnants of ancient fences meander across Island fields and woods in varying design from Pilot's Landing to North Neck. On more than a few slopes and ridges, rough stones emerge through the leaves, marking burials from a more ancient civilization. People have occupied the Vineyard for much of its history. They assumed and earned their place on the land. Their works that remain—from subtle to grand—warrant attention and often deserve our reverence and care.

This consideration that began with a singular beach helps to explain the broadly chronological organization of this volume, the selection of processes and times in Island history it explores, and the way in which ecological and conservation thinking is advanced. Each and every past event and process is imbedded in the present. Henry Thoreau fancied that he could see ancient Indian paths in the woods at Walden. Given his remarkable skills of observation and the land's history, this is not unlikely. Today, we are much more likely to detect the faint lines of ditch and ridge fences made by Thoreau's contemporaries as we walk the south shore and periphery of the Great Plain. Yet the subtle influence of early native people must also linger in our woods. Each process and product from the past has a signal, though the signals manifest through time with varying strength and fade at different rates. To understand how all of this fits together, we will start with ice, water, wind, and native times, and then progress through farmers, sheep, and brickyards to a modern era when nature is reasserting itself but battling human forces that work the Island in novel ways. We finish looking to the future as we attempt to put into practice lessons from the past and desires of the present.

Curving lines of posts are all that remain of Bob Welch's duck-hunting blind on Poucha Pond, Chappaquiddick.

The Setting and Stage

If we look upon [Martha's Vineyard] and its people in this way, we shall find that every nook and cranny will . . . give to those who search it in this spirit a mass of curious things, quite as rich as those we could bring to light out of the darkest corner of Europe, and far more comprehensible.
— Nathaniel Southgate Shaler, 1874, *Atlantic Monthly*

This variety attracts an extraordinary mix of citizens to the Vineyard. One of Massachusetts' two Indian towns is at the west end of the island, one of America's first and most exclusive yacht clubs at the east, and in between, a community frequently cited as the only middle-class black seaside resort on the East Coast. There is as impressive a collection of powerful, important, and influential members of the Establishment here in the summer as can be found in any one place; there is a burgeoning enclave of its better-known young radical counterparts. There are descendants of Portuguese pioneers who came from the Azores and Cape Verde Islands on whale ships. There are new Vineyarders who have migrated to start a business or to retire in one of the three more urban down-island towns, there are New Englanders whose ancestors were the first white settlers here, whalers, fishermen, farmers, whose names still dominate up-island villages as well as the streets and stores of the towns.
— Anne Simon, 1974, *No Island Is an Island*

(*opposite*) Clouds over the Vineyard and beyond the Elizabeth Islands as seen from Woods Hole.

Lying five miles by ferry from the mainland port at Woods Hole, the Vineyard forms a slightly drooping triangle. Its long southern coast stretches twenty miles against the Atlantic Ocean and extends from the town of Aquinnah in the west to Chappaquiddick, the sometimes separate island tenuously connected to the east. The Island's two shorter sides front the quieter waters of Vineyard and Nantucket Sound and converge to the north in two "chops"—landmasses that frame the intervening harbor—and the villages of Vineyard Haven and Oak Bluffs. Overall, the Island contains nearly one hundred square miles or about sixty thousand acres. Though modest in size, the landscape is remarkably complex in physical, cultural, and ecological detail. With six distinctly different towns, many hundreds of miles of convoluted coastline, extraordinarily varied forests, and terrain that ranges from occasionally monotonous to abrupt and baffling, no person has explored it thoroughly. Here I paint a bit of that variety, touching on highlights to set the stage for topics explored later in detail. This brief journey winds and jumps, for it covers deep time, many environments, and more than a few cultural settings. But, all of that is a necessary prerequisite for exploring this highly varied place.

Martha's Vineyard is clearly a part of New England and yet unique within that region. Its location has led a distinct combination of physical, biological, and human processes to sculpt and pattern the land over time. The result is a landmass and landscape that looks, feels, and operates unlike any other. And this, in turn, has led to a different quality of life. While sharing strong affinities with its coastal neighbors—Nantucket, the Elizabeth Islands, and Cape Cod, in particular—Martha's Vineyard stands apart in the region and world.

Most significantly, the Island sits at the southeastern corner of New England, far up the eastern seaboard and yet at the southernmost reach of the great ice sheet that buried everything to the north until some twenty thousand years ago. This great glacier formed a singular and highly confined set of land masses extending from Cape Cod across to Long Island. Each is a pile of glacial debris, devoid of the bedrock that gives structure to the rest of New England. Each consists solely of silt, sand, gravel, rock, and chunks

New England topography showing Martha's Vineyard on the mostly submerged coastal plain. Inland, the highlands comprising the White Mountains of New Hampshire and Maine and the Green Mountains of Vermont are separated by the narrow Connecticut River, which occupies a broad valley in Massachusetts and Connecticut. Further west, the massive highlands of the Adirondacks, Catskills, and highland regions of New Jersey and Pennsylvania are separated from New England by the Champlain and Hudson valleys and bisected by the Mohawk Valley.

of clay, all waiting to be shaped or carried away by the forces of water, wind, or man.

To the north, from Boston to beyond Labrador, the coast is distinctly different. Up there, land's edge is dominated by exposed bedrock that was stripped clean, first by ice and then through submergence beneath the sea. This distinction—bedrock versus glacial debris—is a critical one. The shores and bays of Maine are rock solid, firm against pounding waves, and resistant to the rising ocean. Structures anchored to that bedrock are similarly firm, or at least as firm as structures can be in the face of wind and water. In contrast, the glacial sands and gravels that characterize the Cape and Islands are capricious and deceiving. Dependably providing a solid footing when dry, they seem to melt and flow when saturated or exposed to even the gentlest lap of water. When subject to wave and current, the sand that once seemed so stable collapses entirely and is swept into rippling forms that disappear off into the sea, where they are sculpted into new patterns and forms.

South of Long Island lies another type of coastline. Untouched by the great glacier, the older, worn land of the Mid-Atlantic is formed of an inundated coastal shelf of loose materials deposited over the eons by rivers flowing to the sea and reworked there by water and wind. Although presenting many features similar to the southern coast of New England—including dunes, barrier beaches, brackish ponds, and lagoons—it lacks the dramatic relief and varied rockiness of the younger glacial landscape. The two regions do share the submerged plain, the continental shelf that extends out some fifty miles into the ocean. In earlier periods of lower sea level, this became an exposed expanse of land that provided a welcoming home and avenue for the movements of plants, animals, and people under a distinctly different climate.

While glacial history distinguishes and unifies the southern New England coast, geomorphological variation within that pile of debris helps to characterize the various landmasses and its constituent parts. The Vineyard is clearly distinct from Nantucket, the Elizabeth Islands, and the Cape. At the same time, although the different parts of the Vineyard are linked geographically, the landscape in Chilmark contrasts sharply from that of Edgartown, and the Island's Great Plain shares only a few qualities with Aquinnah or Chappaquiddick. This subtle variation among the many parts of the coastal landscape is also tied to glacial action, and has exerted profound influence over the way that ecological and human activities have played out over time.

The glacier thrust immense quantities of rock, sand, and clay into jumbled ridges of heterogeneous composition and variable heights known as moraines. Spending time here, it is easy to take this familiar landscape for granted, but as early as the 1880s these irregular lines of glacial debris forming the backbone of this coastal terrain—the lower Cape, the hills of the Vineyard, the north and central ridges of Long Island, the bulk of the Elizabeth Islands, and all of Block Island—were recognized by scientists as one of the world's most impressive morainal complexes. As the ice thrust forward to form these ridges, torrents of rushing water were released from the melting ice each summer, carrying sand and gravel outward, where they settled into

Cape Poge with its morainal bluff and lighthouse.

flat deposits of outwash, forming the Vineyard's Great Plain, the Mashpee Plain on the Cape, the level expanses of Long Island, and most of Nantucket. Where these moraines and outwash plains lie at sufficient elevation, they have survived twenty thousand years of sea level rise and remain as land today.

Moraines and outwash differ in more than topography and origin. In general, moraines have finer, more varied, and slightly richer soils that hold water better and support lusher vegetation than the sandy outwash. A substantial amount of the variation that we recognize among these coastal landscapes is tied to aspects of this glacial history, especially: the relative proportion of moraine (lots on Block Island to little on Nantucket); whether it is attached to the mainland (Long Island and Cape Cod versus the Islands); how large it is (compare Penikese, Nantucket, and the Vineyard); and how distant it lies from the mainland (Naushon versus Nantucket). These simple differences established long ago by the idiosyncrasies of ice, climate, and New England physiography conditioned each landscape's ecology, accessibility, land use, and other qualities that we appreciate today.

The control of glacial history on vegetation and land use is clearly seen on the Vineyard. For thousands of years, the undulating

Gay Head cliffs and lighthouse before its relocation, October 2011.

morainal hills of Chilmark and West Tisbury supported diverse forests of oak, beech, beetlebung, hickory, sassafras, and red maple. But, because these relatively moist productive hills also grow grass well and have water at or near their surface, they were widely cleared by early settlers for agriculture. In contrast, the flat and sandy Great Plain is droughty and was largely unclaimed by landowner or town until the twentieth century. Lacking water, only its periphery was settled and cleared for colonial agriculture. The rest remained wooded with oaks, scrub oak, and pitch pine and were used in common as woodlot and hunting ground and otherwise dismissed as a wasteland.

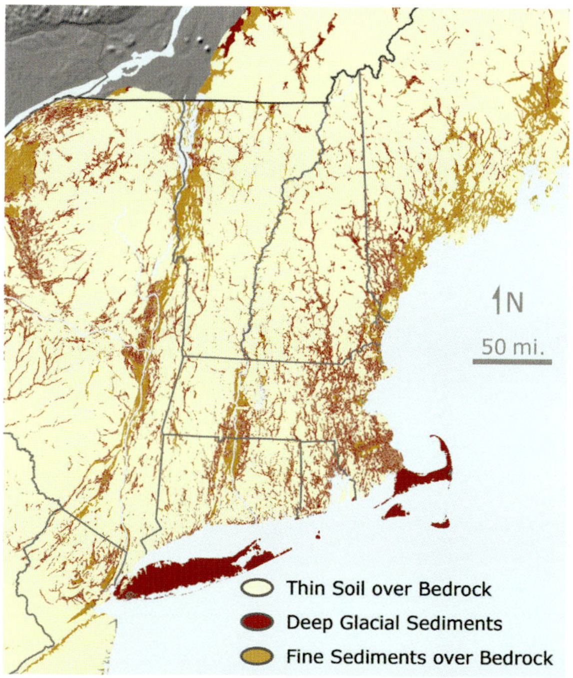

While most of New England is characterized by bedrock overlain by just a few feet of loose soil, the Cape and Islands comprise a distinct region of deep glacial materials. Deposited by ice and its meltwater more than twenty thousand years ago, this debris is hundreds of feet deep and provides the coastal region with its characteristic topography and soils.

Oaks frame Meetinghouse Road as it climbs the southernmost moraine in Chilmark.

The same broad patterns hold elsewhere. Naushon and other Elizabeth Islands are entirely moraine; their outwash plains were modest and trail off, submerged beneath Vineyard Sound to their south. Today, this chain of islands supports a delightful mixture of rolling grazing lands and relatively lush vegetation, including more than a thousand acres of stunning beech forest that dominate the heart of Naushon. Nantucket, with modest moraine and abundant outwash, is a poorer landscape for growing luxuriant forests or crops. Consequently, it was cleared of its oak woods, grazed intensively by sheep for three centuries, and, in the process, gradually converted into an extraordinary mosaic of sandplain grasslands, heathland, moors, and scrub oak thickets. On Long Island and the Cape, the broad patterns seen on the Vineyard are revisited on a larger scale. The moraines comprise high, undulating, and somewhat more fertile terrain, and the flat and expansive outwash plains support pitch pine, oaks, scrub oak, and similar droughty vegetation.

While geomorphology sets the stage, ocean processes have sculpted the margins of these lands over eons, and human activity has reworked the land surface; together these two forces have determined much of the configuration and modern appearance of these various landmasses. Coastal currents and winds have been most active in exposed ocean settings—eroding, smoothing, and rounding the southern and eastern shores exposed to the ocean and then sweeping the eroded sands down-current into graceful features such as the Provincetown spit, Monomoy, Norton Point, and Cape Poge. Sheltered waters such as Buzzards Bay, Long Island Sound, and Vineyard Sound inflict more subtle treatment on the land.

The Vineyard's location and setting have influenced the human use of the land. During prehistoric times, the Northeast lay a continent's width away from the Bering land bridge that brought people through Alaska from Asia. Consequently, humans have occupied the Island and New England for only about ten thousand years, less than western and southern regions. Similarly, the Island's Algonquin people lived thousands of miles north of the Mexican center of North American agriculture, which means they were the last of the native people to receive corn and beans, the cultivated staples of the continent. Though embedded in local lore and legend, these critical cultigens arrived merely a few hundred years before English explorers. As a consequence, the original Vineyarders never adopted the settled horticultural lifestyle and large village populations of southern groups or even of their Iroquois and Huron neighbors in adjoining New York and Canada. Instead, throughout the Cape and Islands, the confluence of a favorable climate with abundant resources from the land, estuaries, and sea allowed these coastal hunter-gatherer-collectors to become the most populous people occupying New England.

While the Vineyard's location delayed the arrival of early cultural influences, it predisposed the region to subsequent ones from Europe. Contact began diffusely with Portuguese, Italian, and other exploring forays (if not with Vikings, who left only tantalizing clues in their sagas), culminating in Bartholomew Gosnold's arrival to Cuttyhunk and the Vineyard in 1602. Gosnold set off from Falmouth, England, to establish a colony in the New World, and though unsuccessful in that endeavor, he explored the Maine coast, sailed into Provincetown Harbor, where he named Cape Cod, and established an encampment on Cuttyhunk, the westernmost of the Elizabeth Islands. In local forays he encountered and named Noman's Land Martha's Vineyard for his young daughter. Gosnold's subsequent efforts at colonization proved more successful, for after returning to England he ultimately led the expedition that settled Jamestown, Virginia.

Following Gosnold's discovery of the Vineyard, another Englishman, Thomas Mayhew from Tisbury in Wiltshire and a Puritan who arrived in Massachusetts in 1631, acquired the area encompassing Martha's Vineyard, Nantucket, and the Elizabeth Islands for forty pounds and two beaver-skin hats. In 1642 his son, Thomas the Younger, and some forty English families established the Island's first settlement at Great Harbor (Edgartown). The pattern of settlement and land disposition gradually brokered by the

Old pastures at Waskosim's Rock Reservation before being grazed by the Land Bank's new goat herd, August 2012.

Vineyard Haven Harbor with the spires of the First Baptist Church and Town Hall.

Mayhew family lent a decidedly New England flavor to the Island's government, organization, architecture, and land-use patterns.

The history and patterns could have played out quite differently. With the Basque and Spanish operating well-established whaling bases in Labrador since the early 1500s, the French settling Quebec and parts of northern New England, and the Dutch rapidly advancing from Manhattan up the Hudson River in the 1600s, other influences could have dominated the early history of the Vineyard. Alternate futures could have played out even within the history of British colonization. What if, for example, Gosnold's men had chosen to stay in 1602 as they intended, and Cuttyhunk rather than Plymouth had become the first settlement and a government-backed trading stronghold in Massachusetts? There may have never been Pilgrims escaping religious persecution to establish a foothold at Plymouth with such modest means that they sought and received native help and enjoyed that first Thanksgiving.

Within the coastal region, the human imprint across four hundred years of intense use has been determined by ownership and access. Under the ownership of only three families over the past four centuries, Naushon remains a delightful mosaic of expansive forests ringed with pasturelands that are conserved and managed by the Forbes family with care and an eye informed by history and

Edgartown Harbor looking up Katama Bay and across Oak Bluffs to Falmouth.

ecology. A mere forty houses are largely clustered around Hadley Harbor at the eastern end of that island chain. Block Island, all moraine, small and fertile, was stripped of its native forest early and then farmed and grazed intensively. Today it supports only a handful of native trees, is gracefully wooded by tall shrubs like shadbush, and boasts beautiful remnant fields and many houses. Long Island and the Cape, sprawling but attached to the mainland, offer easy access and commuting to major metropolitan hubs. As a result, they have been transformed by full-time and seasonal occupants into a dense mixture of expansive residential landscapes, commercial centers, distinctive villages, and gorgeous beaches. The mosaic of moraine and outwash on the Vineyard is fortuitously separated from the mainland and experienced a long history of conservation stimulated by alarm over what was happening elsewhere on the coast. Today, it supports a much lower population density than its mainland neighbors in a landscape of forest, farms, and villages.

The variation in the land governed the pattern of settlement on the Island, which together yielded a set of strikingly different towns, each unique in its setting, history, commerce, and character. Each town's distinctive character prompts strong opinions and preferences, helping to attract and retain the diversity of people that characterize the Island. Early on there were two large towns—Old Town and New Town, soon known as Edgartown and Tisbury—that stretched from the sheltered sounds in the north to great ponds and oceanfront on the south. Both centered on harbors, yet on different shores. And both spawned secondary village centers separated by the intervening plain or water and characterized by growing social, political, and economical differences that eventually led to their division into separate towns in the late nineteenth century.

The long-standing village of West Tisbury shared much of the plain with Edgartown, yet lacked a port and encompassed a sizable chunk of the fertile moraine and so became allied more closely with Chilmark in its agricultural and milling focus and its small New England village character and layout. Oak Bluffs (Cottage City) arose last and *de novo* with the camp revivalist movement and planned residential development in the nineteenth century. Though lacking a harbor and formally part of Edgartown, Oak Bluffs hung a series of wharfs from its exposed shores to accommodate ferries, stubbornly carved a harbor from ponds and wetlands, and through its old Eastville and newer bridge connection became more aligned with Vineyard Haven than any other town. Meanwhile, over time the original Island-wide Wampanoag population contracted

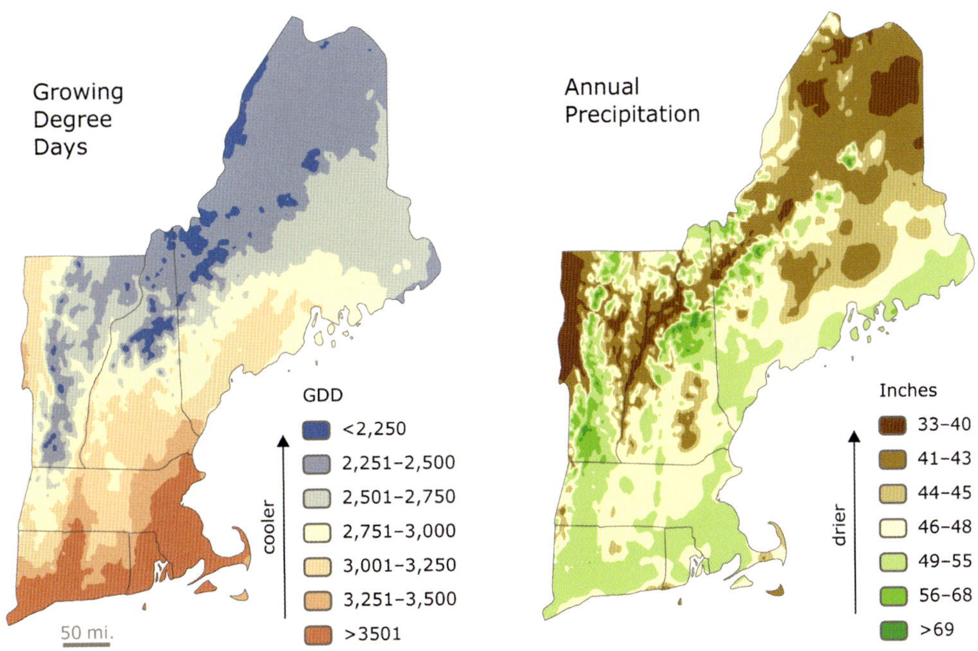

Although precipitation and temperature (shown as growing degree days, a measure of total warmth over the year) broadly decline to the north in New England, those patterns are strongly modified by elevation.

A lighthouse view east across Aquinnah, November 2012.

The quickened pulse of the Oak Bluffs waterfront.

to peripheral settlements at Chappaquiddick, Christiantown, and Gay Head before becoming concentrated in the last area. Formally shifted from its status as a guardianship of the U.S. government to an incorporated town in 1887, Gay Head adopted the ancient name of Aquinnah in 1997.

Although each town shares some qualities with others, they are as contrasting as any six adjoining townships in New England. The nature of this diversity is captured in a community-planning document from Edgartown.

> Each of the Island's towns reflect[s] its origins: Edgartown as the historical home of master seamen during the whaling era and still the seat of County government; Tisbury [Vineyard Haven] as the Island's year-round gateway and market town; Oak Bluffs as the Island's first summer resort and continued concentration of summer activity; West Tisbury and Chilmark as agricultural villages; Aquinnah as the Island's remaining Wampanoag Indian settlement and the site of perhaps the Island's most recognizable feature and only National Natural Landmark—Gay Head Cliffs. Three-quarters of the Island's population is distributed equally among the three "down-island" towns: Tisbury, Oak Bluffs, and Edgartown, each with a busy commercial town center. Vineyard Haven serves as the Island's main port, seconded by Oak Bluffs in the summertime. The three "up-Island" towns, West Tisbury, Chilmark and Aquinnah, are more rural in character.

Of course, the water that isolates the Vineyard provided many resources—from the salt marshes, estuaries, near-shore regions, and ocean. It also facilitated and stymied movement. Coastal people, from prehistory to the present, have always been among the most well-traveled groups, maintaining expansive connections to the larger world even while oftentimes being less familiar with their own backyard. Water is the great medium of trade, and thus the resources of the mainland and Island have always been exchanged freely, if often expensively. And yet the five miles that separate the Vineyard from the nearest mainland pose a significant barrier to some movement. This is particularly true for many plants and animals that arrived to New England after the rising seas isolated the Island five thousand years ago or those that have been locally extirpated since that time. A case in point is the magnificent Atlantic white cedar, a conifer prized for the durability of its wood that provided shingles and shakes for the exteriors of many early homes. White cedar was heavily cut on the Cape, where the Marconi Swamp on Cape Cod National Seashore provides one of the few remaining examples of the great groves of early history. Henry Thoreau reported that oystermen in Wellfleet recounted stumps of cut cedar as large as wagon wheels lying in exposed mudflats beyond Billingsgate Island. Remnants of great cedar stands also grow around the ponds of Naushon Island. Yet cedar arrived in the region as the sea was isolating the Vineyard, so it never became a part of the Island's landscape. A similar, though not bemoaned, fate has fallen to one of the region's recent immigrants. Over the past century the coyote has spread from its native southwestern habitat to nearly every part of the continent. Widespread across New England today, in the 1970s this hybrid canid with a few wolf genes made its way to Naushon, where it curtailed the centuries-old practice of open sheep grazing and culled much of the deer population, leaving larger, more robust survivors. Aside from a solitary animal reported in recent years, the only verified coyotes on the Vineyard have washed ashore dead.

Geography has shaped and influenced the region in other ways. New England's position at the northeasternmost reach of the country places it at the convergence of the major storm tracks that cross the continent, from cold and dry Arctic winds to warm, moist air streaming in from the Gulf coast. The fluctuating influence of maritime versus continental conditions conveyed by these contrasting air masses leads to the highly variable and rapidly changing nature of our weather. This convergence also yields a less glamorous distinction—the region's appellation as "the tailpipe of America." The collective pollutants spewed from power plants,

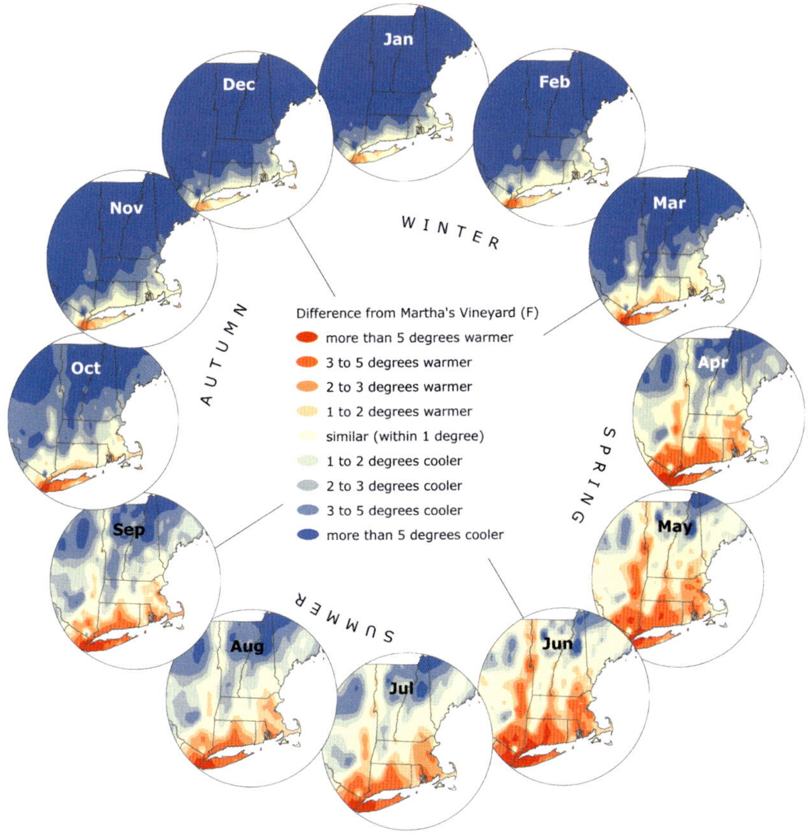

The Vineyard is a seasonally moderate place. Monthly maps depict the temperature difference between Martha's Vineyard and the rest of New England and adjoining New York. From November through February, the Vineyard is substantially warmer than most of the region. Conversely, from June through September the Vineyard remains blissfully cooler than much of southern New England and experiences temperatures similar to southern New Hampshire and the Massachusetts uplands. Map derived from U.S. National Oceanic and Atmospheric Administration thirty-year weather data (1981–2010).

industry, and farm factories of the Midwest and Lake States all trend toward New England, along with the smog of ozone and nitrogen generated by automobiles, cities, and industry up the eastern seaboard. The result is acid rain, nitrogen loading to watersheds, mercury contamination of our waters, and episodes of heavy haze.

Reminiscent of the nineteenth-century whaling fleet engulfed by arctic ice, the *Alabama* and *Shenandoah* weather February 2015 in Vineyard Haven Harbor.

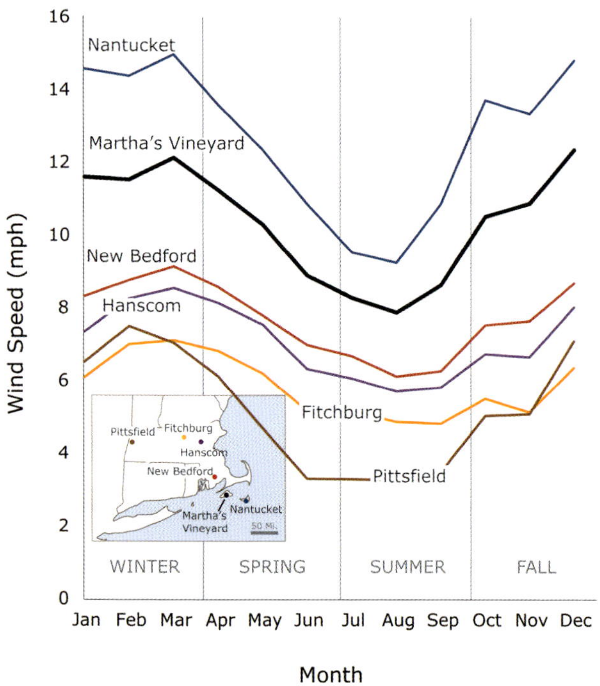

Winds in southern New England display a seasonal pattern of greater speeds in the winter and calmer conditions from June through August. Geographically, wind speeds decline from the coast inland to western Massachusetts. Data from U.S. National Oceanic and Atmospheric Administration's National Climatic Data Center (2001–2010).

As the Vineyard and New England thrust boldly into the Atlantic, they also experience extreme storminess and exposure to two of the most damaging storm systems known to man—hurricanes and nor'easters.

Given this strong coastal influence on the Island's environment, it is surprising how restricted the maritime effect is to the extreme coast. With prevailing winds coming from the west, only the islands and a narrow coastal strip benefit from the moderating influence of the sea. The Vineyard is particularly well situated to benefit from these ocean influences. Lying south of the Cape, it is shielded from the frigid waters of the Labrador Current that originates in the melting ice and great rivers of the high arctic. Instead, the Vineyard bathes in the oftentimes warmer southern waters that arrive with the Gulf Stream, which swings up the east coast and then arcs northeastward to ameliorate Europe's environment as far north as Norway. Less fickle than the major air masses, these currents are nonetheless subject to occasional changes accompanied by dramatic shifts in weather across the North Atlantic and even the globe. Indeed, these great ocean streams are held in a dynamic tension that could play a critical role in our future climate.

The sea's moderating influence on Island climate has been long recognized and was responsible for drawing many early summer visitors. One of these, Nathaniel Shaler, a Harvard dean and geologist who came to forge an indelible mark on the Island by founding Seven Gates Farm, captured these qualities well in an 1880s article in *Atlantic Monthly.* "The island is an oasis of salubrity in our New England bad climate. Its average warmth is at least two degrees above Boston and all the region north of the Cape; and all this gain is on the winter half of the year. The thermometer in the summer heat of the redoubtable hot term of July, 1872, when we had six days of succession above ninety [in Cambridge], never came above eight-four degrees in Tisbury. The winter average is probably at least ten degrees above the mean of Massachusetts Bay."

As Shaler observes, annual averages do not adequately portray the Vineyard's moderate climate. The surrounding ocean, slow

Old fenceline behind the dunes at Quansoo, Chilmark.

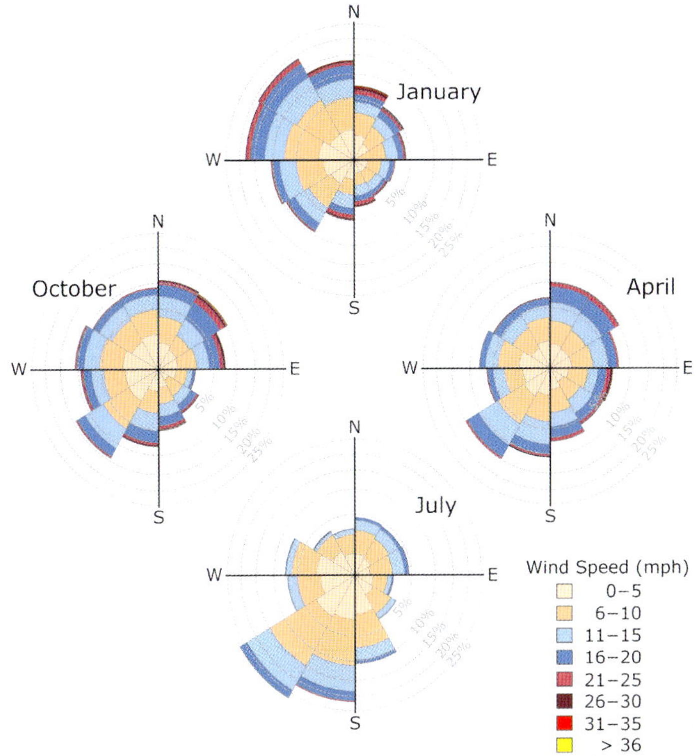

Wind roses for Martha's Vineyard illustrate the seasonally complex nature of the wind environment. Each circle is broken into the twelve major wind directions, with the length of each slice representing the frequency of annual wind from that direction (total equals 100 percent) and the colors representing wind speed. In summer (July) the winds are generally gentle and overwhelmingly from the southwest. In contrast, winter (January) winds shift to a predominant northwestern direction and are stronger from all directions. In fall, winter, and spring, the most intense winds come from the northeast to southeast. Data from U.S. National Oceanic and Atmospheric Administration's National Climatic Data Center (2001–2011).

to warm or cool in relation to the air or even the ground surface, provides this moderating influence. The inertia in this massive oceanic heat reservoir also delays the arrival of seasons and slows the pace at which they unfold. As the chill of fall and winter descend inland, the ocean's warmth helps ward off early frosts and prolongs the fall of most oak leaves until well toward Thanksgiving. The sea's persistent heat also allows the hurricane season to extend into October, as illustrated by the arrival of Hurricane ("Superstorm") Sandy on October 29, 2012. Conversely, although in the Vineyard mild spring weather arrives and the first daffodils generally peek out well before those of much of Massachusetts, warm spring conditions arrive late, and the green-up of the landscape and vegetation is delayed by about two weeks relative to southern New England.

This late spring has ecological implications, most notably a dearth of spring migrating birds. When April finds many of the great flocks wending their way north from southern wintering grounds, the Vineyard remains rather gray-brown and lifeless. The warmer, greener inland lowlands are busy with insects at that time and provide a bountiful route for these passing migrants.

This coastal environment is further shaped by the wind, which consistently exceeds that experienced inland. While delightful and nearly constant breezes moderate the summer heat, a change in

Wind-sculpted oak on Sheriff's Meadow Foundation's Quansoo Farm, Chilmark.

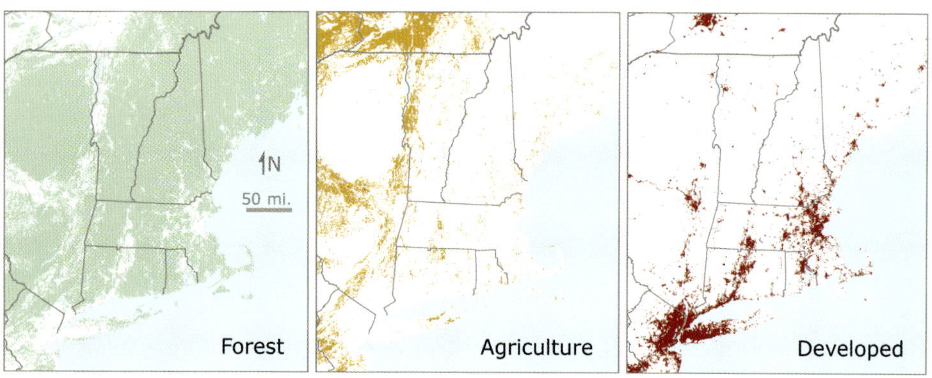

Although densely populated, New England is the most heavily forested region of the United States, with more than 80 percent forest cover. Agriculture is scattered but concentrated in the Champlain and Connecticut valleys. Developed areas spread along the coast from New York, up the Connecticut Valley, and outward from Boston. Data are the MODIS 2008 landcover from NASA.

weather can readily turn these to blustery and biting winds that sculpt and dwarf trees into distinctive shapes near the coast and on exposed sites. Windspeeds increase strongly in the winter, with peak averages in December and January close to double the lows experienced in July.

As any landowner with a windrow of trees can attest, wind exposure is strongly moderated locally by vegetation. As forests have grown back in the past century and a half, the wind at ground level has lessened considerably. Seasonally, the winds shift direction and speed in predictable but complex ways, leading to great variation in coastal dynamics and the exposure of different sites, harbors, and moorings. Winter winds come predominantly from the north and west and, less frequently but significantly, from an easterly direction driven by nor'easters. Summer winds angle more from the south and west, with the strongest hurricane winds coming from the south and east.

With abundant moisture, a temperate climate, a low frequency of natural disturbances, and a native population that subsisted on hunting, gathering, and collecting natural resources rather than farming the land, New England and its coastal landscape were thickly covered with forest before European arrival. These forests varied due to topography, climate, soils, and geomorphology. Once again on the Cape and Islands, glacial geology played a decisive role in giving this region a most distinctive type of vegetation.

Across New England, hardwood species were common. In the north these were beech, maple, and birch, along with spruce and fir in the mountains and hemlock and white pine at lower, warmer elevations. To the south, oaks joined these and other species in a broad transition zone across northern Massachusetts and then dominated further south into Connecticut, Rhode Island, and eastern Massachusetts, where they were accompanied by hickory and chestnut. Throughout the region the soils are generally thin, only a few feet of glacial debris overlying bedrock. Thus the uniquely deep deposits of sands and gravels laid down by the glaciers tweak the regional vegetation pattern substantially.

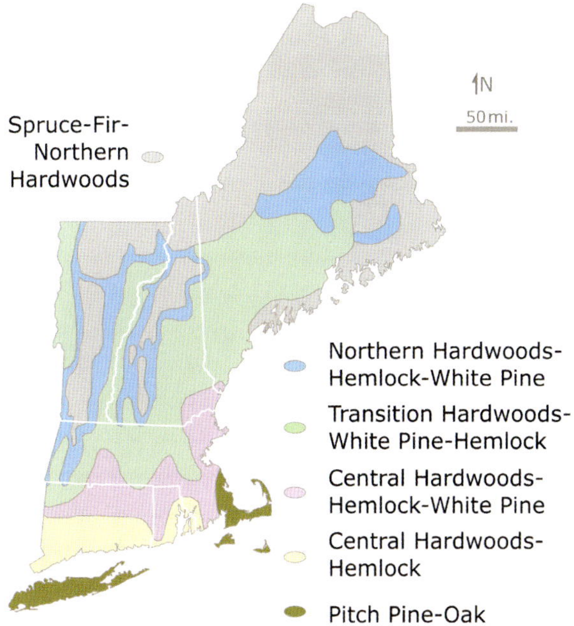

With a climate that varies strongly with elevation and latitude, the vegetation in New England transitions from hardwood trees in the south to mixed hardwoods, white pine, and hemlock centrally, to spruce, fir, and northern hardwoods in the far north and at higher elevations. Although depicted as pitch pine and oak, the sandy southeastern coast supports much oak and other hardwood species. Vegetation types from Westveld (1956).

Pitch pine, an uncommon species across much of New England, becomes abundant on the Cape and many of the islands, while its counterpart, white pine, assumes a minor role, and hemlock is nearly absent. Drought-resistant species, especially many oaks and hickory, assume a predominant role, and the more moisture-demanding hardwoods such as birches, chestnut, and ash are scarce, while beech and red maple are variable. From Long Island to Cape Cod this distinctive assemblage of plants is described as pitch pine and oak, although there is considerable landscape-scale variation within the region as within all of these forest vegetation zones.

Amidst the collection of unusual plant forms, vegetation structures, and human artifacts is an unusual assortment of uncommon plants and animals. About 65 percent of the Island has been designated by the Commonwealth of Massachusetts as priority habitat for rare and endangered species of plants and animals. A large number are southern and coastal species that have established a northern bulwark in the Vineyard's benign environment. A few

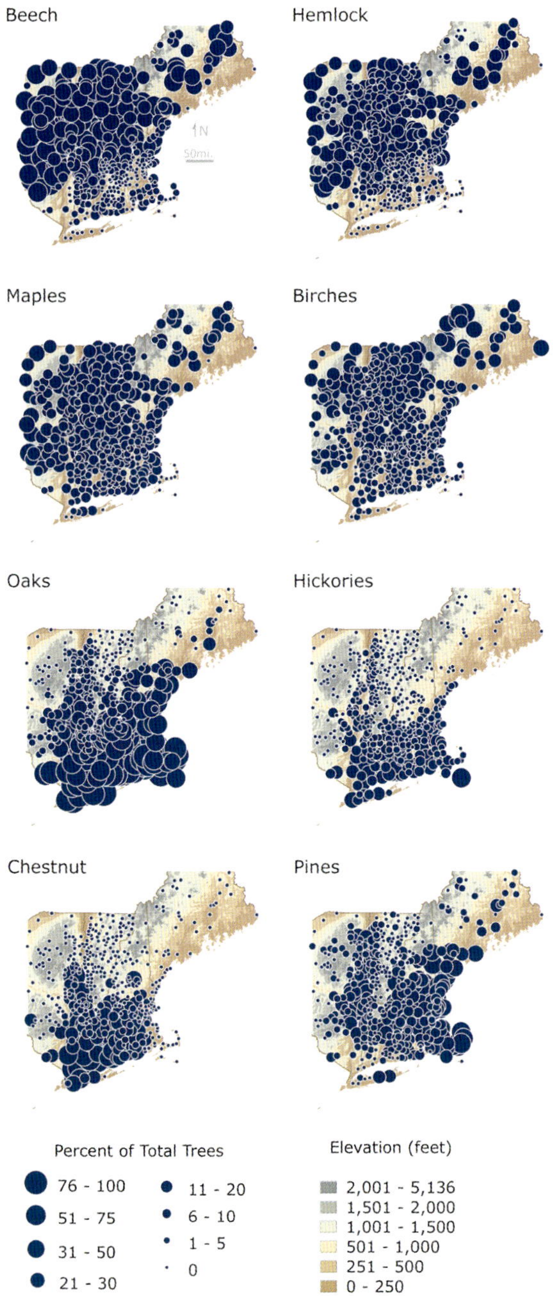

Mapping of land survey data at the time of European settlement documents a regionally varying vegetation of beech, hemlock, and birch in northern New England; oaks, hickories, and chestnut in the south; maple throughout; and abundant pine in the east and west. Data from Charlie Cogbill and the Harvard Forest.

have northern affinities. Many seem dependent on sandy coastal habitats. And quite a few others seem to thrive when the sandy, nutrient-poor landscape is disturbed through farming, erosion, the widespread uprooting of trees, or fire. Collectively, these plants and animals make the coastal landscape a center of biodiversity and a focus for species conservation.

The past few hundred years of New England history bring two remarkable stories of human endeavor and land transformation that are critical to interpreting the Vineyard landscape. The first is the tale of deforestation and recovery; the second of conservation and land protection. In a manner unparalleled across the globe, our region displays a history of resilience. Centuries of relentless deforestation, logging, fire, and grazing commenced with the first arrival of European colonists until the mid-nineteenth century, when New England began a process of natural reforestation. By 1850 forests covered only about 40 percent of the entire region, with much of this cut over and in poor condition. Today, New England is more than 80 percent forested. Author and activist Bill McKibben has called this regreening of our landscape the great environmental story of the United States and perhaps the world. On the Vineyard, forest cover nearly doubled over the past century and a half to equal the Massachusetts average of a little more than 60 percent. Today, with the harvesting of timber and firewood proceeding more slowly than the rate of tree growth, forests are continuing to mature across the region. In many ways, New England and the Vineyard are becoming wilder.

This process has played out with somewhat mixed reception on the Island and across the coastal landscape. On the one hand, the spread of trees has yielded beautiful stands of oak, pine, beech, and beetlebung, surprising many naturalists whose reading of the windswept landscape of contorted and stunted trees in the twentieth century led them to believe that grass, shrubs, and scrubby woods were all this coastal landscape could support. But, at the same time, the expansion of full-grown trees and large forests is transforming and obscuring the pastoral landscape of open fields, fencerows, and

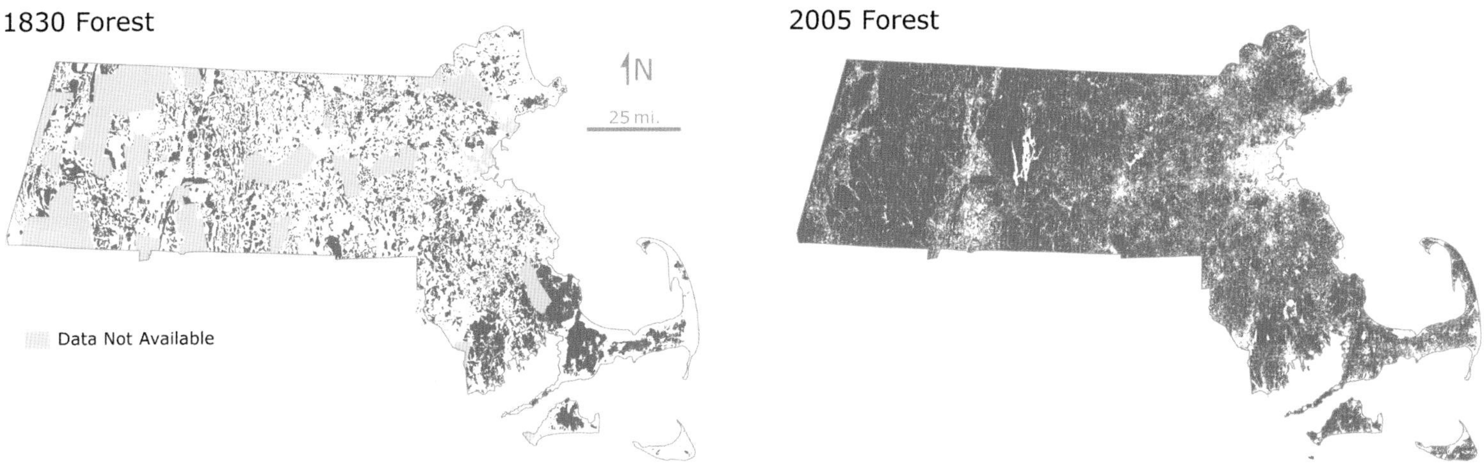

Following broad-scale abandonment of farmland, Massachusetts and New England reforested in the last half of the nineteenth and twentieth centuries. During the peak of deforestation, southeastern Massachusetts was the most heavily forested part of the state. 1830 data from Harvard Forest; 2005 data from MassGIS, Commonwealth of Massachusetts Executive Office of Environmental Affairs.

distant water views and putting species that once thrived in these wide-open habitats in a declining spiral.

Within this distinctly New England history of environmental recovery, the coastal region and Vineyard have a unique story. Across the coast, many farms were abandoned after the mid-nineteenth century, and native forests have come back. But in this windblown landscape where plants endure whipping gusts carrying salt, sand, or snow, progress has been slower than inland. The result is a delightful mixture of farm, field, and forest intermingling with an unusual diversity of successional vegetation—barren grasslands, juniper woodlands, shrub and scrub thickets, tangles of bayberry, catbrier, viburnum, sumac, and scrub oak. Mixed in this vegetation is an array of plant forms and groupings that are unique to our historical period—wind-sculpted trees, broad-spreading and open-grown oaks, beeches and shadbush, and thickets of pitch pine.

Lying amidst the recovering and recovered landscape are the usual artifacts of New England colonial history—stonewalls, cellar holes, gravel pits, farm ponds, and abandoned woods roads. But on the coast we add to these saltworks, windmills, herring ditches, cranberry bogs, and the piers and foundation stones of abandoned commercial wharfs, along with buildings razed or relocated due to the incessant inland creep of the eroding shoreline. The Vineyard adds its own distinctive assortment—the paint mill and brickworks ruins, remnants of World War II military placements, vestiges from futile efforts to contain the shores, and a procession of dams and mill sites arrayed along the most meager of streams.

In the late nineteenth century another process began to shape the land, though lying partially hidden by the new growth of forest—the conversion of farms and forests to houses, other structures, and roads. This process threatens to undercut the return of native forests and the persistence of local agriculture. Fortunately, the second story of remarkable New England progress is that of conservation. On this front, the region is a national leader, and the Vineyard boasts one of the most successful and innovative histories. New England contains the first park in the nation (Boston Common), the first land trust in the world (The Trustees of Reservations), the first Audubon Society, the country's greatest density of and most effective land trusts, and a legacy of leading voices in conservation—Henry

Brick mill chimney with the Menemsha Hills in the background, January 2013.

Grove of beetlebung above Lambert's Cove Beach, West Tisbury.

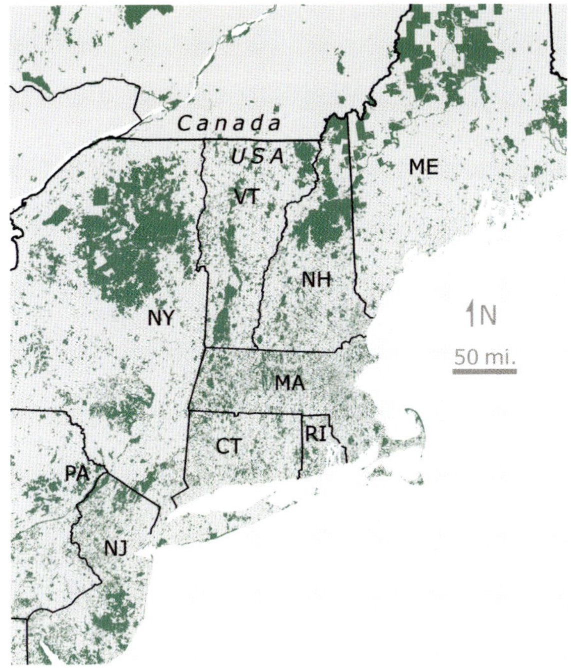

Conserved land composes nearly 25 percent of New England (*above right*) in a distinctive pattern of small to moderate tracts in more settled areas and large blocks of state, federal, and private lands in northern rural areas. The success of New England land trusts in conserving many small private tracts produces a sharp contrast with adjoining New York and Canada. Nonetheless, New England remains far from the goal of an interconnected conservation landscape. Conserved land data from The Nature Conservancy, Harvard Forest, and the National Conservation Easement Database. On Martha's Vineyard (*right*), a mosaic of conservation lands covers 40 percent of the Island. Data from The Martha's Vineyard Commission and the Harvard Forest.

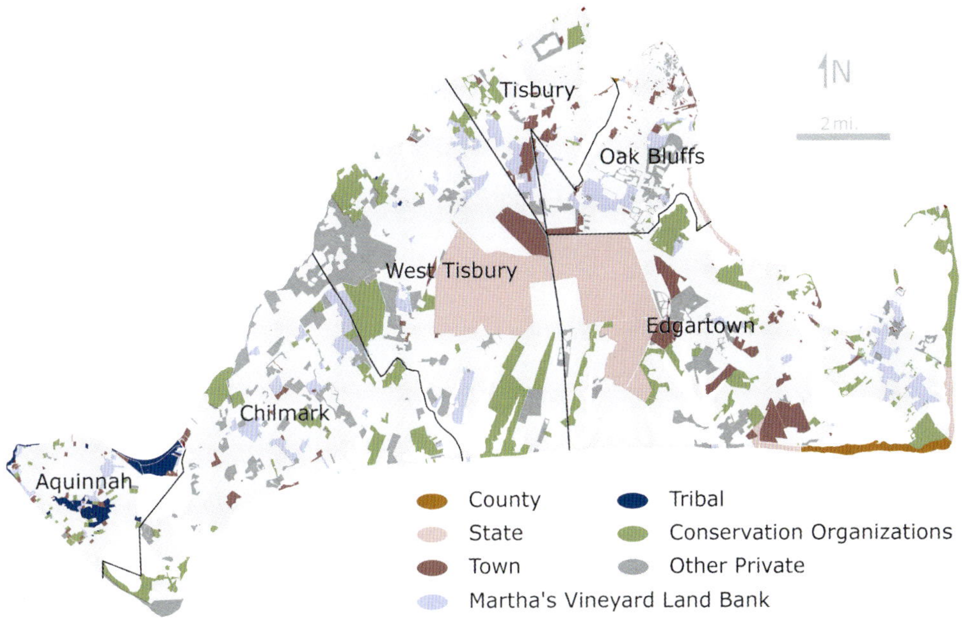

Vineyard Waters: Salty, Brackish, and Fresh

How does it come to be that wells can tap freshwater on a heap of sand and gravel surrounded by the sea? And, what determines whether the large coastal ponds are fresh, brackish, or salty as that sea? The Vineyard supports a remarkable source of groundwater, far more, in fact than the future population is anticipated to require. The secret to this resource is twofold: New England is a well-watered environment where precipitation greatly exceeds evaporation and the excess continually enters the ground. Below the Island surface lies a vast bubble of freshwater floating on the heavier saltwater that intrudes from the surrounding ocean. Across the Island's Great Plain, this groundwater aquifer lies some fifty or so feet below the surface. As water accumulates, the excess flows laterally to the Island's margins, where it enters the great ponds and the sea.

Although vast, this supply is highly vulnerable to contamination. Nitrogen from septic systems and fertilizers, herbicides and pesticides from lawns and farms, and spills of chemicals and solvents like gas, oils, and paints all enter the ground. There they can travel easily through the porous sands and gravels to groundwater and then into our wells or the Island's ponds and streams. Herein lies one great threat and challenge to the Vineyard's environment and one great motivation for conserving vast areas of the Island's landscape and carefully controlling the way we develop the rest.

Meanwhile, the coastal ponds vary in salinity from nearly fresh to extremely salty, largely determined by their degree of separation from the sea. The ponds are fed by groundwater, streams, and precipitation, which are all fresh. Breaching of the barrier beach that separates the pond and sea, either through human effort or a storm, allows saltwater to enter, producing brackish conditions. The result is beneficial for shellfish and many plants, allows saltwater fish to enter, and flushes the pond of contaminants. For these reasons the opening of ponds is often favored by fishermen and many landowners. However, with a connection forged between the pond and sea, the height of the pond drops and fluctuates with the tide, which may be detrimental to the interests of other landowners. As a result, arguments, changing perspectives, and polarized opinions have surrounded the opening of ponds through history.

Thoreau, George Perkins Marsh, Charles Weeks, and many others. To date, New England has conserved about 25 percent of its land, and if those of us who champion the Wildland, Woodland, and Farmland conservation vision for the region's future are successful, that will reach 80 percent over the next half century.

Within New England's history of conservation, the Vineyard and surrounding coastal region provide distinctive and yet unfinished stories. Partly in reaction to intense development pressure, but due equally to a passion for the land's special qualities and leadership by innovative people and organizations, the coast has won major battles in conservation, advanced unique approaches and institutions to protecting lands, and forged a framework that may yet be able to retain many of the grand qualities of this landscape. Cape Cod National Seashore is a unique member of the National Park Service and the sole reason that a substantial chunk of that densely settled region remains accessible to all. The Elizabeth Islands are conserved by dedicated families aided by a few of the region's major conservation organizations.

With 40 to 50 percent of their landscapes protected in conservation land, Martha's Vineyard and Nantucket fortunately lie well beyond the New England average. Both are distinguished by an uncommon conservation institution—land banks that utilize a small tax on real estate transactions to acquire and manage conservation land. Both are reinforced by strong island-wide planning and

Shear Pen Pond
and Cape Poge Bay,
Chappaquiddick.

regulatory agencies and a solid range of local, regional, and national conservation organizations. And yet on these islands, like all of New England, there is much still to accomplish in short order, especially given the rapidity of development and astronomical prices for land. Essential habitat for many species remains unprotected. Exquisite views, varied waterways, and unusual sites, along with many intriguing ruins and artifacts from history, remain in jeopardy of development and impairment. Much remains undone in the quest to link island sites for walking and biking and provide adequate access for recreation. Meanwhile, the broad sweep of unusual forests, fields, and open landscapes that the Island uniquely supports is being whittled away.

And there is the Future. Increasingly we have come to recognize the important role that intact forests, swamps, shorelines, and stream corridors play in buffering human lives and infrastructure in the face of nature's extremes and the stresses that human actions place on the environment. Intact shores absorb the brutal hits of winter storms and hurricanes, and fully functioning streams, wetlands, and ponds accommodate torrential rains. Broad swaths of forests and salt marsh absorb carbon dioxide and pollutants like nitrogen and mercury. Large expanses of woodland rebuff and recover more rapidly from wind, ice, and insect onslaughts than isolated trees or remnants of the perforated whole. By conserving

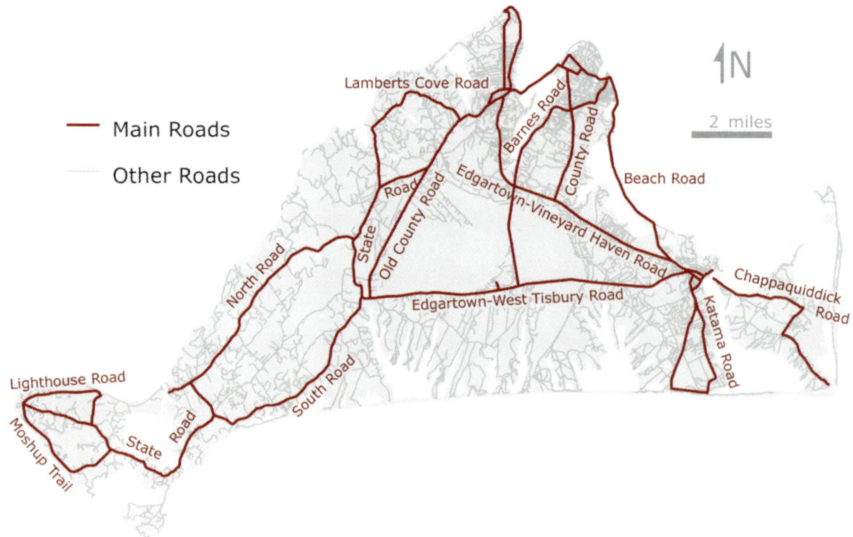

The traffic layout on the Vineyard allows visitors and residents to move among towns and major sites on busy roads while leaving much of the Island peacefully oblivious to the changing tempo of seasonal activity.

nature, leaving large swaths intact, and concentrating human activity distant from the fragile, vulnerable, and most important of nature's infrastructure, we ensure the soundest ride through environmental extremes that are certain to come. This has been a goal of Martha's Vineyard conservation groups, the Land Bank, and the Martha's Vineyard Commission.

All of the Vineyard's preparation through its history of conservation is now essential in the face of the changes coming to our coast. In the future the Island will experience more rapid changes in ages-old processes like sea level rise and coastal erosion. It will confront novel challenges from environmental shifts that our actions are generating, such as extreme droughts, rains, and the emergence of little known native and newly arriving pests and pathogens. It will also be visited by the occasional extreme event known only from long historical and prehistorical records. One likely event? A repeat of the 1635 hurricane—a storm with winds over 130 miles per hour, three times the force of those in 1938, and a surge at least double that experienced in Sandy. All of this will come as we continue to cope with existing environmental issues such as nitrogen eutrophying our ponds and groundwater and the imposition of human activity into some of the most vulnerable and dynamic parts of the landscape.

While there is little that the Vineyard can do directly to combat climate change other than seek to lead by local examples of conservation of resources, energy, and nature, and pressing for political action, there is much to be done to accommodate the impacts that this may bring. First and foremost is to conserve more and develop less in more focused and wise ways. We need to leave as much of nature intact as possible, both in places where nature is unusual and vulnerable and in places where nature can serve to mitigate the inevitable impacts of weather, seas, and insects. This means retaining the landscape much as it is and perforating precious little more with houses, yards, and hard surfaces like asphalt.

History, and the efforts of many people, have given today's Island a precious gift of a remarkable intact landscape. Across this landscape, natural processes are flourishing more strongly than at any time in the past few hundred years. Our collective success in building on past accomplishments in securing this natural infrastructure will largely determine how the future alters this setting and stage.

II Shaping the Island

PREHISTORY TO A.D. 1600

Geology and Sea Level Rise

Is it possible that the whole of Cape Cod is nothing but a vast terminal moraine, produced by a glacier advancing through Massachusetts Bay and scooping out materials . . . ? In this case . . . most of Nantucket and Martha's Vineyard might be regarded as moraines of the same glacier, when it existed further south.
— E. A. Hitchcock, 1841, Presidential Address,
Association of American Geologists

(*opposite*) Sand and rounded cobble on a morainal beach near Gay Head.

We largely know it as a solid, liquid, or gas, but when water falls as snow and comes under immense pressure from its own weight, it transforms into ice that will gradually become plastic and flow outward as an extremely viscous fluid. This insight and others that were needed to conceive of and reconstruct the actions of the great ice sheet that formed the coastal landscape came to Swiss scientist Louis Agassiz in the mid-1800s through his study of much smaller mountain glaciers. High in the Alps, Rockies, and other great mountain chains, permanent fields of snow blanket the summits and form ice that extends down valleys in long tongues. As long as the accumulation of snow continues far upslope, the mass will grow and ice will spread under its own pressure, sliding outward and downslope. At lower elevations, where the ice encounters warmer temperatures and longer summers, the rate of melting increases and eventually equals that of ice advance. At that point, the forward movement of the great ice mass ceases and a seasonal cycle develops. Under summer sun the front of the glacier melts, retreating and getting thinner, releasing torrents of water laden with silt, sand, and bouldery debris. Each winter, heavy snows resupply the glacial mass above with new ice, and the front advances. Though the glacier's front appears to shuttle back and forth, the interior of the mass of ice continues to advance outward relentlessly, a spasmodically surging stream of frozen water and conveyor of rocky debris to the front of the icy mass.

With prolonged cold and adequate moisture, isolated glaciers can expand and coalesce. The result is something that nineteenth-century geologists, in the days before flight and satellite images of Antarctica or Greenland, could only imagine—broad domes of snow and ice that cover expansive northern lands in broad, flowing sheets a mile or more thick. As these glaciers grow, the outward forces drive the ice to cross flat and undulating terrain and even spread up and over elevated areas and obstructions. The flow follows gradients of pressure and gravity, moving along the path of least resistance and guided by the land that it erodes as it travels. Beneath its great weight the ice and its rocky load gouge, abrade, and smooth the

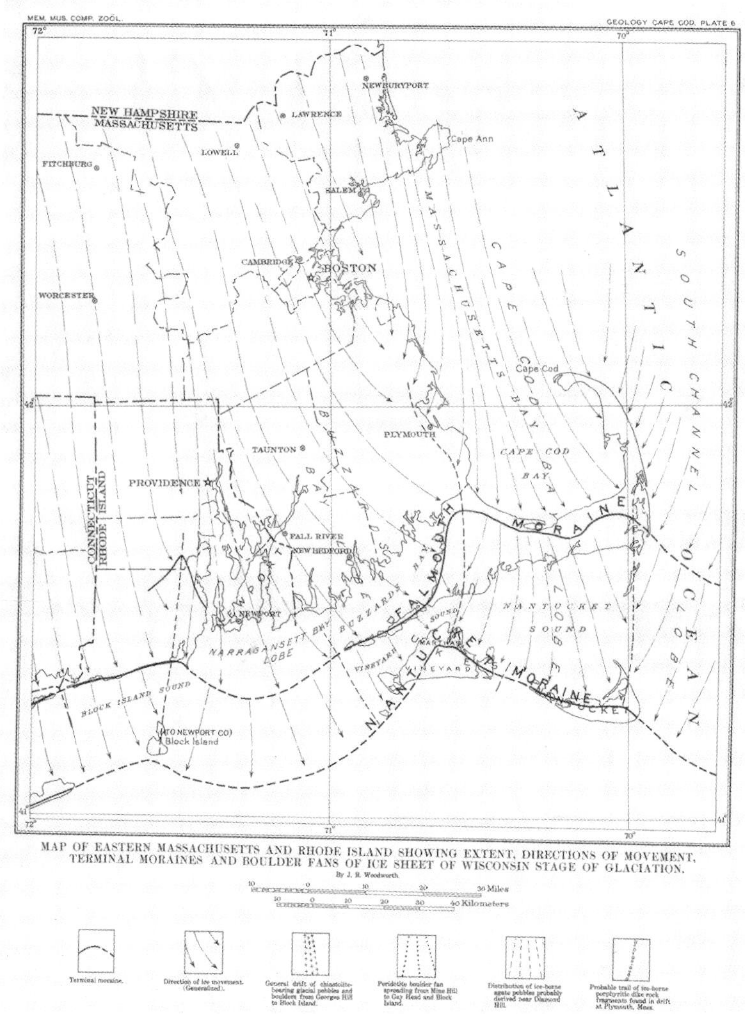

Harvard geologist J. B. Woodworth's 1930s depiction of the converging glacial flows that formed the Cape and Vineyard. The Vineyard and Nantucket developed at the southernmost extent of the ice. The Cape formed after the ice retreated and then readvanced. Map modified from Woodworth and Wigglesworth, 1934.

rock that it passes. The ice deforms and flows around and over resistant obstacles, melting under increased pressure and sliding past, lubricated by the water it releases. The pressure eases as obstructions are passed; water then freezes rock to the mass of ice, and the backsides of mountains may be plucked off and carried along. The sediment that accumulates in and is pushed in advance of the thrusting ice is deposited at the melting front in a jumble of boulders, gravels, and sand. This debris, broken and rough-edged, forms a heterogeneous collection delivered by the largely indiscriminating force of ice.

Meltwater is the other, more discriminating agent associated with glaciers; through its action, very different landforms emerge. Waters released by summer warmth flow along the surface and margins of the ice and plunge into fissures, where they erode the underlying ground before being released at the front or sides of the glacier. There and through streams that extend out beyond the ice, they deposit vast quantities of sediment in a highly predictable manner. The discriminating quality of water is a product of the fact that its ability to transport material is determined by its velocity. To move boulders requires powerful torrents, which glaciers can readily deliver. In contrast, silt and clay are suspended and ultimately deposited in calm water. Water thereby transports and deposits sediment in an orderly fashion that geologists refer to as "sorting." The largest boulders come to rest as glacial rivers first weaken, cobbles and gravels are dropped by streams, and fine sands and silt may be transported great distances by small rivulets and sheet flow. Similar-sized sediments tend to be deposited together in beds of fairly consistent composition. Therefore, in a cliff face at the ocean's edge or in any excavation, the alternating layers of different-sized sediment define periods of alternating water velocity. Meanwhile, the constant jostling and grinding of boulders, cobbles, pebbles, and sand in the watery flow blunt corners and round edges. In contrast to the jumbles of variously sized and rough-edged material known as *till,* which is dumped in place by the ice, water's hallmark is level *stratified deposits,* sorted into distinctive layers of silts, sands, and gravels, all rounded through their flowing travels.

Applying these basic observations from mountain glaciers, geologists have pieced together the story of the formation of the northern landscapes of North America. Under the bitterly cold climate that prevailed during the last glaciation, an immense dome of ice grew and spread out of northern Canada as far south as Long Island, the Vineyard, and Nantucket. As the ice approached this region, its front was made up of abutting lobes that were linked overall in a harmony controlled by climate and continental ice dynamics. Two lobes spread down through Vermont and New Hampshire carrying forward debris that came to form the triangular head of the Vineyard. A third eastern lobe slid along the large basin of the Gulf of Maine and came to a final rest over Nantucket. The ice front persisted there for decades or likely centuries—accumulating, thrusting, depositing, and releasing jumbles of sediment and torrents of debris-laden water. East and northward, the ice front extended to Georges Bank and on to Newfoundland. To the west it looped to Long Island, then on through Pennsylvania to the Midwest.

Never still, the tongues of ice advanced and retreated. Over

Immense erratic atop the moraine on the Allen Farm, Chilmark.

In the coastal landscape, the maximum extent of the glacier is marked by a series of ridges and hills (moraines) extending from Nantucket and Martha's Vineyard through Block Island to central Long Island. A second moraine reaches from Cape Cod and the Elizabeth Islands through southern Rhode Island, Fishers Island, and northern Long Island. Level outwash plains lie to the south of each moraine. As the sea has submerged and eroded these deposits, extensive spits, barrier beaches, and shoals have formed.

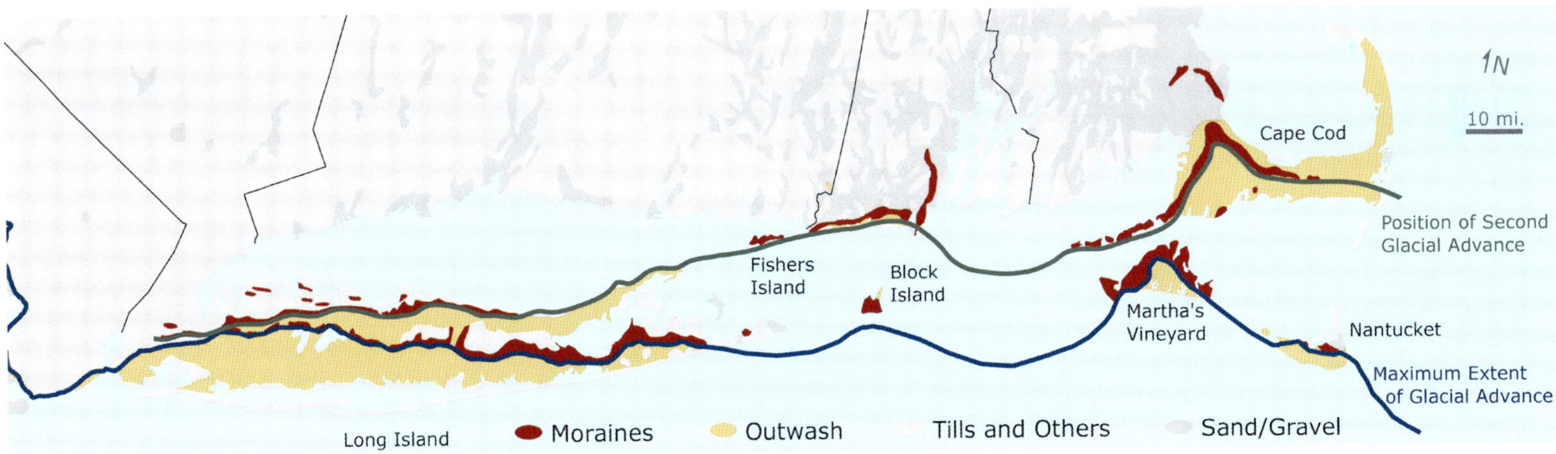

Edward Hitchcock (1793–1864): Geomorphological Genius

"While reading [Agassiz's] work . . . on the evidence of ancient glaciers . . . I seemed to be acquiring a new geological sense; and I look upon our smoothed and striated rocks, our ancient accumulations of gravel . . . with new eyes."
— E. A. Hitchcock, 1841, Presidential Address, American Association of Geology

Imagine the enormity of change in worldview, fundamentally shifting your understanding of a landscape you have sought to interpret your entire intellectual life. Your training and all established science have reinforced the view that the earth's surface was shaped by biblical floods. Then one day a colleague places a paper in your hands proposing that a global ice age was the real culprit. With an openly inquisitive mind you reexamine the evidence, and your life's observations fall into place.

The year is 1841. New England, at its peak of agrarian splendor, is largely bared of trees and primed for geological interpretation. Nantucket is a thriving whaling port of five thousand people and four times as many sheep, while the more varied Vineyard landscape supports rolling grasslands, farms, scrubby woods, and competing maritime enterprises. A hundred miles away in Amherst, Massachusetts, Edward Hitchcock, one of the world's great authorities on earth processes, was revisiting the final draft of his three-hundred-page opus, "The Geology of Massachusetts," with eyes newly opened by an audacious theory penned by a Swiss academic. Grounded in lessons from the Alps, Louis Agassiz's slim volume *Etudes sur le Glaciers* asserted that a sea of ice had once covered much of the northern hemisphere. He portrayed this continental glacier as sharing the qualities of both solids and liquids. It flowed over the earth's surface yet scoured and transported staggering quantities of rocky debris.

Edward Hitchcock, 1793–1864.

Theories explaining northern landscapes had changed little through Hitchcock's decades of training students at Amherst College. The prevailing interpretation invoked a great flood to explain the streamlining of landscape features, the movement of boulders tens to hundreds of miles from their bedrock source, and the deposits of sandy, gravelly residues or "drift" across the land surface.

Yet for Hitchcock, "the theory of drift has for some years been the most unsettled part of geology." While water explained many features, it failed to account for deep grooves cut into bedrock, ridges of debris, and pressures that crushed rocks. Moreover, a torrent that moved house-sized rocks should have swept the land clear of the soil that so ably supported New England's forests and farms. Equally perplexing to Hitchcock was the absence of organic matter in the drift that blanketed the land. Floods should have left deposits teeming with algae, fish, and mollusks. But thirty years of study had revealed only sterile, gravelly material.

Now, Agassiz provided a solution to Hitchcock's dilemma. Through a brilliant leap of logic, the Swiss scientist extrapolated from modern glaciers in the Alps to staggering images never witnessed on earth: domes of ice burying parts of continents. Hitchcock marveled, "The object of Agassiz is, first, to describe the miniature, and then to enlarge the picture till it reaches around the globe."

When Professor Benjamin Silliman at Yale passed Agassiz's book on to Hitchcock, he sought an evaluation of the audacious theory by a revered colleague uniquely positioned to test it against field evidence. Hitchcock was chairman of the American Association of Geology (AAG) and legendary for work documenting ancient dinosaur footprints and great lakes in New England and interpreting the formation of the oxbows in rivers like the Connecticut in his backyard. When Hitchcock tested Agassiz's assertions against his experience, he was immediately sold.

Thus, as he finalized his book for delivery to John Wesley Powell, director of the U.S. Geological Survey, Hitchcock took the unusual step of composing a postscript comprehensively amending the work. He asserted that the power of Agassiz's theory was that it explained and allowed disparate processes to work simultaneously in harmony and in close proximity: abrasion, thrusting, deposition, erosion, channeling, and collapse. In every setting the theory and field evidence meshed. "It explains satisfactorily the origins of those singular accumulations of gravel . . . the smoothing, polishing and furrowing of the rocks at different altitudes . . . the transportation of bowlders, and their lodgment upon the crests and narrow summits of mountains; and that often without having their angles rounded . . . the existence of deposits of clay and sand above the drift . . . the source of water to fill valleys . . . why these deposits of clay and sand are almost completely destitute of organic remains . . . that the northern slopes of some mountains in New England . . . exhibit scratches and furrows."

In one year and two capstone works—his presidential address and a major opus—a great scientist embraced a novel theory and provided the first satisfactory interpretation of the New England landscape and the enigmatic landforms that compose the Cape and Islands. In classic scientific tradition, Hitchcock supported his points with local examples and open acknowledgment of the source of his inspiration. At the same time, he defined and sought to address limitations to this brilliant theory. How, he questioned, could the glaciers spread so far across the flat coastal landscape? And what mechanisms triggered the frigid temperatures that grew the great sheets of ice?

time the conveyor action combined with the thrusting of the ice front to produce ridges and hills of heterogeneous sediment—from boulders to clay—that marked the southernmost reach of the ice. Across much of the land affected by the glacier, the thin soils and hard rock yielded little debris, and the resulting morainal ridges are discontinuous and challenging to trace. But at the position of the present southern New England coast, the advancing ice encountered great depths of clay, sand, and gravel deposited by previous glaciers and the deltas of rivers formed over hundreds of thousands of years. These deep sediments were plowed and piled by the ice into moraines—ridges, hills, and valleys displaying hundreds of feet of relief. The extent and depth of this reworked material is substantial. Across the Cape the deposits range to 600 feet in depth. The Vineyard sits on upwards of 150 feet of unconsolidated material.

The Vineyard's two opposing moraines derive from different ice lobes and sources of material and therefore differ sharply. The western moraine stretches from Vineyard Haven to Gay Head and is most readily apparent as three parallel ridges running north of South, Middle, and North Roads in Chilmark and West Tisbury. These ridges are bold in relief and varied; they comprise mixtures of different-sized materials, sporting immense boulders and in many places incorporating massive chunks of clay plowed up from deposits lying just to the north. At the time of their formation, they also engulfed immense blocks of ice that subsequently melted to form bowl-like, kettle-hole depressions in the undulating terrain. Where these depressions intercept the groundwater or are underlain by dense layers of clay, they support scattered lakes and ponds, including Occoch, Black, Harlock, Seth's and Ice House Ponds, many wetlands, and the historic cranberry bogs in the Lambert's Cove area from Bog House Road over to Blackwater and Duarte Ponds.

The eastern moraine reaching from Oak Bluffs to Edgartown and Chappaquiddick contains less material and little clay; it is correspondingly subtle in relief and covered with drier vegetation. At some point its surface was blanketed with sands from the

The Wequobsque cliffs, Chilmark, a morainal landscape of gravel, sand, and clay, January 2012.

Morainal sand, gravel, and stone capped by a layer of blown sand along the beach at Red Gate Farm, Aquinnah.

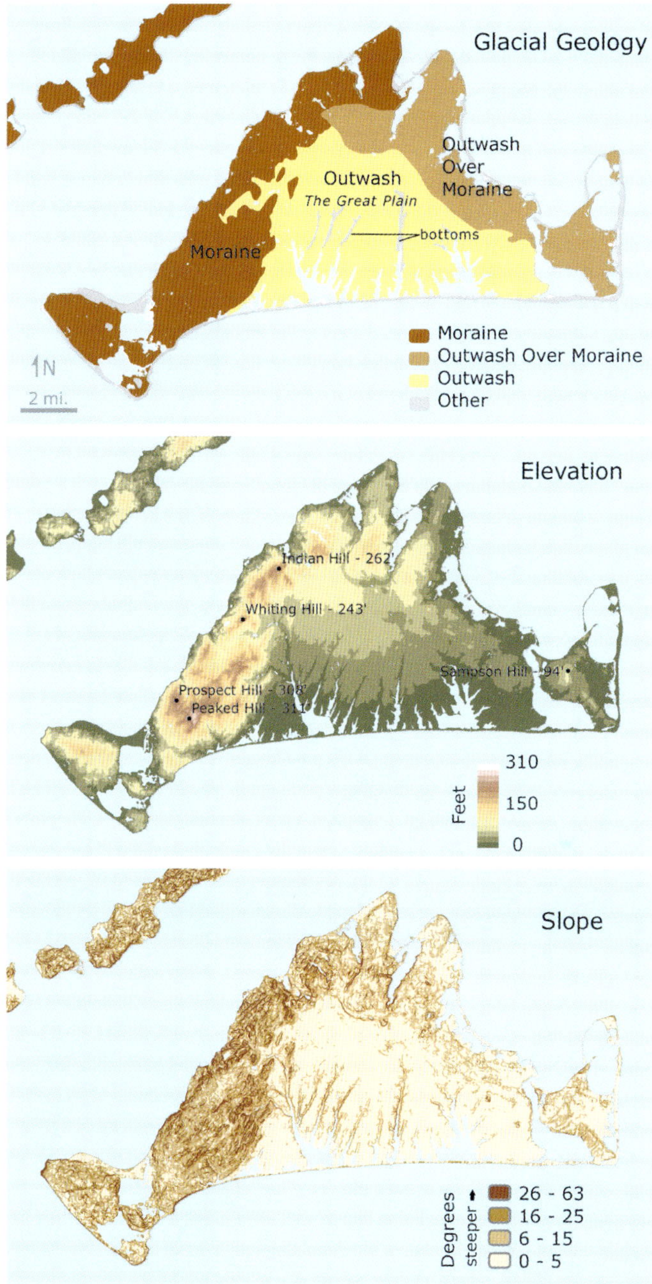

Robert Oldale and Roger Barlow's research (1986) has shown that glacial history left a legacy of striking topographic variation across Martha's Vineyard (*right*). The Great Plain is a largely level expanse of glacial outwash sloping to the south and interrupted by narrow valleys called "bottoms" or "hollows"; the down-Island region from Vineyard Haven to Chappaquiddick is gently rolling deposits of moraine overlain by outwash; and the up-Island region of moraine that runs from West Tisbury to Gay Head varies in elevation and slope.

melting glacier, a process that muted its features further and left its landscape more similar to the large outwash plain in the center of the Island than to the western moraine.

In summer months, the melting ice spawned rivulets, streams, and rivers. These carried relentless loads of sand and silt southward to form an immense and nearly level outwash plain that thickened and expanded over time. These great waters also created substantial but transient lakes where topography, ice, or recently deposited debris formed dams. An immense shallow body of freshwater once lay south of the Vineyard's moraines, where the meltwater fed a broad delta of sand—flat and dozens of miles wide. The surface of the resulting delta and outwash deposit forms the Great Plain, which

Playing with Coastal Geomorphology: The Cape and Vineyard Compared

The interplay between geomorphology—the origins and configuration of the earth's surface—and land-use history is illuminated beautifully through comparison of the Vineyard and base of Cape Cod. Though easily overlooked, the physical structure and origins of the two landscapes are nearly identical save for one key feature—the presence of surface water. That difference resulted from a geological quirk that, in turn, led to distinct colonial and modern histories. The result today is profound: the center of the Vineyard is dominated by a large conservation area, whereas the Cape is developed.

Focus on the thick triangular shoulder of the Cape—from the apex at the canal down to Falmouth and over to Hyannis and Barnstable. Rotate that triangle clockwise about a quarter turn and set it above the Vineyard. The two form nearly identical land masses, with the Cape roughly 25 percent larger. Structurally, both are bounded by moraines formed by the confluence of two ice lobes. To the south of the triangles are broad, flat outwash plains—the Vineyard's Great Plain and the Mashpee Plain on the Cape.

Other uncannily similar details are attributable to analogous processes. The south and north coasts of both regions are sculpted smooth by long-shore currents into barrier beaches that enclose extensive salt ponds or infilled marshes. Along both south shores, coastal ponds reach northward in fingerlike depressions, known locally

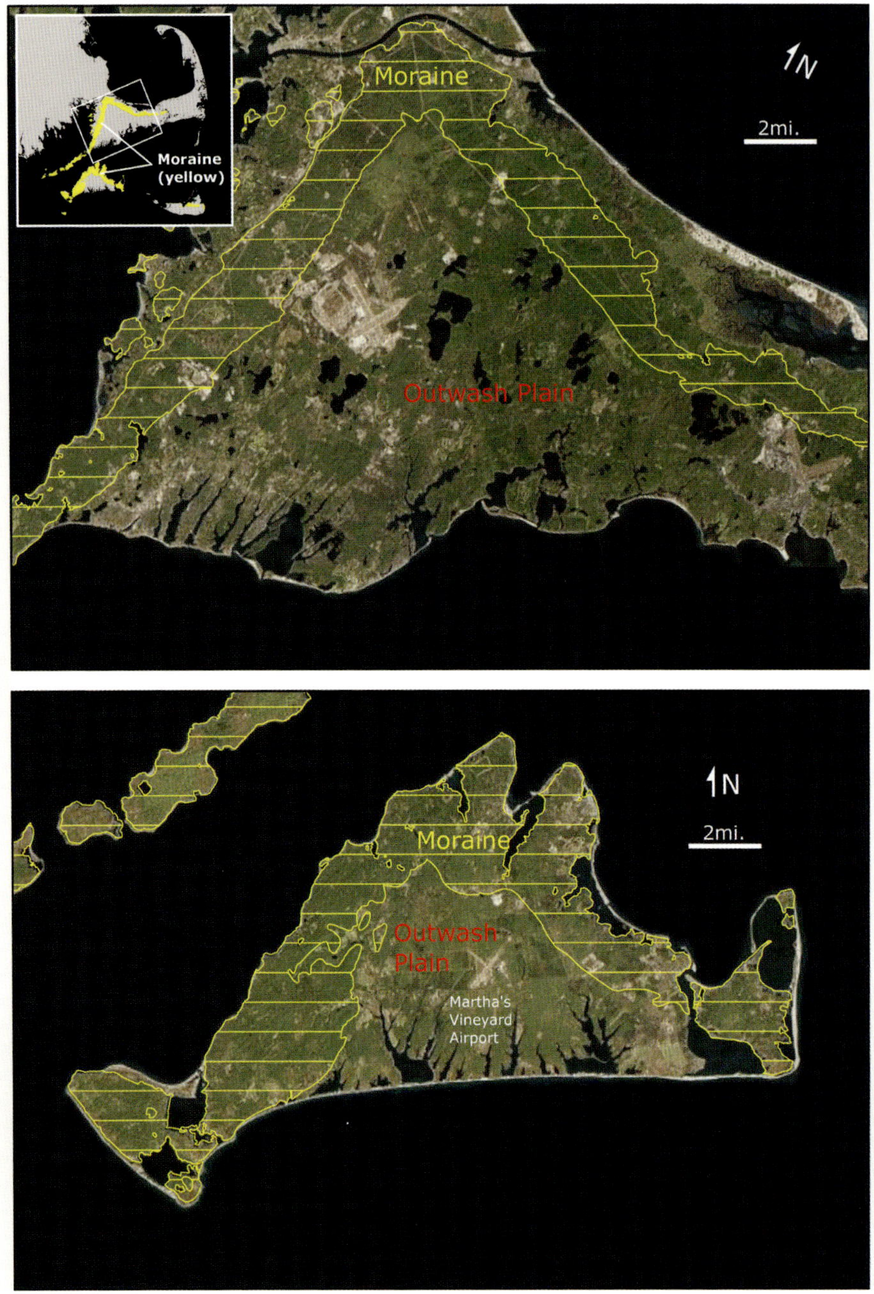

Playing with coastal geology: the Cape and Vineyard compared. Aerial photographs from ESRI and the USDA National Agriculture Imagery Program.

as bottoms or hollows that extend far into the plains. In contrast, the western coastlines are rougher and composed of steep hills, cliffs, or undulating shores.

There are also striking parallels in human activity. Long connector roads wind through both moraines. Each plain supports a large and conspicuous airport with a military history—Martha's Vineyard Airport and Otis Air Force Base. Both plains also contain smaller airfields, large gravel pits, and golf courses, along with housing developments that post-date the 1950s.

But a key difference between the plains caused their histories to diverge. The Cape supports large deep ponds, whereas the Vineyard is flat and dry. Lacking the ability to extract water from great depths, colonial farmers settled where ponds, streams, or shallow wells could supply water for their families and livestock. Consequently, through the centuries, while the Great Plain earned the label "a barren plain of no town," the Mashpee Plain became crisscrossed by roads converging on houses and the small villages of Greenville, South Sandwich, and Newtown, with adjacent ponds named Snake, Peters, Triangle, Great, and Wakeby. While the coast remained more desirable, surface water made an outwash plain usable.

The source of the Cape's water is glacial ponds—kettle holes or ice block depressions. On the Vineyard the great ice lobes extended only as far as the moraines. The outwash plain was formed as meltwater carried sand and gravel further south across the coastal plain. As the ice melted back to the Cape, it left immense scattered blocks of ice. Then, when outwash streamed from the Cape's moraines, it engulfed these ice blocks. As these melted, the gently sloping Mashpee Plain became pitted by depressions and then ponds. This slight quirk in glacial history of the two plains determined the difference between the settled landscape on the Cape and the neglected one on the Vineyard.

With its farms and small villages, the Mashpee Plain was progressively divided into more private ownerships. Lacking water, the Vineyard's Great Plain continued to be ignored as little beyond a source of firewood, blueberries, and hunting and was ripe for purchase when state wildlife officials focused on protecting the heath hen. Following the bird's extinction, state ownership expanded to form the largest conservation property on the Island. It was only in the twentieth century with the deep drilling of wells that the barren plain attracted the airport, housing, and commercial uses that now crowd the State Forest from all directions.

was largely featureless save for the distinctive narrow valleys—bottoms or hollows—that run to the southern coast. Beneath this surface lie countless layers of alternating sands and gravels.

Indirectly, the great continental ice sheets controlled two major processes that were critical in shaping the history of the Vineyard and New England coast. The first was fluctuations in global sea level. The storage of vast quantities of ice on land stole water from the world's oceans. Given the fixed volume of water on earth, this generated up to a four-hundred-foot drop in sea level and substantially expanded the global land area as shallow coastal regions became exposed. New connections were forged among distant lands, facilitating the movement of plants, animals, and humans into new regions as the temperature warmed and before the oceans returned to their current levels. Across the eastern United States, a vast continental shelf extended from New England to Florida and back up into Canada; England became part of the European continent; and the continents of Asia and North America were joined across the Bering Strait. One major consequence for New England: as the glacial ice spread southward and approached the modern coastline, it was greeted not by water but by a dry stretch of gentle terrain—the coastal plain and continental shelf that extended fifty miles or more to the east and south.

Isostasy, the equilibrium inherent in the earth's crust, was the second major process influenced by the glacier; the weight of the mile-thick ice mass depressed the malleable crust of the earth. Since the earth surface rebounded from this depression more slowly than the sea level rose with the melting of ice, many coastal regions to the north were first inundated by the sea and then subsequently reemerged. The region best known for such a history lies around Stockholm and northward in Sweden and Finland; here old fishing villages have had to relocate their harbors and moorings seaward periodically to accommodate the receding ocean. But a large coastal strip from Boston to beyond Maine also followed this pattern. This inundation is one reason that the rocky ocean-washed shoreline of northern New England differs so distinctly from that of the Cape and Islands.

The changing configuration and balance of land, water, and ice exerted other, more subtle effects on climate and ocean conditions. Glaciers ringing the North Atlantic combined with cold temperatures and arctic rivers to fill the ocean with ice and frigid water. As a consequence, the Gulf Stream, the warm and salty current that sweeps north and east from the Mid-Atlantic region and New England to bathe the British and Norwegian coast with moist, temperate weather, was diverted far to the south. With this warm current trending across the Mid-Atlantic toward Spain, New England and maritime Canada experienced arctic conditions characterized by extremely long, dry, and bitter winters. Barren tundra long dominated these areas as well as vast ice-free parts of Europe.

Long Cove and other south-shore coastal ponds extend inland toward the distant airport.

Outwash deposits of alternating sands and gravels at Wasque, Chappaquiddick.

Retreat of the Ice

The ice reached its maximum extent more than 26,000 years ago, coinciding with the low stand in global sea level, and then, quite abruptly, reversed course and pursued a nearly unilateral and complete retreat for the next 20,000 years. Following 100,000 years of relentless cold and icy conditions, the global climate switched,

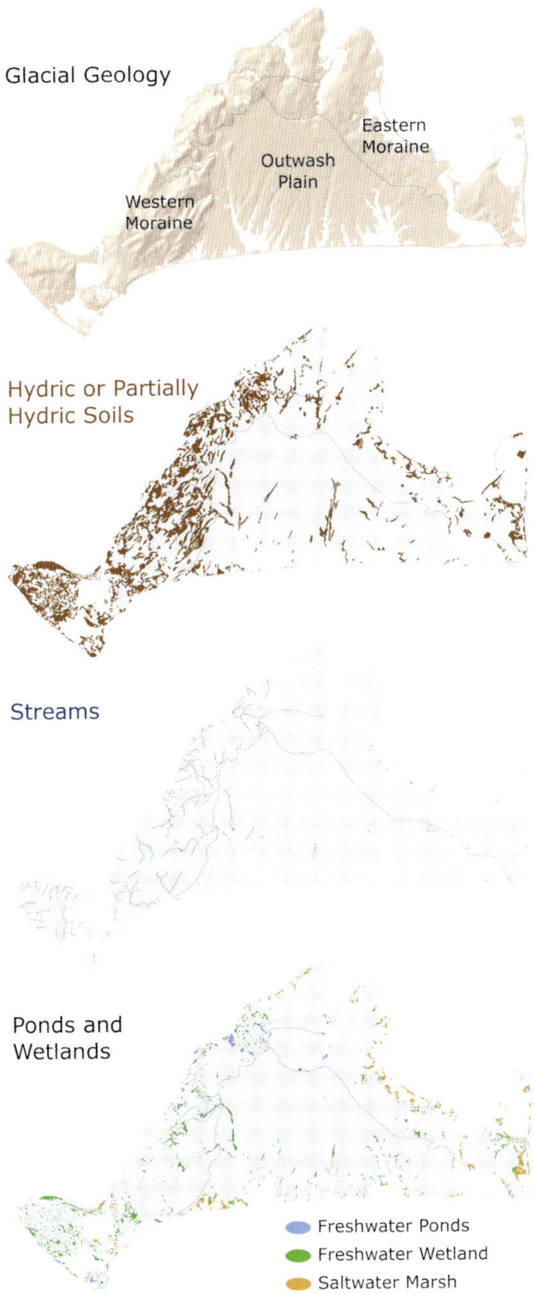

Glacial history strongly controls the distribution of freshwater on and below the Island's surface. The western moraine is underlain by clay deposits that hold rainfall near the surface and support most of the Island's wetlands, ponds, streams, and moist or hydric soils. In contrast, the sandy outwash plain allows water to percolate readily into the freshwater aquifer that lies deeply below its surface. The sandy eastern area contains just a few ponds (e.g., Duarte, Fresh, Lily, and Jernegan Ponds) and wetlands (e.g., Dodger's Hole, Beetle Swamp, and Airport Bog). The Island's relatively few salt marshes are concentrated on Chappaquiddick and around the great ponds.

ice shrunk rapidly (on a geological timescale) to nearly its modern extent, and the world's ecosystems were transformed in response. This climate shift was driven largely by a cyclical increase in summer solar heating that was reinforced by subsequent changes in the oceans, atmosphere, land surface, and ice mass itself.

As temperatures warmed, snow depths diminished, the glacial front halted its advance and began dropping more of its stony load, and the water volume in summer torrents escalated. Metaphorically, and from the perspective of a distant viewer, the glacier began to "retreat" northward at the astonishing rate of tens to hundreds of meters a year. By 16,000 years ago it was well north of Boston, and 3,000 years later it was gone from New England.

The last remnants of the continental glacier melted in western Labrador 7,000 years ago. With that final act, the great North American ice sheet became geological history, and the modern coastal setting was complete save for subsequent flooding and sculpting by the rising oceans. Since then continental ice sheets have been restricted to the vast islands of Antarctica and Greenland. And by then the Cape, coast, and islands were broadly configured and experiencing an environment that would be recognizable to us today.

Though largely unidirectional, the warming period of ice retreat was marked by at least one large reversal and many minor fluctuations. When the active ice front had receded less than ten miles from its coastal maximum at the Vineyard and Nantucket, the climate cooled sharply for a period and the ice became active once more. The ice lobes advanced each winter and then withdrew the subsequent summer, maintaining a largely stationary front overall. Carrying debris forward, the two ice lobes bulldozed two joined ridges and flooded the intervening land strewn with scattered blocks of melting ice with sandy water to form a second parallel landscape of triangular moraine embracing outwash. We call it Cape Cod.

A transient arch of water-sculpted rock at Lucy Vincent Beach, March 2013.

Since the Glacier

As the glaciers diminished, the rapid loss of ice reversed the two geological processes it had initiated, with great consequences for the coastal landscape. First, water locked up for millennia in ice on land returned to the ocean and sea levels rose. Then, unburdened by its heavy load, the land surface rebounded, while the sea floor beneath the increasingly deeper oceans subsided under the additional weight of water. The first process operated in tandem with the melting of ice. But the adjustment of the earth's crust to its shifting load lagged and is best described as operating at a geological pace that varied gradually through time.

The global increase in ocean waters and the isostatic rebound of the earth surface interacted over the thousands of years of glacial retreat to determine the local level of the sea. Because these two processes operate at different rates and are individually dependent on many factors, including the regional thickness of ice, timing of ice melt, and the increasing ocean depths, it is impossible to describe a single curve for sea level globally or even along the New England coast. Each landscape within and surrounding the region affected by the glaciers has a unique historical curve. Southern coastal areas near the Vineyard sank about thirty meters under the weight of the ice, while northern areas were depressed much further. As the sea rose, areas north of Boston were initially inundated by marine water, depositing the famous Boston Blue Clay and Maine's Presumpscot clays, before the land surface rose and emerged from the sea. The presence of sharp topographic features on the Cape and Islands confirm that these areas of easily eroded sands and gravels were never submerged.

To determine the absolute change in sea level we need to go to the tropics, far from the influence of any glacier's weight on the earth's crust. Barbados has become a global benchmark for sea level, the height of the ocean as controlled by the relative amount of water in ice on land. Here the record confirms that sea level reached its low of nearly 400 feet below modern levels about 26,000 years ago. It then rose twenty to thirty feet every thousand years, until

The Gut, Cape Poge Bay, and North Neck on Chappaquiddick were formed by the glacier and coastal currents.

The sea floor has been shaped by ice and water. As dry land deposited by the glaciers became submerged by the rising ocean, shorelines were eroded and the ocean bottom was modified by strong coastal currents. In this view subtle local variation has been accentuated to make the overall patterns more visible. Bathymetry data from U.S. National Oceanic and Atmospheric Administration.

Sandy beaches characterize the outwash plain at Quansoo and east along the south shore.

7,000 years ago, when the remnants of the North American ice sheet finally disappeared. Since then sea level has continued to rise at a slower rate to the present. It was five meters lower 3,000 years ago and has been rising about two meters every thousand years since.

As a consequence of changing sea level, most of the action during the past twenty millennia since the ice produced the Island's basic structure has occurred right at the coastline. Ever since the Vineyard and Cape emerged from the ice, their extent and shapes have been progressively reconfigured by water. Immediately after the ice melted, the distinctive coastal landforms that are fixed in our minds were irrelevant topographic irregularities on a broad coastal plain that extended dozens of miles toward the ocean and hundreds of miles down the coast. As the plain was gradually submerged under

Long Point, looking west across coastal ponds and outwash plain to the western moraine.

Small pond, marsh, and red maple swamp perched high on the moraine in the Wompesket Preserve, West Tisbury.

advancing waters, the slight elevational irregularities took form as a series of progressively changing and distinct capes, bays, and islands. Waves and powerful coastal currents eroded, transported, and deposited sediment from and to these lands and the floors of the surrounding sea. Landmasses were rounded; exposed southern shores were eroded smooth and progressively straightened; spits were shaped and further extended into lengthy capes at Poge, Monomoy, and Provincetown; and barrier beaches and bars formed from sands swept along the shore enclosed coastal ponds and secured quiet embayments and harbors. Most of the land area eventually slipped beneath the sea, but the slightly more elevated areas endured as the Cape, Vineyard, Nantucket, and the Elizabeth Islands (themselves an extension of the Cape Cod moraine that was partially inundated).

With rising seas, the coastline eroded and migrated inland, and modern landforms began to emerge. The details of this history are both fascinating and critical to the interpretation of past and future changes in this region's environment, vegetation, and people. Nonetheless, despite two centuries of intense research, many details remain elusive. Sea level curves for the New England coast remain generalized. Maps seeking to show the changing configuration of the coast bear many question marks. Three factors frustrate the reconstruction of the coastal landscape over time. First is the complex interaction between sea level rise and crustal rebound. Sorting out either is extraordinarily difficult. Second, the sea is continually altering the evidence for the past distribution of land. Even if we could create a perfect timeline for sea level, the landscape beneath the water that we want to apply this to no longer exists as it did before it was inundated. The chains of morainic hills that once extended from the Vineyard to Long Island and from Woods Hole to Block Island are not simply submerged. Through the action of powerful currents, underwater ridges have been partially erased, valleys filled, shoals built, and sediments carved by deep channels.

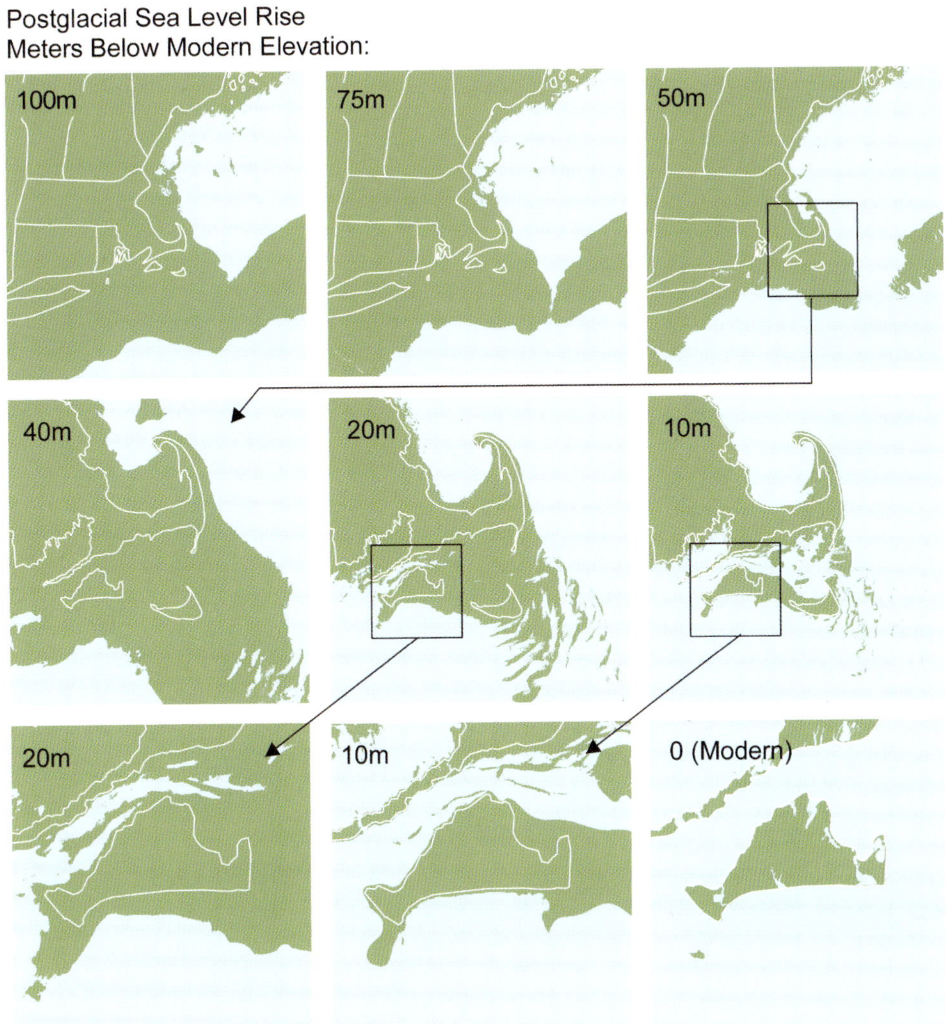

Using the history of sea level rise and modern bathymetric maps, we can reconstruct the changing configuration of the coast over time. Martha's Vineyard and Nantucket became isolated from the mainland between five and six thousand years ago. Since that time, slight changes in sea level have produced remarkably profound alterations in the shape of the major landforms. Bathymetry data from U.S. National Oceanic and Atmospheric Administration.

Finally, this is a landscape of subtle relief in which a matter of inches may determine the difference between dry land and submergence for sizable areas. Defining the past shapes and dates of isolation of the many landmasses requires a degree of precision in sea level and original topography that remains elusive.

Major gaps remain, but we can illustrate the broad details using modern bathymetric maps, recognizing that submerged features have been modified. When ice began its rapid retreat, the dry continental shelf extended out fifty miles from the Vineyard. Flat and vast, this expanse was largely featureless save for the pattern of moraines, bottoms, and a few deeper valleys. Broad canyons like Stellwagen Basin and the Gulf of Maine were inundated by

Mill Brook, meandering through the Chilmark Town Forest.

12,000 years ago; Vineyard Sound, the 150-foot deep channel that separates the Island from the Elizabeth Islands, was initially flooded 10,000 years ago; Muskeget Channel's 100-foot depth was filled a thousand years later; and Nantucket Sound was inundated by about 7,500 years ago. Approximately 6,000 years ago the islands—the Vineyard, Nantucket, and the Elizabeths—were isolated and the great fishing grounds of Georges Bank were completely submerged. In all, it took nearly two-thirds of the postglacial period—more than 15,000 years—for the land and seascape that we know today to be reasonably well defined.

The relentless rising of the sea continues. The sixty acres of Billingsgate Island, just west of Wellfleet in Cape Cod Bay, supported thirty houses, a schoolhouse, and a small fishing community in the early 1800s; by 1940 the entire landscape had disappeared beneath the sea. Skiff's Island, marked on nautical charts for over three centuries and once supporting grass and a few sheep, has disappeared on and off the Vineyard's southeast shore for the past century. And, the south shore of the Vineyard progressively recedes northward.

Once waters capture the land, it appears to dissolve. The sea bottom morphs as the powerful coastal currents reconfigure landforms that once appeared stable, but are simply a mass of loose sand and gravel. Billingsgate provides a poignant model for the process. Today, a vague hint of the island's prior extent is revealed by a smoothed shoal four by eight miles across. No clues remain of the cellar holes, house foundations, or baseball field that characterized this once-thriving landscape. Studies based on the coastal survey maps have concluded that between 1850 and 1935 more than ten million cubic meters of that once distinctive landscape have been redeposited across the ocean floor.

Though the pace declined over time, the sea never ceased to rise. As it did, the lands changed progressively, "adjusting," as geologists broadly describe the transformations that accompany inundation by the ocean. Coastal ponds extended inland, lapping up valleys, preceded by marshes that reach farther inland and are progressively submerged themselves. Elsewhere, isolated dry depressions become wooded swamps, then grassy marshes, and finally ponds as the groundwater table rises upward through the underlying sands and gravel on top of the heavier saltwater.

In the face of all this change, the remaining landforms laid down more than twenty thousand years ago lie distinct today. The basic pattern set by the ice continues to define the Vineyard. Morainic hills guard the northward coast facing Vineyard Sound and form the Island's highlands, including Peaked Hill, Prospect Hill, and Indian Hill, along with spectacular cliffs at Menemsha Hills, Gay Head, Squibnocket, and Wequobsque. Along the morainal shores that face Woods Hole, the Elizabeth Islands, and Noman's Land, the heterogeneous jumble of glacial debris has given rise to many rocky beaches, littered with boulders and cobble and framed by steep slopes that loom above. Most striking is Stonewall Beach on the southwest coast, where rounded stones deeply cover wide stretches. Boulder beaches abound from Gay Head to Cape Higgon and Cedar Tree Neck and beyond. In contrast, along most of the southern and entire eastern shore, and out on Chappaquiddick, long stretches of fine-grained sandy beach bear testimony to the paucity

Bottoms, Hollows, and Pamets

These deep linear valleys provide the major relief on the Great Plain and distinctive landmarks that figure strongly in local history. They also add variation to a globally valuable sandplain landscape and have become focal areas for the conservation of rare species. Known as bottoms or hollows on the Vineyard and hollows and pamets on the Cape, these unusual features have been erroneously described as streambeds for centuries. Early on they were cited as evidence for the biblical deluge. As the role of glaciers emerged, they were interpreted as relict valleys cut by streams issuing from melting ice. This intuitive explanation sits comfortably in natural history guides but collapses under close scrutiny.

The basic problem is hydrology. Pour water on the plain and it doesn't flow; it disappears. In contrast to arroyos, the bottoms don't support flash floods, even with torrential rains. The shape of the valleys also belies the stream explanation. None of the valleys reach the moraines and the purported source of water. After traversing the plain, the bottoms terminate abruptly, rising to the flat surface with no trace of an active or ancient source of water. Finally, in contrast to the meandering pattern of streams in gentle terrain, the hollows are straight, with few short branches and an undulating floor.

The bottoms remained enigmatic until the 1990s, when their formation was interpreted simultaneously in Scandinavia and the United States. The explanation is unexpected, for the process worked up the slope, in the opposite direction than intuition would suggest, and involves processes more common in the desert west and limestone bedrock.

These distinctive features were produced during deglaciation by an extremely modest process, the grain-by-grain "sapping" of sand. Spring sapping is a process whereby material from the face of an exposed bank that is oozing water erodes, causing the adjoining and overlying materials to slump and the face to retreat. The process can be readily viewed in miniature along a sandy beach. Just beneath the surface of the sand lies a water table revealed by scooping out a handful of sand. Although the sand initially seems firm along the excavation, as water oozes through the wall on the upslope side, particles of sand will loosen and fall. As this process continues, the adjoining and overlying sand slumps. Progressively, the process extends in a focused way upslope, in the direction of subsurface water flow and slightly higher water table. The process of sapping is thereby controlled by water flow below the surface rather than on top of it and in the process creates a fairly straight channel. The erosion occurs through the undermining of the surface rather than through overland flow and downcutting of the surface, which is how most watercourses are created.

The bottoms formed through this process at a time when a high groundwater table was supported under the Great Plain by large glacial lakes dammed up against the moraines on the north side of the Vineyard. In locations where this sloping groundwater table intersected depressions or the downslope front of the plain, seepage springs emerged from the exposed face. In a modest but progressive process, the seeping water eroded sand from these faces, undermined the material above, and caused it to collapse. Over time the springs eroded up the gentle groundwater slope toward the moraine and glacial lakes that supplied the water. The resulting valley patterns—their distribution, orientation, depth, and branching—were controlled by the lines of groundwater movement through the plain. This, in turn, was determined by the permeability of the layers of gravels, sands, silts, and clays that the glacial meltwater had laid down. It is a relict process, operative thousands of years ago. The bottoms have endured ever since.

Spring sapping continues as long as the water table and subsurface flow are sustained. Thus, when the glacial lake level dropped more than fifteen thousand years ago, the process ceased. Today these valleys lie as stranded relicts of the ice age and legacies of once active processes, separate from the glacier but very much dependent upon it.

Vineyard Conservation Society walk to the boulder-strewn slope and old forest at Goat Rocks on Seven Gates Farm, West Tisbury.

of large rocks, the sandy nature of the glacial outwash, and the steady replenishment of the receding shorelines by sands carried by coastal currents.

Above all, the story of its glacial history reveals a paradox of dynamism that defines the Vineyard. On geological scales, the Vineyard is extremely active and continually changing. But on human scales of daily life, the landmass is largely static. Much of the Island conveys a strong sense of permanence and security. But at its margins the land is episodically dynamic. Here lessons from the past portend the future. Permanence is perception, transience is reality.

Climate and Vegetation Dynamics

It is difficult for us to conceive of Noman's Land, which is now as innocent of any foliage as an infant's poll, once supporting great forests and a thicket of undergrowth, yet in two centuries the whites, without forethought or wisdom, had despoiled it of its verdure and rendered it an unproductive, barren isle, where for centuries the aboriginal occupants had preserved its fertility and the productiveness of the Vineyard, sparing the trees as a part of nature's household economy. On Noman's Land, in the swamp, may be seen the trunks and stumps of huge cedars, the decaying remains of a noble forest growth that existed two hundred years ago.
— Charles Banks, 1911, *History of Martha's Vineyard*

In the mid-1850s, Henry Thoreau began to show a fascination with swamps. He had long enjoyed the boggy forests of Concord, Massachusetts, for their wild qualities that transported his imagination to Labrador and other boreal environs. But as he sought to unravel his own landscape's history, he came to appreciate these wetlands for the clues to a long forgotten past that emerged from their depths. In his daily saunters, he took to probing woody swamps, grassy marshes, and mossy bogs with long, thin poles that he cut from nearby woods. In places he reached twenty feet down through layers of peat and muck. Snagging material on a branch stub or extending his arm through the cold, stringy mess to his shoulder, he recovered parts of plants that often differed in striking ways from those surrounding him. Pursuing this quest further, he began to explore the ditches that farmers had cut through drained wetlands. Startling and thought-provoking patterns appeared in the exposed depths. In places the firm surface was underlain by oozy mud, suggesting that swamps stood on the sites of ancient ponds. Elsewhere, cattails appeared in the peat beneath a thick growth of tall shrubs, indicating a succession from wet to somewhat drier vegetation. And, on at least one occasion, deep in the peat he extracted the wood and cones of spruce and larch, trees absent from his landscape but common to those northern lands that filled his boggy reveries.

Thoreau recognized these buried plants as fossils, and he came to think of wetlands and ponds as repositories of the ancient past. Familiar with other geological pursuits, he appreciated that his probings took him back through successive layers of time, revealing changes in his landscape. The materials encountered at an elbow's or arm's length were fairly recent; the deep layers and evidence for ponds beneath firm swamplands were much older. In combination with his knowledge of native and colonial history, this evidence provoked questions in his mind that resonate with our own quest to understand the New England countryside. What were the earliest forests like? What had the first Europeans witnessed? How had this land changed with the forces of climate, fire, ax, and plow? The

Pollen analyses show that Noman's Land and treeless parts of the Vineyard were heavily forested by oaks, beech, hickory, and beetlebung until cleared by European colonists.

simple clues that lay beneath his feet opened Thoreau's eyes to a past world that he anticipated would tell him much about the forces that had shaped his own.

Thoreau also grasped that these soft fossils filled a looming gap in the hard rock geological story being unraveled by his contemporary Edward Hitchcock. The Amherst professor's studies of mountaintops, gravel pits, and shorelines revealed glacial details and powerful climatic and physical forces that were nearly as riveting as that other scientific breakthrough just entering Thoreau's awareness—evolution. But, what of the time between the glaciers and the early records of Verrazano, Gosnold, and other explorers that Thoreau pored through in his trips to Harvard's libraries? Geologists were severely limited by the tools and science of their day in addressing that critical time. Their understanding leapfrogged from the era of mountain building to the ice age and then on to the narratives of European exploration, leaving a significant gap. Perhaps the soggy peats and muds might help bridge that void.

Thoreau fixated on a period of time that is critical for our understanding of the coastal landscape. So much had happened since the ice retreated. The islands appeared; the climate, sea, and land were transformed; and two distinct civilizations—Native American and European—arrived and flourished. To understand these processes, we can follow Thoreau's lead and explore those soft sediments that have accumulated since the glaciers formed the Vineyard—the muck beneath wetlands and the layers of mud extending tens of feet below the oozy bottoms of the Island's ponds. Using stainless-steel tubes and extension rods that improve on Thoreau's knobby pole and fistfuls of peat, we recover samples at progressively greater depths going back in time. We then dissect these samples for an array of fossils from plants, animals, and the environment. The materials we recover have persisted for thousands of years and include the seeds, leaves, and especially the pollen and spores of many plants; the hard parts of insects; charcoal from fires; and sands and silts delivered to the pond's bed through erosion by water or wind. Each provides part of the story, but collectively, and in combination with records of temperature and precipitation derived through accompanying analyses and ages from radiocarbon dating, they paint a picture of the changing landscape that can extend back 15,000 years. By taking Vineyard samples and extending our efforts and those of our colleagues and predecessors to Naushon, the Cape, and Nantucket, we can weave together a broad picture of how this coastal tapestry evolved.

But what is the relevance of these ancient stories today? Why should we care?

When I speak with students about possible directions for their research, I often suggest that they envision an audience less receptive than their professors and fellow students, who are

How to Get Mud

It was not until the 1960s that technology emerged that allowed scientists to tap one of the most valuable repositories of ancient ecological information—the muddy sediment filling the bottom of ponds and lakes. Swimmers avoid this ooze, but its value is unrivaled for insights into the past. Pond sediments tend to be much older than wetlands and yield a more consistent record of change. They also receive diverse materials like dust, pollen, eroded soils, and algae from the air, water, and surrounding land that yield insights into the larger landscape.

But while obtaining pond mud may sound like a simple proposition, it is quite challenging to retrieve an intact and continuous sample of many feet of ooze from below ten feet or more of open water. The consistency of "mud" ranges from watery at the top to compact, clay-like, or gravelly at the bottom. There may be layers of sand, charcoal, peat, wood, or roots anywhere in between. All is fragile, easily disturbed, and runny in contact with water.

The simple but ingenious solution to this conundrum is the Livingston corer, named for a professor from Duke University who famously tangled with a giant African crocodile while deploying a prototype of his invention. The corer was substantially modified and improved by another colorful academic—Herb Wright from Minnesota—with a reputation for subsisting on granola and yogurt for weeks on wilderness forays with devoted students.

The corer is brilliantly simple: a meter-long, stainless-steel tube with detachable aluminum rods and a rubber piston that snugly fits inside the tube and is attached to a long cable. From the ice surface or a boat secured in three directions, the tube, with the piston held at the bottom by the rods to keep material out, is lowered to the desired depth in the mud. The rod is raised a meter to a cocked position and then, while the piston is held by the firmly secured wire, the tube is pushed down, smoothly capturing one meter of mud. The rods, cable, and tube are then pulled to the surface, while the piston maintains suction to retain the mud. On the surface, the corer is laid horizontal and the piston is pushed with the rod to extrude the sediment onto plastic wrap and aluminum foil for storage and future sampling. Coring is then repeated downward in meter-long intervals until an impenetrable sandy or gravel bottom is reached. With mud accumulating less than a centimeter per decade, each meter contains more than a thousand years of ecological history.

Paleoecological studies are therefore conveniently divided into distinct efforts: planning and the selection of sites to address specific questions; strenuous but memorable field excursions to collect samples; years processing and analyzing the sediments for various chemical, physical, or biotic materials in centimeter detail; and the interpretation, writing, and publishing of results.

academically inclined to lend a sympathetic ear to arcane studies. I recommend something I call the holiday test. Imagine arriving back home just in time for the Thanksgiving gathering of your extended family. Filled with the conflicting emotions that accompany such holiday events, you enter the living room of relatives and conversation and bump into Uncle Bob and Aunt Hilary, the building contractor and investment banker. A hearty handshake and hug are followed by the inevitable "So, what are you up to these days?" After you stumble through a few sentences laced with terms like "lake sediments," "postglacial," and "pollen analysis," Hilary is peering over your shoulder toward the shrimp cocktail and Bob interrupts to ask exactly who is funding this work and how it will lead to a real job. Based on my own flat-footed experience, I admonish the students to test their projects and pitches against real-world concerns. How will your work help the tax-paying public? What intriguing questions or societal challenges will your studies

Moshup's and Lily Ponds adjacent Squibnocket Pond on Red Gate Farm.

address? And, how can you convey the results in compelling ways to your relatives or other diverse crowds? If our work isn't relevant and captivating to people whose opinions we value, then perhaps it needs rethinking and we need to improve our storytelling. Ancient archives reveal tales that fascinated Henry Thoreau, but of what real use are they?

Today, our world faces substantial challenges. Human activity is transforming the environment and earth's surface at a breathtaking pace. We expect temperatures to rise, rain and snowfall patterns to shift, glaciers to melt, oceans to expand, coastal erosion to accelerate, and the frequency of extreme weather—drought, hurricanes, and more—to increase. The ultimate driver of this is alarming: humans are destroying the natural infrastructure of forests and other ecosystems that support all life and replacing these with human enterprises that consume vast quantities of natural resources and release pollutants that jeopardize our health and alter the environment. How substantial are these changes relative to those of the past, and how has nature responded to these transformations? How might we conserve our landscapes to help them accommodate ongoing impacts and continue to support the diversity of life, including humans?

Modern science provides much relevant information to address such issues. But the time span of most studies is fleetingly brief—generally a few decades at most. The magnitude and rates of changes we anticipate for the near future lie well outside the realm of what has been experienced in even our longest historical records. To extend these observations, to evaluate combinations and magnitudes of changes in climate and other processes beyond historical experience, and to study how nature and people have coped with these in the past, we need to delve back further.

Using historical records, tree rings, and maps we can go back three to four centuries, which helps immensely. Henry Thoreau pursued such studies, as did Islanders Henry Whiting, Charles Banks, and Pete Ogden. Over that period, forces beyond recent experience have stressed nature and human infrastructure. Great hurricanes have struck, intense fires have swept the landscape, and important species have declined or even gone extinct. In the past few hundred years, human activity transformed the Vineyard through deforestation and reforestation. Since Gosnold's arrival in 1602, shorelines have retreated more than a thousand feet.

And yet even this record only nods toward the immensity of changes since the great ice sheet formed this coastal landscape. Pond and wetland sediments capture tremendous shifts in vegetation and climate over a period in which sea level has risen hundreds of feet and the vegetation has shifted from tundra-like to temperate. From these records come insights into the rate and ways in which forests and other ecosystems have absorbed many impacts. We can witness the environments that confronted early native people and visualize the land as the colonists arrived. We can also assess how much the landscape was transformed in the past few hundred years and how it continues to change. In these records we

Each Coring Site a Story Itself

To assess the long-term evolution of the coastal landscape, we explored sites across the region while focusing in greatest detail on the Vineyard. With its varied glacial deposits, the Island presents substantial challenges. The central Great Plain, lacking permanent ponds, remains the biggest frustration; its most detailed record comes from Duarte Pond south of Lagoon Pond, an intriguing area with a long history of Indian occupation, and adjacent "Airport Bog," a red maple swamp at Thimble Farm studied by Pete Ogden in the 1950s. Shorter records came from Lily and Jernegan Ponds on the Edgartown outskirts. In contrast, numerous sites provide long records on the western moraine with its moist soils, rich forests, and agricultural history. Our efforts to investigate parts of the Island with detailed native histories led us to explore a group of ponds on Red Gate Farm and Squibnocket Ridge as well as Ben Luce Pond adjacent to Skiff Avenue and the Lagoon in Vineyard Haven. Many sites in treeless landscapes on Squibnocket Ridge, behind the dunes near Squibnocket Pond, on Noman's Land, and on Chappaquiddick were explored to test the notion stated by Charles Banks and many others that these areas supported thriving forests until European times.

Each pond yielded a different story of challenge, surprise, frustration, or revelation. Inexplicably, Ice House Pond, an ice-block depression created by the glacier, has no sediment, an enigma not previously encountered in my four decades of coring. A few hundred yards away at Uncle Seth's Pond, we retrieved four meters of mud and a stellar 13,000-year-old record. The sediments at Harlock Pond and sites on Naushon contained only 5,000- to 7,000-year sequences, presumably because these basins lay dry until either increased precipitation or rising sea levels

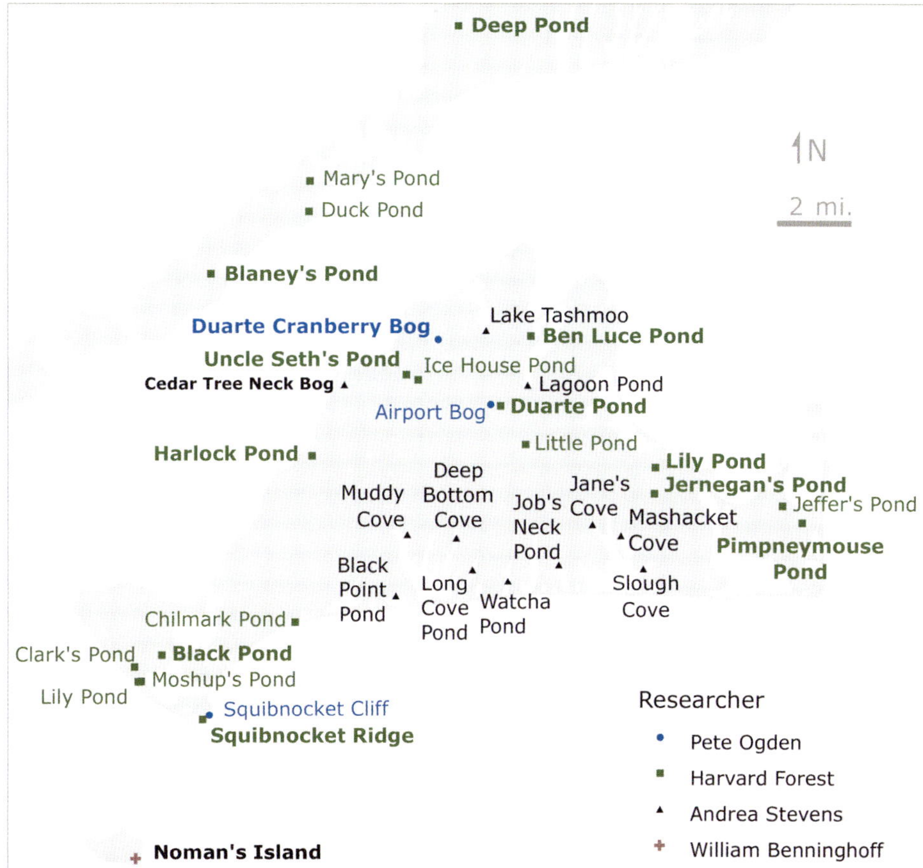

Over the past fifty years, paleoecological studies have been pursued by Pete Ogden, William Benninghoff, Andrea Stevens, and our Harvard Forest group. The quality of the records differs by site: intact ponds (longest and best record), swamps/bogs or other wetlands (variable in quality and duration), and the upper ends of coastal pond (shortest, most subject to disturbance, and challenging to interpret). Our work draws from the best long records, including those for Deep (Falmouth), Blaney's (Naushon), Black, Harlock, Seth's, and Duarte Ponds and Cedar Tree Neck Bog, but valuable short-term perspectives come from Jernegan and Lily Ponds, Squibnocket Ridge, Ben Luce Pond, and Noman's Island.

caused groundwater to fill them. Surprisingly, tiny, shallow basins like Moshup Pool (our informal name), a twenty-meter-wide pool behind Squibnocket Beach, yielded good but discontinuous sediments extending back 10,000 years. Meanwhile, Chilmark Pond was filled with sand from the overwash of past hurricanes, frustrating efforts to explore native history near Lucy Vincent Beach. Finally, glacial history limits ponds on the extremities of the Island. Though we've cored Jeffers Pond and a tiny wetland pool at Pimpneymouse Farm, we still seek better records from Chappaquiddick.

We augment our work with that at Cedar Tree Neck Bog by Andrea Stephens, a Nature Conservancy scientist who focused her Ph.D. work on the Vineyard, and Pete Ogden's study at Duarte Cranberry Bog near Lambert's Cove. Stephens's study also embraces the grasslands, shrublands, and maritime forest along the south coast and suggests that this landscape remained open in prehistory due to Indian fires. Yet the sites studied are the heads of coastal ponds and are dynamic with rising sea level, breachings, and hurricane impacts, and land use, and the resulting records are short, often disturbed, and challenging to interpret due to the high levels of grass and sedge pollen from surrounding marshes. We are reluctant to draw conclusions from these sites.

To place our Vineyard studies in broader perspective, we examine records from other coastal sites. These include Blaney's, Duck, and Mary's Ponds on Naushon; Deep and Fresh Ponds in Falmouth, which we explored with Bryan Shuman from the University of Wyoming; Peter Dunwiddie's sites from Nantucket and Block Island; and Bill Benninghoff's study from Noman's Island.

uncover a trove of natural experiments that far outstrip the lessons available on earth today. By examining this past, we advance basic science while deriving insights that may capture the interest of our relatives and many nonscientists because they can inform our conservation and management of the land. They may also alert us to the nature of future change.

One assumption that underlies many discussions of global change is that it will bring about conditions that are simple variations of ones experienced previously or elsewhere in the country. We've all seen the maps—the state of Massachusetts is superimposed on the eastern United States and shifts southward over time, experiencing first the environment of New Jersey, then Virginia, and, in a century, that of North Carolina. The assumption is that the environment will change and nature will simply adapt. Our climate will warm to be southern, and in response, plants, animals, and ecosystems will adjust progressively and predictably. Forest zones will shift neatly as, for example, loblolly pine from Virginia begins to replace pitch pine in New Jersey and then on Cape Cod. We expect reliable extrapolations of what we experience today. Intuitively, we do not expect major surprises. Yet, deep history teaches us otherwise. And, interestingly, recent events on the Vineyard give us a taste of what we might expect if the ancient past is any guide.

The scene comes from traveling up-Island on North Road late on a summer day in 2009. As you reach the top of the rise just a bit beyond the Southgate entrance at Seven Gates Farm, gaze off to the left in an easterly direction across the Woods Preserve with its fields, barn, and view of Fisher Pond. As the sun sinks behind you, a brightly vivid scene is illuminated on the distant ridge. Beyond the shadow of the pond and Mill Brook, the glowing horizon contrasts sharply with the deep green of the trees and grass below. As far as the eye can follow the ridge and beyond, for more than a mile from Priester's Pond to Waskosim's Rock, the starkly illuminated gray hulks of countless trees signal that the forest is largely dead. Nearly every tree is leafless and barkless. Reaching up forty to sixty feet, the naked branches form an open canopy of amputated upstretched arms. The oaks—white, black, and scarlet—are so dominant and so uniformly dead that the few beech, sassafras, and red maple that survive stand out, deeply green, unscathed, and thriving. Singly and

Dead oaks above surviving forests of beech and oaks at the Woods Preserve, viewed from North Road, August 2009.

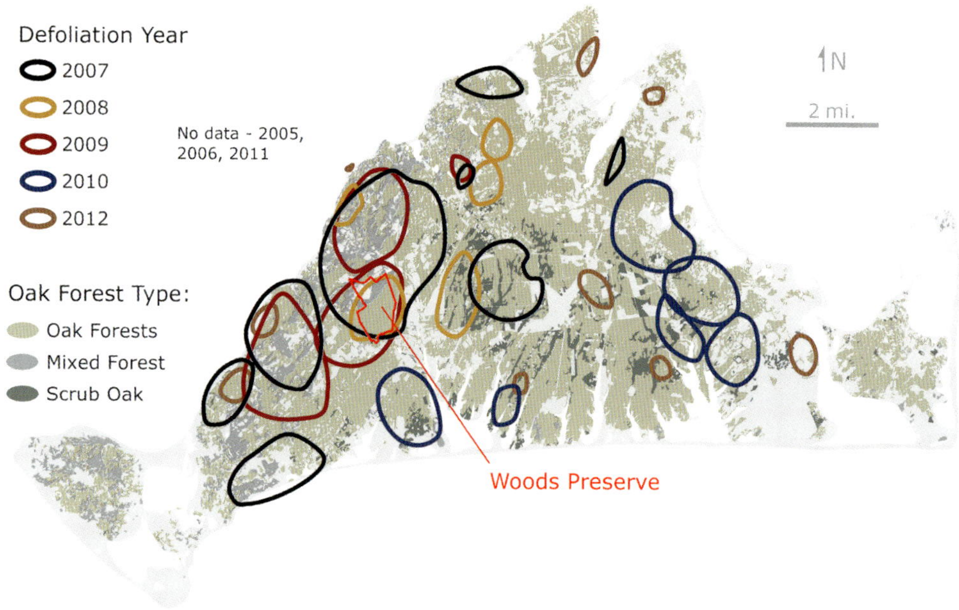

Oak mortality on the western moraine resulted from repeated caterpillar defoliation in 2005–12. The heaviest mortality on the Woods Preserve followed three years of defoliation (2007–09) and a lengthy rainless period. Data from Massachusetts Department of Conservation and Recreation.

in small clumps, these survivors are leafy beacons of growth in the luminescent ridge of gray.

The climate is warming, and oak is a southern tree predicted to thrive and multiply northward like a great wave on the heels of the fleeing sugar maple. Beech, the most abundant survivor on this Vineyard ridge, has a northern affinity and so might be expected to decline and shrink toward Canada with its northern compatriots. So what is happening? What has so dramatically tipped the balance the wrong way here, in one of the mildest climates in New England? The answer is partially weather, but mostly bugs.

For three years straight, the fall cankerworm (so not really a bug) defoliated the oaks. The normal activities of this native insect are generally benign, but it exploded in a series of outbreaks. Seldom seen in any number, the cankerworm suddenly went crazy. Millions of its bright white moths filled the evening skies, clouding Vineyard roads under illuminating headlights, flooding lit windows and doorways, disrupting evening ballgames and chores, and crawling into bark crevices and under shingles, where the eggs they laid yielded millions of caterpillars the following spring that devoured the leaves of oaks. Other defoliating caterpillars, such as winter moth and gypsy moth, joined the feeding frenzy. The oaks recovered, but the following year a similar outbreak occurred. Then the caterpillars appeared a third summer. That year, as the oaks struggled to cope, a drought struck. In previous years scattered oaks had succumbed in a salt-and-pepper pattern that was scarcely noticeable save in one's own backyard. But in this third year, eight rainless weeks tipped the balance. Entire groves of majestic trees died. The center of this seemingly cataclysmic event was Middle Ridge above Mill Brook, stretching from the Polly Hill Arboretum and Woods Preserve to Panhandle Road and up and over to the Land Bank property at Waskosim's Rock and parts of Seven Gates Farm. A solid chunk of the largest block of conserved forest on the moraine died that summer.

Freed from competition, the surviving trees and shrubs relish

Pete Ogden, 1928–1996

Half a century ago and historically situated between this volume and the late nineteenth-century studies of Nathaniel Shaler and Henry Whiting, James Gordon Ogden III made major strides in characterizing the history, ecology, and dynamics of the Island. A lifelong Vineyarder, Pete headed to Yale for a Ph.D. in ecology that benefited from association with the best minds in the field—Edward Deevey, Paul Sears, and Harold Lutz. His thesis centered on vegetation history in relationship to climate change and rising sea levels and revealed great details on the transitions from tundra and boreal forest to our temperate forest today. Although Ogden retrieved only a few meters of sediment from Airport Bog and Duarte's Cranberry Bog, these records remain among New England's oldest. And, like many great scientists with a large worldview, he sought to use the specialized tools of pollen analysis to answer big questions concerning the relative role of climate, people, and natural disturbance processes in shaping our world.

Pete's summer trips back home from Yale, and faculty positions at Ohio Wesleyan and Dalhousie University, allowed him to develop comprehensive studies on the modern landscape and contribute greatly to local conservation efforts. Lamenting the "bedraggled" state of the Island's woodlands, he sought out the best remaining forests for study—stands at Seven Gates, at the juncture of Middle Road and Tea Lane, and in Priester's Woods off North Road. In the process he located many old trees, including two-hundred-year-old red maples at Airport Bog and stumps of black and white oak adjacent Priester's Pond that he aged to 150 to 250 years and 285 to 330 years, respectively.

It is Ogden's efforts to interpret coastal dynamics that are cited most frequently today. In the 1970s, he penned a classic paper on coastal erosion that drew from Henry Whiting's work and his own observations on the barrier beach connecting Chappaquiddick to the Vineyard. His most poignant story came from that work at Norton Point during Hurricane Edna in August 1954. Seeking to understand breach formation, Ogden got more than he bargained for when the barrier beach began to collapse beneath his feet. "With deep fright," he scrambled back to solid ground to observe the tumultuous draining of Katama Bay. With the instincts of a great scientist and writer, he turned that near-death experience into an anecdote in his journal article that is regularly cited in stories about coastal erosion today.

their newfound glory. Fortunately, despite all appearances and the initial consternation of many onlookers, the forest ecosystem is perfectly intact. Rather than "destroyed," the expansive oak woods have transformed into an entirely new and quite different forest. In many ways it remains majestic. Flowers and fruits crowd the branches of shrubs—huckleberries, catbrier, blueberries, viburnum, blackberries, and poison ivy—as they soak up nutrients liberated from the dying trees and revel in a flood of sunlight after years of struggling in dense shade. The beech, maple, sassafras, hickory, and beetlebung sport lush leaves and new branches. Every appearance signals that these species are capitalizing on the loss of the oaks and expanding their previously limited role. Oak seedlings and small saplings also abound; in fact, small oaks are so abundant that they will surely dominate in future decades. But many other species, especially beech, are ratcheting up their abundance, muscling a larger share of the canopy, and using the opportunity to infiltrate new sites.

The widespread oak mortality was abrupt and startling. It bears no resemblance to the incremental change that we expect in our forests. Much of the ridge transformed from a tall, dark forest to an open shrubland or impenetrable tangle of catbrier, raspberries, blackberry, and huckleberry in a single season. In focused locations, the woods converted from oak to beech. Elsewhere, on adjoining parts of Seven Gates Farm, and from Aquinnah to the Lagoon, scattered clumps of trees encompassing 5 to 50 percent of the canopy died.

If oak suffers additional waves of mortality, beech will ratchet up its abundance a notch or two further. What a bizarre twist: a species dominant in northern New England replacing oak, the quintessential tree of the Cape and Island and benighted champion of New England's global change future. A world with so much beech would be unfamiliar. With its dense foliage and unrivaled capacity to intercept light and generate shade, beech would transform the Vineyard. Duck into the groves of smooth, gray-barked trees beyond the Polly Hill visitor center, or along trails on Cedar Tree Neck or the Presbury Norton Reserve, to witness beech's effect. The forest is strangely still and locally empty. The thick canopy of paper-thin leaves obscures the view above, while dried leaves, bare soil, and shadows lie below. Beech controls the environment in a manner unrivaled by any coastal species. On Naushon Island, where more than a thousand acres of ancient beeches blanket the interior, the consequences of a complete shift from oak to beech are clear. From towering groves in deep hollows to short, dense thickets on the hilltops and guarding woodland margins, the forests are dark and quiet. Gone is the waist-high layer of huckleberry and blueberry of the oak woods. Beech forms a magical yet depauperate scene that harkens images of Sherwood Forest.

Such abrupt shifts in ecological condition—from woods to shrubland or oak to beech—are what ecologists call nonlinearities. They are rare yet important. Take the mortality of these Vineyard forests, for example. For the squirrels, mice, and deer that thrive on acorns, the loss of oak certainly seems locally catastrophic, forcing many to relocate. But just the next summer the "catastrophe" brought an unaccustomed bonanza of flowers that insects descend on, followed by a feast of fruits that attract birds and people. The thriving new growth now supports a larger number of deer, yet squirrels may be uncommon for decades. The new ecosystem is completely different.

These shifts may arise under a number of different scenarios:

Duarte Pond on the north edge of the Great Plain fluctuates in water depth with changes in precipitation, October 2014. Beyond the red-leafed beetlebung trees lies Airport Bog on Thimble Farm, one of Pete Ogden's primary sites.

when the environment passes some critical threshold, the climate changes abruptly, a sudden disturbance occurs, or two or more individually modest stresses combine to yield a novel impact. In the case of the Vineyard oaks, the combination of caterpillars and drought provided the trigger. It also may be that some unusual climatic conditions allowed the fall cankerworm and other insects to become uncharacteristically abundant and create major infestations in the first place.

But does the collapse of one oak forest provide much of a lesson? Is the scene across the ridge in West Tisbury and Chilmark an aberrant incident to be ignored, or a glimpse of what greater environment change may bring? Is the specter of oak forests becoming beech possible, or a one-off event? Deep in the mud of many coastal ponds lies a resounding answer.

Five thousand years ago, after reigning for thousands of years as one of the preeminent trees in the Northeast, coastal oak populations crashed abruptly. Within decades, across a region from the base of Cape Cod through the Vineyard and Nantucket, oak plummeted to half its former abundance or less. In its place came beech. Other species also benefited—hickory, beetlebung, and maple on many Island sites, and pitch pine and white pine on the Cape. But beech was the big winner. For nearly two thousand years, this long-lived shade-tolerant tree rivaled oak on the moraines, while oak and pine continued to dominate the dry, sandy outwash.

Early paleoecologists missed recognizing this ancient die-off of oaks, which was only identified about ten years ago when Wyatt Oswald, Bryan Shuman, and I were struck by the magnitude and abruptness of the shift at Deep Pond in Falmouth and then recognized its broader occurrence on the islands. Though questions remain, we have considerable clues as to its cause. Most significantly for us today, this abrupt shift coincided with a period of rapidly changing temperature and drought. We also know that another dominant species elsewhere—hemlock—declined simultaneously across the northeast. When the abundances of hemlock in Vermont and oak on the Vineyard are aligned, they match perfectly.

Drought, extreme weather, climate change, and insect outbreaks—these are familiar themes of our day and the recent oak dieback. They may well become regular occurrences in the future. What we learn from examining ancient records emerging from the muds and muck of the Vineyard is that plants and animals cannot always adjust readily to rapid changes in the environment. While the vegetation seamlessly accommodated many major but gradual shifts

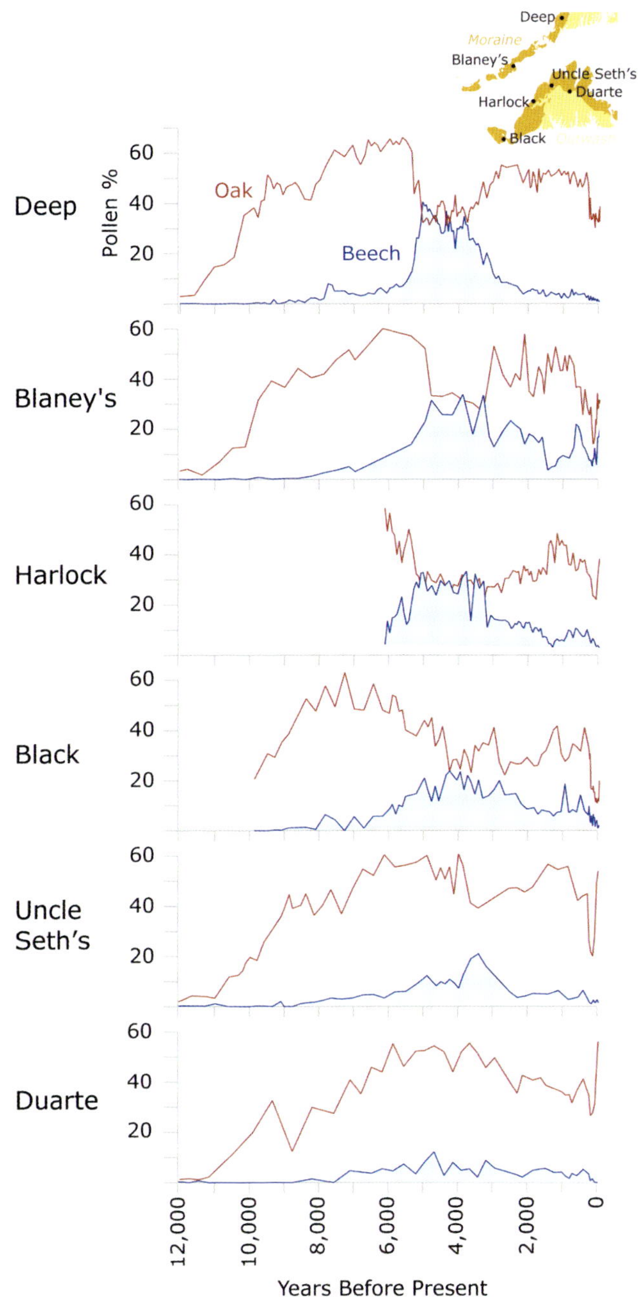

While beech is increasing today, it once played a much larger role on the coast, as shown by pollen diagrams from the Cape, Naushon Island, and the Vineyard's western moraine and Great Plain. Arriving eight thousand years ago, beech increased slowly and then expanded abruptly with climate change and a precipitous decline in oak. The oak decline is most pronounced on the moraine, where beech came to dominate. Near the moraine-outwash border (Seth's Pond) oak declined less, whereas on the Great Plain (Duarte Pond) it remained dominant.

with climate change since the ice retreated, it also experienced a few astonishingly abrupt transitions.

The take-home message for us and the Thanksgiving gathering is that shifting environments do not necessarily initiate predictable responses. Climate change can generate complete surprises, including dramatic shifts in one species or the transformation of entire ecological systems into unfamiliar territory. As witnessed on the Vineyard five thousand years ago, these surprises can occur because the environment assumes a novel set of conditions or because extreme stresses weaken or kill trees. They can also occur because the environment triggers other unexpected ecological responses, such as a geometric increase in an apparently innocuous native insect.

Just as Vineyard forests were recovering from the fall cankerworm outbreaks of 2005–12, more oaks began to die in patches across the Island. The new culprit was another seemingly benign native insect—the cynipid wasp. An insect so inconspicuous that it has received little study, the wasp is known to have caused major problems only once before, on Long Island in the mid-1980s. The target of this insect was not all oaks—just black and scarlet oak. As these abundant species suffered or died, beech, hickory, and sassafras benefited, joined by white oak and the less common post oak. Island forests will persist and recover from the wasp and future episodes, but with each unexpected shock, they experience meaningful changes.

As the climate continues its rapid change, we should expect other native insects and diseases to operate differently and new ones to appear on our shores. Already, as the Island is experiencing more winter moths, it is bracing to deal with the recent arrival of the southern pine beetle, a native that is spreading northward under warming conditions and may greatly reduce populations of pitch pines. We should also expect new combinations of temperature and precipitation coupled with sea level rise, variable weather, and human activity to challenge our ecosystems in novel ways with unexpected consequences. While it may come to pass that sugar maples will migrate neatly toward Canada and oak and pitch pine will increase on the Vineyard as predicted, we may be surprised. As a local forester from Chilmark remarked to me in a discussion at the Agricultural Society one evening, “After all these years worrying about fires and deer damaging the trees, some insect that no one ever heard of rose up and killed off the entire forest.”

The core from Jernegan Pond in Edgartown displays highly organic brown mud transitioning to light-colored silts and sand.

The Larger Context of Postglacial Change

It is difficult to comprehend the enormity of change that has taken place across New England since ice loosened its control on the land. Along the coast this task is doubly challenging because it involved so much more than just the climate. The entire landscape was reconfigured. As the seas rose, the balance between land and water shifted in major ways, the groundwater table rose within the Island, and the ocean and its currents—warm Gulf Stream waters from the south and frigid ones associated with the Labrador Current from the north—played an increasing role in modulating the local environment.

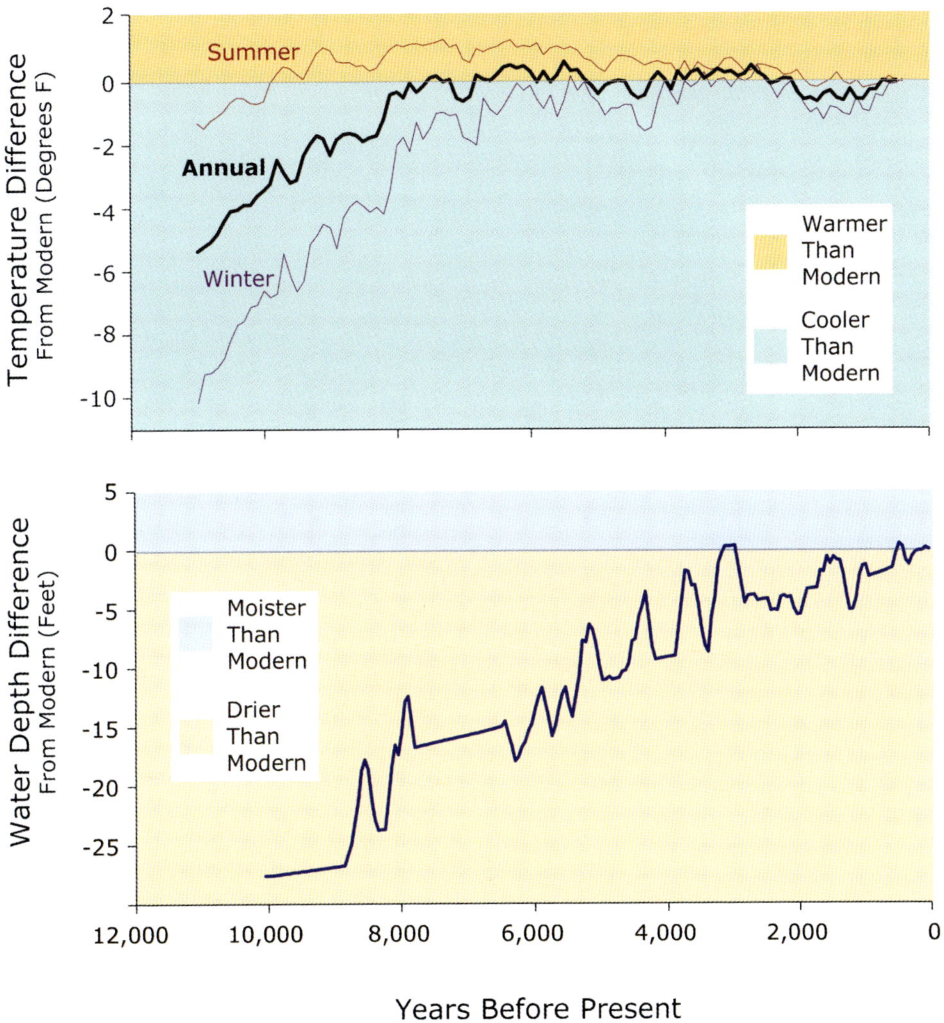

Since the retreat of the glacier, temperatures rose rapidly to about eight thousand years ago, then varied until two thousand years ago, when they declined and then rose to the present temperatures (*top*). Until the past few thousand years, seasonal differences have been greater than today. The depth of ponds (and water available to plants and groundwater) has increased progressively, interrupted by notable fluctuations (*bottom*). The past thousand years represent the wettest period since deglaciation. At no point was the climate (combination of temperature and moisture) similar to conditions today or anticipated in the future. Data from Bryan Shuman, University of Wyoming.

For millennia, as the ice withdrew from New England, the Vineyard was an unremarkable spot on a vast tundra steppe that stretched across the coastal plain dozens of miles to the east and hundreds of miles south and north. The expanse was flat, dominated by low sedges, grasses, moss, and shrubs and interrupted by low ridges of moraine, stream valleys that meandered to the sea, and broad plains. In the enormity of this expanse, the distinctive place we call the Vineyard would have been unrecognizable. The characteristic triangle that graces bumper stickers was lost in an expansive landscape akin to the North Slope of Alaska, a land where space is great, relief is slight, and the views are far-reaching.

Early on, the climate was harsh, with the ocean distant and

dominated by polar currents and winter ice. Strong winds descended from the giant ice sheet and mingled with arctic air masses to reinforce the long, bitter winters and dry conditions. Lacking trees or substantial relief, the land offered little shelter from the constant wind. Water lay deep in the ground, precipitation was slight, and most depressions that hold ponds, familiar streams, and wetlands today were dry.

But the rapid warming that drove the ice northward brought changes. The growing season lengthened, rain and snowfall gradually increased, and treeless tundra gave way to a boreal landscape of spruce, larch, and paper birch that extended with modest variation across the center of the continent. Still dominated by arctic air, the climate remained much drier than today; as it warmed, more temperate but drought-resistant pines prevailed. First came jack and red pine, and as more moisture accompanied the warming climate, they were joined and replaced by white pine, and then the first oaks. An approximation of this landscape can be found today in the Lake Country stretching from northern Minnesota to Michigan. There a dry continental climate of warm summers and long, frigid winters supports a landscape mosaic of conifers with

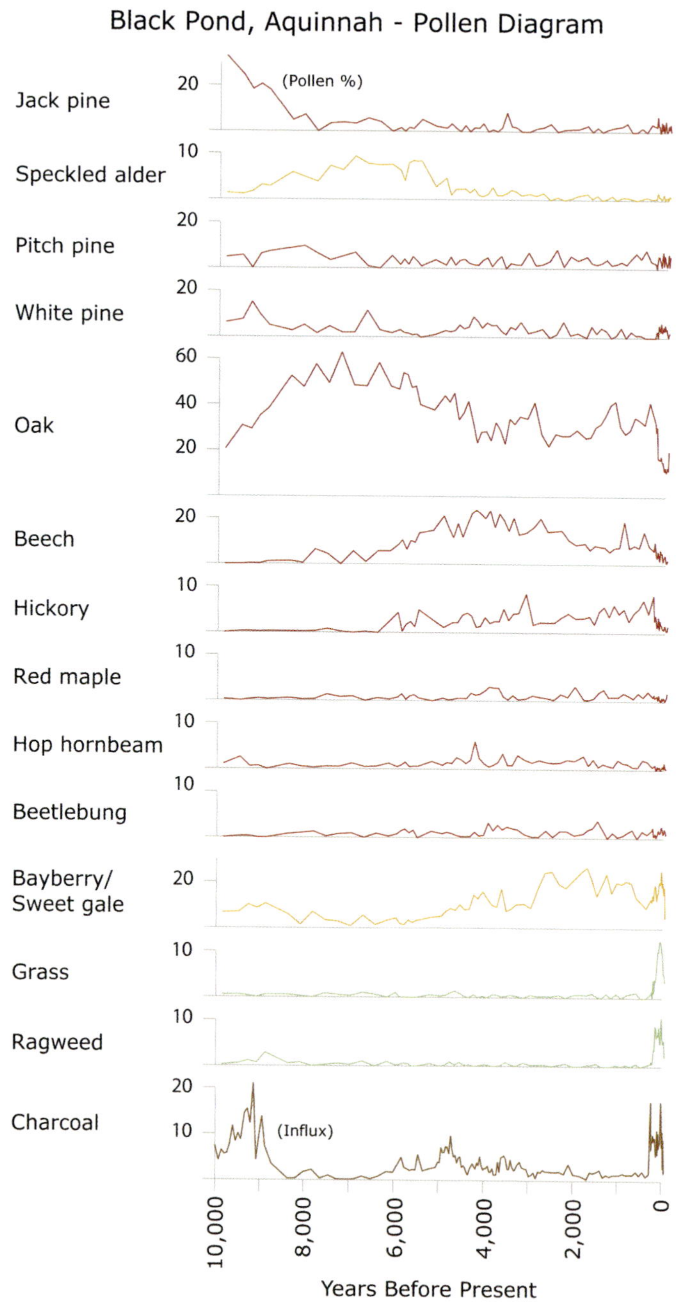

The landscape around Black Pond in Aquinnah was open farmland in the nineteenth century and now supports a mixture of fields, shrubs, forests, and swamps. The long-term record indicates that the area was heavily forested for the past ten thousand years. Following boreal forests of spruce (not shown) and then jack pine, the area was dominated by oaks for most of the past nine thousand years. Beech arrived around eight thousand years ago and sustained a three-thousand-year peak when it formed a diverse forest with oaks, while maple and beetlebung occupied moist sites. *Myrica* shrubs include sweetgale immediately around Black Pond, and bayberry on the surrounding uplands. A diverse forest continued until European deforestation, marked by the sharp decline in tree pollen and increases in grass and ragweed.

Given that Black Pond is situated in an area of intensive native occupation and use over many thousands of years this record confirms that native people did not clear large areas of forest or have major impacts on the native vegetation. Charcoal values do not parallel the abundance of people on the landscape, but are greatest during the early pine period and oak-to-beech transition and were low over the past few thousand years until colonists cleared the landscape. Fire and charcoal decline in the most recent decades.

birch and aspen that varies kaleidoscopically with lightning-set wildfires. On and around the Vineyard we can see this vegetation clearly in cores. At Black Pond, we saw a major peak in fire activity as the pines gave way to the oaks.

Temperatures continued to rise and then peaked between eight and six thousand years ago, when it was a degree or so warmer than today. For thousands of years leading up to and embracing this period of maximum warmth, oaks increasingly dominated the shrinking coastal plain. There was a bit of variation—beech was present at many sites, hemlock reached Naushon and the base of the Cape, and pitch pine was a substantial part of the forest on the plain around Duarte Pond—but oaks composed the solid majority under the warm and gradually moister conditions. All the while, ocean waters encroached more closely on the coastal region. Across our Island sites, Duarte Pond positioned at the northern edge of the dry Great Plain to Black Pond on the moister morainal landscape at Aquinnah, we witness a range of variation in vegetation and fire.

Through the years of oak dominance under warm, dry conditions that might reasonably appear to be fire-prone, the evidence for fire is surprising and variable. At Duarte Pond, charcoal, a good indicator of burning in the surrounding landscape, paralleled the increase in oak and stayed at high levels for six thousand years. At Black Pond the lowest levels of charcoal coincide with the warmest two thousand years. Elsewhere the evidence for fire registered between these extremes. This variation is fascinating, both for the picture of heterogeneity it provides for the Vineyard landscape and for what it says about the relationship between oak and fire. While many ecologists and foresters assume that oak is a pyrogenic species that requires fire to persist at great abundance, the reality is more complex. Clearly, oak can accommodate fire, as it did on the Plain, but records from the other sites confirm that it can thrive without it.

Through the end of this period around six thousand years ago, many changes in the physical landscape and environment triggered the oak-beech dynamics that we discussed previously. Broadly, the great ice sheet had lost its hold on the continent and climate as it

Wyatt Oswald and Lindsay Day probe the sediments at Pimpneymouse Pond, Chappaquiddick.

disintegrated in western Labrador. As the ice shrank and cold arctic currents declined in magnitude, the Gulf Stream pushed farther north and began to bathe the coast with warmth. The disappearance of the great ice sheet encouraged a northward shift of the jet stream, which in turn allowed the incursion of moist southern air masses. The increasing moisture and milder winters yielded the moderate temperate conditions that we know today and favored the arrival and proliferation of many of our common trees.

Locally, the rising seas and increasing moisture began to shape the Vineyard landscape into a semblance of what we recognize around us. With both forces driving up groundwater, existing ponds filled further and new wetlands, ponds, and streams began appearing. A number of our pollen records began during this period, when the sites first became ponds accumulating sediment. Below this level the mud terminates in sands and gravels deposited by the glacier.

With the ocean rising, minor differences in elevation determined the fate of the land and configuration of the coast. The pattern of inundation looms as a critical factor controlling the extent to which plants and animals were able to access the land in their journey northward after deglaciation. This migration depended on the timing of separation of the islands, the size of the remaining landmasses, and their distance from the mainland. Block Island was the first of the southern New England islands to separate, some ten thousand years ago, and its flora and fauna lacked most species arriving subsequently. Between eight and six thousand years ago the other islands—Nantucket, the Vineyard, and the Elizabeths—separated from the mainland progressively, and the broad configuration of the coast became well defined. As the sea encroached on all sides, the mass of water surrounding the Island increasingly moderated summer and winter temperatures, bringing more frequent fogs, delaying the full onset of spring, and extending the frost-free fall. The coast became increasingly exposed to wind, salt spray, and extreme weather. A climate reigned that would be familiar to us, and we would recognize the species that began to take control.

Right up to the time of European settlement, we see substantial amounts of beech, hickory, red maple, hop hornbeam, beetlebung,

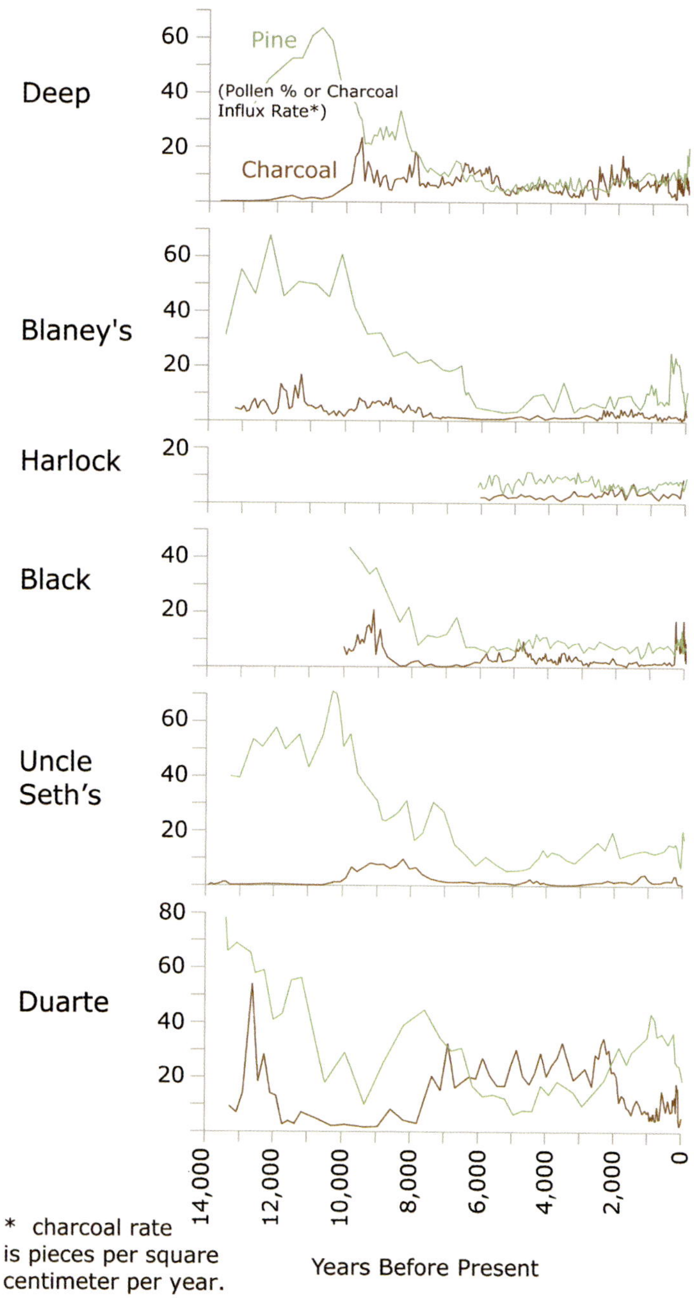

Long-standing interpretations of the coastal region have suggested that fire played an important role in prehistory and was responsible for maintaining the abundance of oaks and pitch pine versus other species. However, the paleoecological records of pollen and charcoal from the Vineyard and surrounding sites provide a contrasting, more complicated picture. Fire activity, represented by charcoal deposited in the bottom of the ponds, is low on the western moraine for the past eight thousand years and shows little relationship to the abundance of pitch pine. Around Duarte Pond on the Great Plain, charcoal is two to three times greater than at other sites and generally parallels the abundance of oak. Notably, however, charcoal falls sharply and stays low during the more recent time of pitch pine abundance, when the Native American populations also increased. At all sites, charcoal rises with European settlement and the clearing of forests for agriculture.

and pine variously in the landscape along with oak. Locally, we see variations of this diverse forest across the western part of the Vineyard—on Noman's Land, at Gay Head around Black Pond, up through Harlock Pond in Seven Gates Farm, and on to Cedar Tree Neck and Seth's Pond. Lesser amounts of these diverse forests occur toward the center of the Island around Duarte Pond and Pete Ogden's old site at Airport Bog on Thimble Farm, adjacent Jernegan and Lily Ponds in Edgartown, and over on Chappaquiddick at Pimpneymouse and Jeffers Ponds. Though oak dominated the outwash plain and eastern moraine, we see solid evidence for a mixture of broad-leaf trees across large areas, including the heads of the coastal ponds.

During the past few thousand years, somewhat cooler and moister conditions prevailed. This intriguing period, characterized by a highly variable pattern of changing vegetation on the Vineyard, is an important part of our story for two reasons. First, it embraces a period of flourishing Native American populations, characterized by the greatest density of human sites. It also ends in the landscape transformations associated with European arrival and so establishes the most recent baseline for the natural range of conditions and vegetation on the Island.

On the western moraine, the changes in vegetation through this period are subtle and broadly similar from Aquinnah north. Around Black Pond the landscape supported a diverse hardwood forest, fire was infrequent, and oak increased gradually over time as lake levels rose and temperatures cooled. Out on the northern plain, the changes were more striking. At Duarte Pond, oak persisted but pitch pine increased from low levels to become a substantial part of the forest in the last centuries before European arrival. During this period, in which a seemingly fire-prone type of conifer forest coincides with the most rapid growth in the native population, we see a sharp unanticipated decline in charcoal. Once again, the simple relationships that we expect between climate change, human activity, vegetation, and fire fail to materialize.

With this final shift of forest conditions in prehistory, the scene is nearly set for the arrival of Europeans who proved to be a force in vegetation change that rivaled all others outside the glacier. Exploring this scene at the height of the Late Woodland Indian period brings our pollen and historical records together.

Black Pond on Red Gate Farm. The open landscape was heavily forested until European settlement.

When Gosnold plied Vineyard waters in 1602, the voyage's chronicler John Brereton described the coastal landscape as covered with "high timbered oaks . . . Cedars, straight and tall; Beech, Elme, hollie, Walnut [hickory] trees in abundance . . . [and] Sassafras trees." Though the details of individual species may vary, the paleoecological evidence is unequivocal in its broad support for this account. As had been true for well more than ten thousand years, at the time of European arrival the Vineyard and entire coastal landscape were heavily forested.

Evidence suggests that the extensive forest groves may well have included many stands with tall timbers. Forest continuity over thousands of years had led to the growth of large and in many places ancient trees, providing shelter from the chronic winds that shape and deform trees on exposed sites today, and allowing expansive

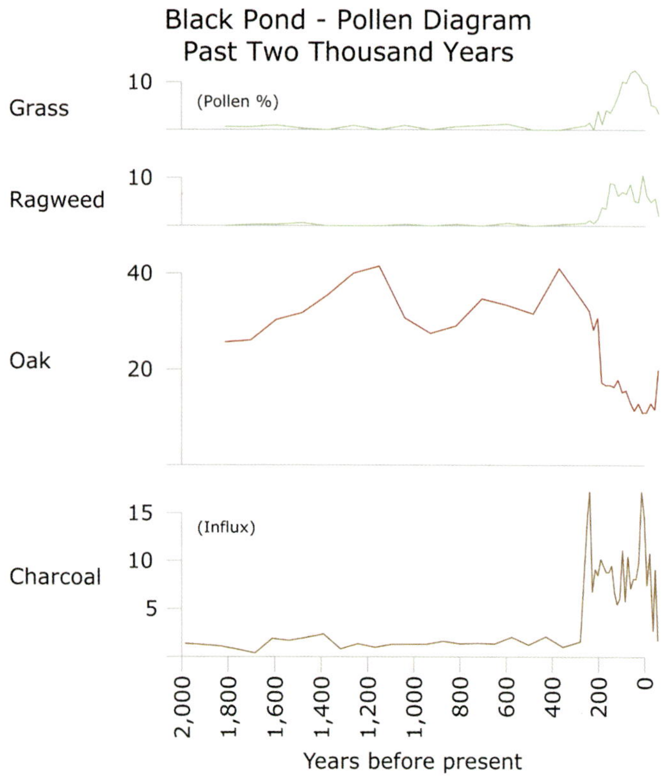

Despite the archaeological history of intensive occupation, the pollen diagram from Black Pond in Aquinnah indicates widespread forest with low levels of charcoal, grass, and weeds during prehistory. Following European settlement, oak and forest cover decline abruptly, and open-land grasses and herbs increase with land clearance and farming. Charcoal displays peaks corresponding to seventeenth-century forest clearance and late nineteenth-century farm abandonment. In recent times trees have increased, and ragweed, grass, and charcoal have declined.

groves to develop. These conditions encouraged a diverse group of long-lived species: oaks, pines, beech, and hickories broadly across the uplands; beetlebung and red maple on finer soils and in the lowlands; and red cedar on bluffs at the coast (Cedar Tree Neck, for example) and along barrier beaches like Cape Poge, where it persists today. Species that we often think of as reaching moderate size and age in our landscape, such as pitch pine, likely achieved surprising stature and longevity in those woods. We get a hint of these stands in the Island's oldest forests, places like West Chop Woods, where many pines are closing in on two hundred years in age, reach two feet in girth, and display the plate-like bark of ancient growth. Across the coastal region, old buildings with floor boards and planks of pitch pine eighteen or more inches wide, tight grained, and with few knots signal the presence of even more majestic trees in earlier days. Such big pine likely grew along with massive oaks in awe-inspiring stands of old-growth timber.

Of course there were disturbances that disrupted this continuity of growth—infrequent hurricanes in late summer, fierce nor'easters with horizontal winds in the winter, occasional fires, droughts, insects, and the myriad activities of a thriving native population. The powerful storms would have toppled many trees, and even entire stands. Smaller disturbances would have punched holes in the forest canopy, allowing streams of light to flood the ground and tree seedlings to establish among the downed boles and upturned root mounds of fallen trees. Each event would maintain the shifting forest mosaic.

The pollen and historical records show that the landscape in A.D. 1600 differed in significant ways that would nonetheless resonate with us today. The hills of the western moraine, with their finer soils, deposits of clay, and varied relief, supported a rich and often moist range of forests. In many spots there would have been sharp breaks from one forest type to the other. Beetlebung stands perched on a clay-rich hillside contrasted with adjoining tracts of oak and hickory on surrounding sands, and loams. The most majestic forests likely occurred in the protected valleys and coves on rich soils along Mill Brook, Tiasquam River, Witch Brook, and broadly through the valleys and varied terrain off Middle and North Roads.

The outwash plain and eastern moraine presented a contrasting environment and vegetation. With deep sandy soils, the land was considerably drier and more hostile to the growth of moisture-demanding herbs and trees. Here oaks, pitch pine, and huckleberry

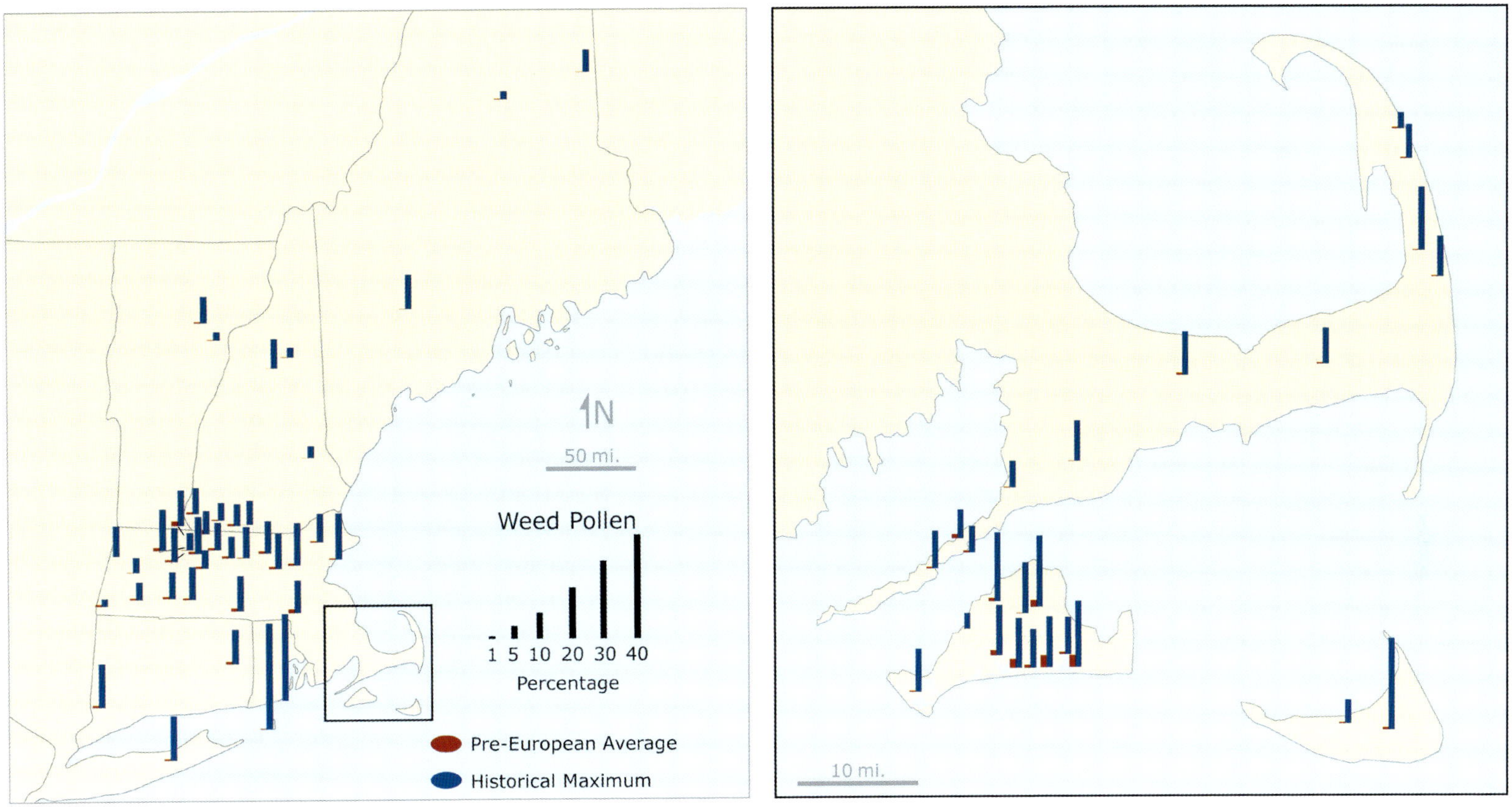

Comparison of weed abundance (ragweed, plantain, sorrel/dock, knotweed, and other herbs) in the centuries immediately before European settlement and the peak amount afterward indicates that New England was heavily forested until colonists initiated widespread deforestation and farming.

(and scrub oak on the droughtier sites) would have flourished, as seen around Duarte Pond. But that site at the north edge of the Great Plain also hints at landscape variation. Moister forests with beech, hickory, and maple occurred where the flat expanse intercepted wetlands like Airport Bog or steeper terrain such as the head of Lagoon Pond and Lake Tashmoo or the margins of the moraine. Around Seth's Pond, for example, a surprising amount of hickory was present.

The center of the Great Plain will likely always remain a hole in our story. Without water, peats, mucks, or mud this twenty- to thirty-thousand-acre expanse remains a void to be filled by whatever narrative the particular storyteller's intuition and biases dictate. In the version that our data help craft, drawn from sites like Jernegan, Lily, and Duarte Ponds, the plain is solidly forested with oaks and scattered pitch pine over huckleberry and scraggly scrub oak. This moderate-stature forest was visited by occasional surface fire, drought, and wind but remained intact for millennia.

Around the Island's margin, the forest cover was broken. The

Burned white oak forest in the Wapatequa Woods, Tisbury, June 2012.

long arms of great coastal ponds, marshy wetlands, and narrow streams supported aquatic, wetland, and sun-loving species. Actively eroding bluffs, the inland margins of barrier beaches, and stretches of expansive dunes provided habitats for successional species, hardy pioneers, low shrubs, grasses, and other plants that thrive with adversity and opportunity. Here and on sites where humans returned seasonally and concentrated their activity over the centuries lay many of the plants of open and chronically disturbed sites that were poised to expand when the forests were removed.

And What About Native People?

As we link the paleoecological and historical records to interpret the forest landscape before it was transformed by waves of colonists, we need to address one last and central issue: the evidence for native activity. Thousands of people occupied New England's coast for thousands of years. Perhaps three thousand resided on the Vineyard. What role did they play in the vegetation dynamics we reconstruct? The answer is surprising but consistent across the coast and all of New England. When we scour the three dozen or so coastal records and more than one hundred pollen diagrams across the region, the hands of native people are completely invisible. We detect evidence of changes in lake levels with drought, charcoal from fires, erosion from hurricanes, and shifts in species with climate and European activity, but see no evidence of early human impacts. The varied activities of native people blended seamlessly into the larger picture of natural change.

Although many modern interpretations promote the image of a landscape tamed by native hands, of forests closely managed with fire, and of sizable corn fields and villages, the independent evidence from the bottom of ponds and swamps supports accounts like Brereton's of a wild and heavily forested land. To understand the reasons for this and to explore the subtle role of people over thousands of years, we will need to turn to additional sources of information.

People on the Coast

A knowledge of local archaeology and history should be part of the ecologist's equipment.
— Gordon Day, 1953

The truth is, that [Martha's Vineyard's] harbours, coves, lagunes, and ponds afford an inexhaustible supply of food. . . . To an Indian they were the most fertile parts of America.
— Reverend James Freeman, 1807

A while back I stopped for a hike on the David H. Smith area on the Pohogonot tract, whose 350 acres extend the Correllus State Forest south toward the coast in a superlative expanse of sandplain vegetation. As I walked the area, I recalled my first visit with state and Nature Conservancy staff and how I was struck by its distinct features—flat with areas of large oaks, some scattered pines, and broad stretches with impassably thick tangles of scrub oak. Now, five years later, the landscape was transformed and, most notably, scorched. The dense tangles were gone, and vigorous young shoots of scrub oak and white oak sprang from the bases of gnarly blackened stems. At the trailhead, I opened a waterproof box atop a wooden post and extracted a few discolored laminated sheets, one of which caught my eye. A photo displayed tall flames and a crew in yellow hard hats. The text explained that the fire on this tract had been set for conservation: "Part of the great sandplain ecosystem on Martha's Vineyard, this area has been burned regularly and purposefully by humans for thousands of years. . . . We are burning this landscape again to maintain critical processes, create plant and animal habitat, and rejuvenate this globally rare ecosystem."

On conservation lands throughout New England, similar activities are being advanced with the same rationale. Across the Vineyard and region, our most rapidly declining habitat is open land—grasslands, heathlands, shrublands, and young woods—which is being overgrown by a dense growth of shrubs and trees. As these areas reforest, the plants and animals shift and the diversity of the broader landscape declines. Lost are distinctive birds—woodcock, meadowlark, bobolink, upland sandpiper, whippoorwill, and many unusual sparrows. These losses are accompanied by huge declines in the herbs, insects, and small mammals such as the New England cottontail on open habitats conserved at great expense to protect them.

To reverse this trajectory, conservationists have turned from land protectors to land managers—stewards fighting the relentless encroachment of trees and restoring open vistas that once made these landscapes distinctive. Such efforts abound across the

One perspective: prehistory shaped by humans; an extreme interpretation of native impacts on the landscape. This painting depicts pre-European New England (A.D. 1000–1600) with large grasslands, scattered woodlands, and cornfields actively managed by native people with fire. From Richard M. DeGraaf and Ronald I. Miller, eds., *Conservation of Faunal Diversity in Forested Landscapes* (Berlin: Springer, 1996), p. 236. Copyright 1996 Chapman & Hall. Reprinted with permission of Springer Science + Business Media.

Vineyard. On the south shore at Long Point, The Trustees of Reservations has applied fire and chainsaws to create savannas and maintain shrubby and grassy expanses that sweep to the sea. At Wasque on Chappaquiddick, the Trustees regularly check the spread of pitch pine through exhaustive mowing and occasional burning. South of Edgartown, The Nature Conservancy (TNC) applies fire to Katama Airfield and adjoining tracts, which collectively form the "largest native sandplain grassland in New England." Elsewhere, the Massachusetts Audubon Society has burned its grasslands at Felix Neck Preserve on Sengekontacket Pond, as has Sheriff's Meadow Foundation down at Quansoo Farm.

Maintaining distinctive ecological habitats is a natural role for conservation organizations, but one might logically ask, "why so much fire?" At Long Point we find a detailed answer. The sign reads, "For thousands of years, practices of the Native Americans on Martha's Vineyard, known as the Wampanoag, had profound effects on species composition and landscape structure. The Wampanoag created drastic changes on their surroundings predominantly through large-scale burning, which assisted in hunting game, maintaining berry patches, and creating fertile soils for cultivation. . . . These fires had a massive influence on the landscape. . . . A substantial portion of Martha's Vineyard was open."

As an ecologist and amateur historian who pursued his Ph.D. thesis studying wildfires in the wilderness of Labrador, I have long been intrigued by these interpretations and the management of landscapes I hike with such a primitive tool. Across the Vineyard and coast, fire appears to link the history of an ancient race and modern conservation.

Another perspective: prehistory shaped by nature. In Albert Bierstadt's (ca. 1858) rendition of Gosnold's encampment on Cuttyhunk, the landscape appears wild and fully forested by ancient trees, supporting diverse wildlife, and exhibiting no obvious native influence. Courtesy of the New Bedford Whaling Museum.

But what do we know of native history in New England? Two iconic but distinct sets of images fill textbooks and often enter our minds as we seek to characterize these earliest Americans. Though these images are often found together, they represent quite contrasting interpretations of Native American livelihoods and the ecological processes that shaped the pre-European landscape.

The first set of images, admittedly an old-fashioned view that populated the textbooks of my youth as well as my grandfather's, captures the event scholars call "contact," the first encounter between civilizations from the New and Old Worlds. In a typical scene, Indians attired in loose skins and adorned with shells and a few feathers stand at the ocean shore exchanging greetings with bearded Englishmen in formal military attire. Other images from this same narrative depict Americans and Europeans straining anxiously and warily for their first glimpse of each other. The natives stand atop a craggy promontory peering seaward through dense trees at a majestic ship with sails and colors unfurled, while others are paddling toward the ship. For their part, the sailors gaze from the deck or yardarms with hand-shielded eyes, fixed intently on the coast, the approaching natives, and thin columns of smoke emerging above the treetops from cooking fires.

In these images of "discovery of the New World," the emphasis is on the event—the first meeting between different cultures—and the contrast between the two civilizations in customs, dress, and technology. The Indians, and landscape that frames them, are wild. They are portrayed as greased, dyed, leathered, and occasionally tattooed. Their savage mien is reinforced by the untamed land of menacing woods, gnarled trees, craggy cliffs, and unbroken coast. The scenes capture a native race at one with wild nature. In sharp contrast, the English (or Dutch or French, depending on the location), with their armor and uniforms, appear uncomfortably formal in this wild land. Clashing metaphorically in their civilized attire with these native people, the newcomers stand poised to unleash the energy of weapons, modern tools, and farm animals to clear and ultimately domesticate this unbroken landscape.

A second, quite distinct image depicts a most iconic American scene—the first Thanksgiving. Here benign Indian families join Pilgrim men and women in a common feast. The outdoor setting is strangely domestic with its tables and platters, and the food is distinctly American: wild meats, corn, squash, and pumpkin. Implicit in the bounty and spirit of sharing is the sense of thanks, for survival, for Indian aid through harsh winters, and for food from native and Colonial gardens. Not far from this illustration in many an old schoolbook is the tale of Squanto teaching the Pilgrims the uniquely American approach to agriculture. In fields tilled with wooden and stone tools, a triad of crops—corn, beans, and squash, known as the "three sisters" in Indian lore—is grown on low earthen mounds fertilized with fish. In these images lies an enduring interpretation of early American life that contrasts with that of the savage and first encounter. These are agricultural Indians, living in open villages

Dianna Doucette and other archaeologists associated with the study by Harvard University and the University of Massachusetts working on the bluff above Lucy Vincent Beach in 1999. Courtesy of Dianna Doucette.

surrounded by productive fields of crops. This is the story of a well-populated, comfortable, and humanized landscape filled with family groups who are actively shaping their land. This second narrative has in recent decades given rise to the view of a strikingly open landscape of cultivated gardens, grassy fields, and breezy savannas.

While these two views of the first Americans are often linked without a second thought, they actually represent sharply divergent interpretations of native people and of nature itself. Both cannot be correct, or at least both cannot represent the same time or place. The scene of wild natives and equally wild landscapes has recently fallen out of vogue. It is a story of an aboriginal people who lived with the land and who hunted, fished, and collected its bounty while slipping fairly benignly through it. Limited in numbers and technology, and reverential of the earth that supported them, these people exerted little impact on nature, for they were part of it, drawing from its bounty while enduring its grim challenges. This is the view espoused by writers and artists of the romantic era—Longfellow, Frank Church, and many Hudson River painters, including Albert Bierstadt, who captured the iconic view of Gosnold's arrival at Cuttyhunk. It is also the perspective of early American historians and a 1960s crowd who embraced this perspective as congruent with their own ecological inclinations.

In this view the world engulfing the Indian was shaped by the environment and its physical forces—wind, water, geology, and climate. The raw qualities of nature pervade the scene in art and prose: dark and foreboding forests, towering clouds, and unbroken coastlines. Variation in the landscape was continuous, molded by physical processes that interacted with the shape of the land and its underlying bedrock. This thickly forested scene provided the home for wild creatures, which included a moccasined and lightly treading people. This is the view that inspired Longfellow's murmuring bough, Henry Thoreau's concept of primitive America, and the description of squirrels passing across the eastern forest without ever touching the ground.

In contrast, the scenes of horticultural bounty, Squanto, and open villages with fields convey something absent from the life of the wild native: the inclination of humans to manage the land for their own purposes. By producing food rather than solely harvesting nature's bounty, the agricultural Indian works to shape nature to his (and likely more often her) own needs; in the process, he shifts fundamentally from passive resident to active manager of the landscape. By adopting agriculture, humans launch down the path from being an incidental to becoming a major force on the land, from subservient to natural processes to increasingly dominant over them. In the inclination to work land locally for food comes the growing motivation and ability to manage the land more broadly for greater objectives. This management may unfold in diverse ways: through local forest clearing, planting, and settling the land, and by applying fire broadly to control the habitats and the abundance of many species—deer, moose, bear, beaver, many birds, and others.

The belief in the noble savage and wilderness held sway well into the twentieth century, when it briefly engendered popular images of the native ecologist. But, today an alternative view of a humanized, managed world pervades the ecological, conservation, and popular literature. For evidence, witness the popularity of the books *1491, Changes in the Land,* and *Fire in America;* scan the displays at National Park or Forest Service visitor centers; or read conservation literature on the Vineyard. In popular and scholarly circles, the light-handed ecological Indian is dismissed. While advertisers used to employ a tearful Indian to decry environmental destruction, today's message is of natives engineering their own version of a finely managed paradise. Major indigenous impacts are claimed for early postglacial times, including the slaughter and extinction of ice age mammals—mastodons, mammoths, giant sloths, beaver, and cave bears. But the most expansive alteration of the land is interpreted for a more recent era. In the millennia preceding European contact, agriculture swept across the Americas from its origins in Mexico. Coincident with the arrival of the three sisters and adoption of a settled agricultural lifestyle, the land began to change fundamentally through forest clearance and wide application of fire.

It is worth recognizing that the intellectual roots for much of this new thinking originated in New England and interpretations of the Cape and Islands. For the conservationists who develop management plans and interpretative materials at Pohogonot, Long Point, and Katama, the most compelling interpretation of Indian life and practices is espoused in two New England writings: a 1953 article by Gordon Day, "The Indian as an Ecological Factor in Northeastern Forests," and Bill Cronon's transformative 1983 environmental history, *Changes in the Land.* Drawing from the writing of explorers, travelers, and colonists, these scholars paint a picture of Indians living in agricultural villages and applying sophisticated stewardship to a varied landscape of open woodland, brush, and fields. Day wrote, "We must conclude . . . that the Indians of the Northeast cleared land for villages and fields, cut fuelwood and set fires beyond these clearings, exercised a wide indirect influence on vegetation."

Thirty years later, Cronon wrote, "When Verrazano described 25–30 leagues of treeless land in Narragansett Bay and Higginson observed thousands of treeless acres near Boston . . . they were observing the effects of agricultural Indians returning to fixed village sites." And, "Selective Indian burning thus promoted the mosaic quality of New England ecosystems, creating forests in many different states of ecological succession. . . . By encouraging the growth of extensive regions which resembled the boundary areas between forests and grasslands, Indians created ideal habitats for a host of wildlife species. . . . In short, Indians who hunted game animals were not just taking the 'unplanned bounties of nature'; in an important sense, they were harvesting a foodstuff which they had consciously been instrumental in creating."

These ecological histories persuasively advance a view of native people widely applying fire and other tools to accentuate the native diversity and productivity of New England for their own purposes while transforming it into a humanized landscape. In this view, man dominates and natural processes recede to the background. The scorched landscapes that appear across the Vineyard today are an attempt to replicate this hand in support of conservation.

In this interpretation, farming is seen as a pivotal step in altering the human relationship with nature. For many native people in the Americas, and some Northeast groups like the Huron and Iroquois, life *did* center around corn. A dependable supply of calories meant that children were weaned earlier, survival increased, and populations grew. The shift to horticultural village life is argued to have triggered a fundamental change in society and its impacts on nature. Once native farmers became empowered by the knowledge that nature could be improved for human benefit, they sought to enhance the land's resources more broadly. Over time, the populated cluster of houses became centered in a rich landscape mosaic of fields in various stages of care and abandonment surrounded by intensively used woodlands and a more continuous expanse of burned and open forest beyond. For Day and Cronon, the skillful application of fire fit neatly into a comprehensive explanation of the pre-European landscape.

Lucy Vincent Beach in 1998 with the archaeological study in progress. Courtesy of Dianna Doucette.

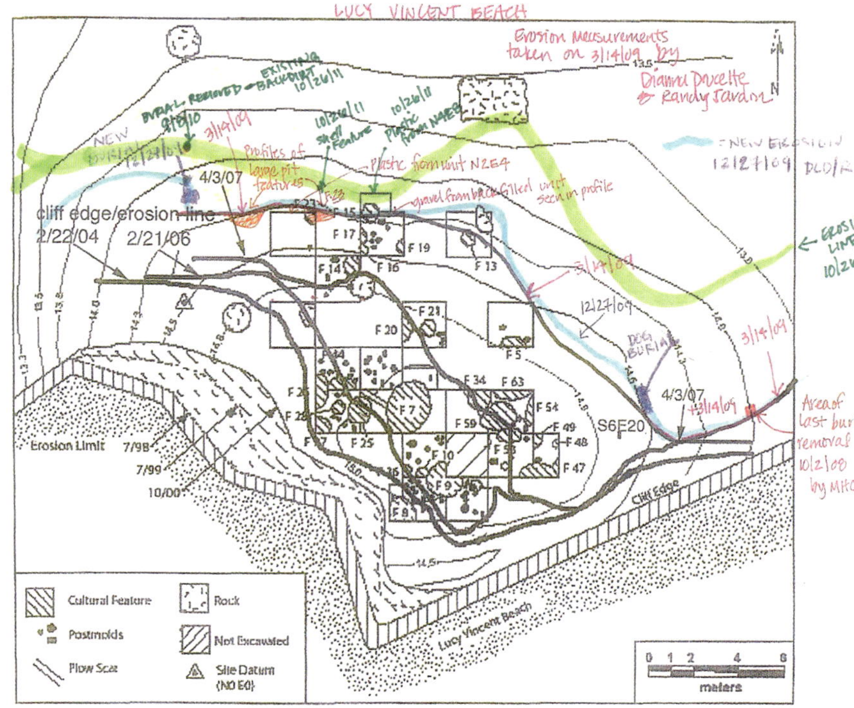

The knoll at Lucy Vincent Beach, a lost cultural site. Annotated sketch map of the archaeological features studied by the groups under the direction of Elizabeth Chilton and Dianna Doucette depicts the progress of erosion from 1998 through 2011 in black and colored lines. Today, all of these features are gone. Map and annotations by Elizabeth Chilton and Dianna Doucette.

Lucy Vincent Beach, 2011. In comparison with the previous photograph, the location of the two boulders near the cliff edge indicates that the entire area studied in 1998 has eroded away.

The Massachusetts-based observations of Thomas Morton (1579–1647) provide an oft-cited part of this intellectual foundation: "The Salvages [*sic*] are accustomed to set fire of the Country in all places where they come, and to burn it twice a yeare, viz: at the Spring, and in the fall of the leafe. The reason that moves them to doe so, is because it would other wise be so overgrown with under weeds that it would be all a coppice wood."

The implications of this cultural landscape interpretation are profound. Agriculturally based Indians are interpreted as having forged a new balance of nature and created habitats that were entirely dependent on people. These activities encouraged many desirable species to flourish that would otherwise have been scarce. If true, the lessons for conservation are clear. If people have maintained nature's diversity for thousands of years, then many species and habitats require a human hand to survive. It is also clear what tools are needed to accomplish this. This thesis was advanced for the Vineyard by Shepard Krech in his provocative book *The Ecological Indian.* The last heath hen disappeared in 1932 about a mile from the Pohogonot tract. According to Krech, the path toward this iconic species' extinction was initiated with European arrival, but not for the usually assumed reasons. At that point, when native hands were no longer able to burn the open scrub oak and grasslands on the Great Plain, the habitat got woodier and the distinctive bird began a downward spiral. In Krech's view the heath hen, and many other species, depended on native people for their abundance.

The Resilient Yet Invisible New England Indian

In applying the concept of agricultural Indians actively managing the land, proponents ignore one critical fact: the past few decades of New England archaeology and paleoecology paint a completely different picture. When ancient Indian sites and the mud from the region's ponds are scoured, there is no evidence for agricultural villages, scant traces of horticulture, and no hint of grasslands or landscapes transformed by aboriginal land use or the tools required to accomplish this. From the Cape to the Canadian border, all evidence indicates that New England was heavily forest throughout prehistory. Thousands of people did occupy the region (75,000 or so across New England's 42 million acres), but their activities blended seamlessly into the natural background; they are invisible to the lens of paleoecology. In contrast to the landscape of the Iroquois and Huron, or many populations to the south where corn farming did arise and evidence abounds for intense human activity, the New England story is quite different. On the coast and across the region, we see a continuity in lifestyle over thousands of years based on hunting, fishing, and collecting of terrestrial, freshwater, and marine resources.

A second lesson is that the region's archaeological and historical records convey complementary but distinctly different stories. Following European contact (ca. 1520–1620), New England Indians did adopt agriculture and a sedentary lifestyle. Native practices changed rapidly and substantively with European encounters and their mixture of positive (trade, new materials) and negative (disease, warfare, cultural change) influences. Written documents capture this departure from the prehistoric archaeological record. Therefore, we must weigh the sources of information carefully. It is surprising just how early in the sixteenth and seventeenth centuries and in how many different ways native people were influenced by exposure to European culture.

This subject of early encounters and impacts is so consequential that we need to explore it a bit. "Contact" was not a singular event that could be captured simply by art or pen. Rather it was an extended process that spread largely invisibly, initiated by direct sightings and meetings and propagated by the indirect exchanges of materials, customs, and disease among thousands of people. The Pilgrims' first Thanksgiving was actually a rather late episode in a lengthy process that had been unfolding for more than a century and rapidly transforming natives toward a settled agricultural society.

After all, the *Mayflower* arrived some six centuries after Vikings

East Beach, Tom's Neck, and Cape Poge Bay area, Chappaquiddick. Native encampments on landscapes like this accessed the rich resources of the ocean, salt marshes, freshwater wetlands, brackish ponds, and upland forests.

engaged in alternating bloody and prosperous exchanges with native "skraelings" along the Newfoundland coast and a full century after the first good descriptions of Indian groups in New England. It is easy for us to confuse William Bradford, Miles Standish, and other Pilgrims who arrived in 1620 with explorers and mistakenly interpret their text or that of later arrivals like Thomas Morton as accounts of an unblemished world. Yet the Pilgrims touched down at Plymouth well after Martha's Vineyard, the Hudson River, and the New England coastline had been mapped and two decades after Gosnold had encountered Indians near Cuttyhunk, paddling a Spanish boat, dressed in pantaloons, speaking broken English, and offering furs in trade.

Indeed, by the time the *Mayflower* limped into Cape Cod Bay, thousands of ships of English, French, Spanish, Portuguese, and Dutch origins had plied the coastal waters of North America for commerce and trade. More than a few of these had direct contact with New Englanders, though we will never know their details. And all were part of a gradual (but archaeologically abrupt) process of transforming native societies through the spread of goods, customs, power, and disease.

By the time Massachusetts was colonized in 1620, the region and its native people were already strongly influenced by Old World forces. After all, Squanto, who taught the English "native" practices, was a transatlantic traveler (eventually crossing the ocean six times) who had already spent two years in Britain and Canada. Captured in 1605 and transported by Captain George Weymouth to Europe, he returned in 1618 to find his village and coastal New England devastated by European diseases that had swept the landscape four years before the *Mayflower* arrived. In some ways he was more worldly in travels and experience than many of the Pilgrims, leading us to wonder: whose agricultural practices did he actually teach?*

Plymouth Harbor (Port St. Louis), 1606. Samuel Champlain's stylized depiction of scattered Indian wigwams with small gardens and woodlands. The scene fits the archaeological interpretation of seasonal and highly mobile family encampments in a productive landscape of forests, streams, estuaries, and maritime habitats.

Recognizing that the historical accounts depict native people in rapid transition, I will base my discussion primarily on archaeological records as I seek to understand native lifestyle and influence on the landscape. It turns out that one of the best places to explore these issues is the Vineyard. Here we find some of the densest native populations and a long history of archaeological

* One remarkable conduit of Old World influences to North America was the movement of kidnapped Indians to Europe and back. Once familiar with European languages, many returned as guides and translators. While the full consequences of this international exchange program remain unclear, the repatriation of Indians and growing influence of European goods, practices, and diseases rapidly rendered historical accounts less indicative of long-standing practice.

studies. If agricultural lifestyles or evidence for fire and land clearance should appear anywhere, it is here.

But as we begin to explore the role of native people on the Vineyard, we should clarify why these issues are relevant to this predominantly ecological work. Pre-contact native people and their landscapes are often the stuff of academic debate, far removed from modern concerns. Why should we focus on the livelihood of people who lived so long ago?

Foremost is the often-ignored fact that the Wampanoag continue to thrive on the Vineyard. Remarkably, this coastal people was among the few eastern groups to survive the centuries of conflict, disease, and displacement wrought by Old World discovery. Their history is the compelling story of the first Americans. To ignore it would be to miss critical elements of Island life and New England history.

Then, if we are going to conserve nature using insights from the past, we must understand people as an integral piece of that puzzle. Native people clearly influenced the land to some degree, but how and where? Looking at Katama, Long Point, and Quansoo, who is responsible for the open habitats that once dominated New England and supported the species that are disappearing today? Was it Indians with hoes and fire, or colonial farmers with axes, plows, and sheep? For our inquiry into ecology and conservation, these are critical questions that must be addressed through archaeology.

Finally, we as a society need to cope with environmental change. Native societies in New England clearly adapted successfully for millennia. We may be enlightened, and even benefit, from examining how this civilization coped with increasingly familiar challenges—increases in sea level, drought, catastrophic wind, rapid climate change, and accompanying shifts in vegetation and long-established foods. Did native people undergo major shifts in lifestyle, behavior, and technology? Did they suffer population collapses? Or did they assume a broadly adaptable mode of living that enabled them to cope with unanticipated changes? The answers to these questions may have broader relevance to issues of sustainability and adaptation.

Who Were the New England Indians? The View from the Pre-Contact Period

In recent years . . . the most interesting question has been whether or not coastal algonquins were raising and consuming maize in significant quantities prior to the initial european arrival. . . . The archaeological evidence suggests that the answer to this question is no. . . . Late Woodland food production was a very minor part of a subsistence strategy best described as a broad-based hunting and gathering subsistence system emphasizing marine and estuary resources. . . . It is clear that the large, semi-permanent settlements characteristic of the Late Woodland period elsewhere in the deciduous Woodlands [like the Iroquois and Huron] are lacking.

— John Hart, 1999

Archaeological and paleoecological records tell us a fascinating story about native people, their lifestyle, and their impact on nature. The story varies surprisingly little from the coast into interior river valleys, and it holds, with some consistency, far back in time.

For thousands of years, the settlement pattern among New England Indians was one of dispersed small bands and family groups that were highly mobile and organized around natural resources that varied in availability and location throughout the year. Small groups and mobility were key adaptations that enabled their flexible responses to changes in seasonal and longer-term food availability. The centerpiece of this lifestyle was the *wetu*, or wigwam. Circular to oval in outline and fifteen to twenty feet across, these sturdy yet temporary structures were constructed with a frame of flexible poles just a few inches in diameter. When covered with one to many layers of woven mat, skin, or bark, the wigwam provided effective protection that warded off bitter winter winds while retaining the warmth from a central fireplace. Adjustable openings in the side and top provided access, light, and vents for air circulation and smoke.

Gay Head Community Baptist Church, Aquinnah.

These structures were well described by the Pilgrims in the booklet *Mourt's Relation:* "The houses were made with long yong Sapling trees, bended and both ends stucke into the ground; they were made round . . . and covered downe to the ground with thicke and well wrought matts, and the doore was not over a yard high, made of a matt to open; the chimney was a wide open hole in the top, for which they had a matt to cover it close when they pleased. . . . Round about the fire they lay on matts, which are their beds."

Though the wigwam was disparaged as flimsy by some colonists and is often dismissed today as rudimentary by connoisseurs of native architecture, its virtue lay in its functionality. The outer coverings could be readily transported to another location where a set of saplings were cut and rapidly framed into a new shelter. Consequently, an entire encampment could be relocated in short order, a phenomenon that astonished explorers and colonists alike. Verrazano wrote in 1524, "They change their habitations from place to place as circumstances of situation and season may require, and this is easily done, as they have only to take with them their mats and they have other houses prepared at once." A century later in Rhode Island, Roger Williams wrote, "In the middle of summer . . . they will flie and remove on a sudden from one part . . . to a fresh place . . . in a halfe a day, yea, sometimes a few houres warning to be gone and the house up elsewhere."

Although the poles and mats disappeared long ago, the widespread use and characteristics of wigwams can be ascertained archaeologically. Soils develop at a geological pace. Once the surface mineral layers are disturbed, they will retain these scars as legacies for hundreds or thousands of years. Among such evidence are "post molds," soil imprints of the poles that supported the shelters. When wigwams were disassembled, the poles were either removed or left to rot. The resulting holes filled with debris that decomposed into a dark organic-rich soil. When excavated carefully, these sharply defined "molds" contrast with the sandy hue of the surrounding soil. The diameter and depth of each mold indicates the size of its timbers, while their distribution across the ground helps define the structure's dimensions. Through painstaking excavation and mapping, it is possible to reconstruct the size, number, and arrangement of structures as well as the domestic use of the site.

Across New England we see a largely unchanging picture for shelter over the past ten thousand years. Post molds are small, from two to four inches in diameter, and indicative of saplings. The outline of individual structures is often challenging to reconstruct due to the long history of reuse of the site, producing overlapping patterns of many molds. From this and other evidence, including radiocarbon dates, it appears that sites were repeatedly used for relatively short duration over long periods of time. Where the outlines have been reconstructed, most form circles from ten to twenty feet across. The interpretation is one of small local clusters of individuals moving around and reoccupying sites for generations. Mobility and flexibility were key. In New England there was nothing like a large permanent agricultural village, a fact backed up by several observers. In 1628, John Winthrop wrote, "They enclose noe

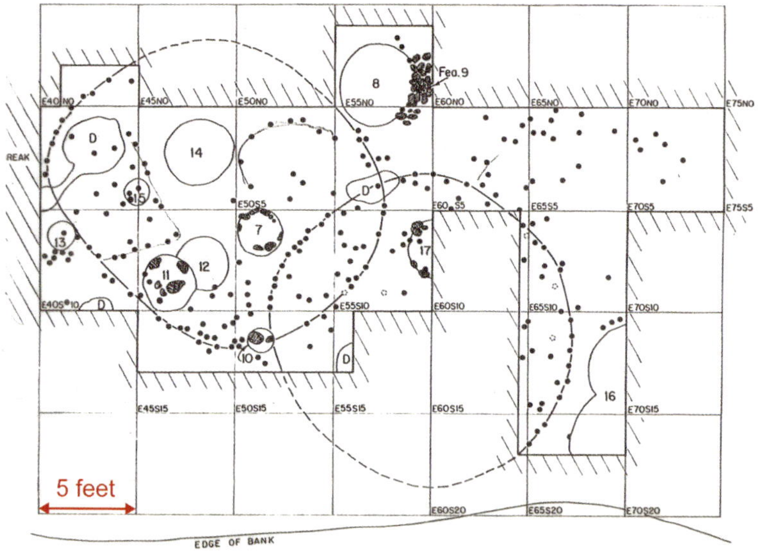

William Ritchie's Cunningham archaeology site. Located along the protected area near Vineyard Haven, the site provided access to the harbor and adjoining Lagoon Pond, as well as the sea, upland forests, salt- and freshwater wetlands, and tiny Ben Luce Pond. Many small post molds are scattered in a largely haphazard fashion along with shell and fire-broken stone, but other post molds outline circular wigwams approximately fifteen feet across. The site is interpreted as a seasonal encampment reused over centuries or millennia. Reprinted with permission from Ritchie (1969).

Shell midden exposed by erosion in the bluff at Lucy Vincent Beach, November 2011.

land, neither have they any settled habitation." His contemporary Frances Higginson observed, "Neither have [the Indians] any settled places, as towns to dwell in . . . but change their habitation from place to place." Forty years later, John Josselyn wrote, "Towns they have none, being always removing from one place to another for conveniency of food, sometimes to those places where one sort of fish is most plentiful, other whiles where others are. I have seen half a hundred of their wigwams together in a piece of ground and they shew prettily, within a day or two, or a week they have been all dispersed. They live for the most part by the Sea-side, especially in the spring and summer quarters."

If New Englanders were not agriculturalists, then what was their diet? Dozens of sites provide evidence for hunting, gathering, and collecting of diverse resources in an opportunistic and seasonally varying way. In the vocabulary of archaeology, this diet was "broad spectrum." People utilized essentially every native vegetable and animal product imaginable: fruits, leaves, tubers, nuts, bark, and stalks from more than five dozen plants; meat, hides, fur, teeth, and sinews of more than one hundred mammals, fish, birds, reptiles, and amphibians; more than a dozen species of mollusks; and even insects.

The broad food base represents an intriguing melding of conservative and opportunistic strategies. By tapping a huge array of items, Indians reduced the risk of starvation. At the same time, they capitalized on the varying abundance and distribution of foods across the seasons and landscape. Annually, the centers of opportunity for food and shelter shifted between water and wetlands, from forest interiors to their margins, and from the shoreline and mudflats to deeper coastal waters. Over thousands of years there was consistency in sites and major staples—deer, mast, berries, and bounty from the coast.

Through time and from coastal islands to northern Vermont, deer

Drift Whales—The Legend of Moshup Revisited

In the early 1980s, Betty Little, an innovative archaeologist from Nantucket, proposed that one factor leading to the coast's large native population was their ability to tap a dependable bounty of food and materials from stranded "drift whales." This assertion meshes nicely with historical observations and modern data.

From the earliest historical times the Cape and Islands have produced riveting scenes of stranded whales. On his arrival in 1602, Bartholomew Gosnold discovered "many huge bones and ribbes of whale" on Cuttyhunk. In like fashion, an early scene on Wellfleet Bay encountered by the Pilgrims was natives butchering dead blackfish or pilot whales.

> We found also a great fish, called a Grampus, dead on the sands. . . . They were some five or sixe paces long, and about two inches thicke of fat, and fleshed like a Swine; they would haue yeelded a great deale of oyle. . . . The Indians . . . cut it into long rands or peeces.

From the coordinated industry that the natives applied to their task, this process was familiar and thoroughly relished.

Some two hundred years later, with a mixture of fascination, scientific intrigue, and revulsion, Henry David Thoreau related a similar scene at Wellfleet during his walk across outer Cape Cod. The value of the resource is clear.

> I could see about a mile south some large black masses on the sand, which I knew must be blackfish. . . . I soon came to a huge carcass whose head was gone and whose blubber had been stripped off some weeks before . . . the stench compelled me to go a long way round. When I came to Great Hollow I found a fisherman and some boys on the watch, and counted about thirty blackfish, just killed, with many lance wounds, and the water was more or less bloody around. . . . The fisherman slashed one with his jackknife, to show me how thick the blubber was,—about three inches; and as I passed my finger through the cut it was covered thick with oil. The blubber looked like pork, and this man said that when they were trying it the boys would sometimes come round with a piece of bread in one hand, and take a piece of blubber in the other to eat with it, preferring it to pork scraps. He also cut into the flesh beneath, which was firm and red like beef, and he said that for his part he preferred it when fresh to beef. . . . They get commonly a barrel of oil, worth fifteen or twenty dollars, to a fish. . . .
>
> I learned that a few days before this one

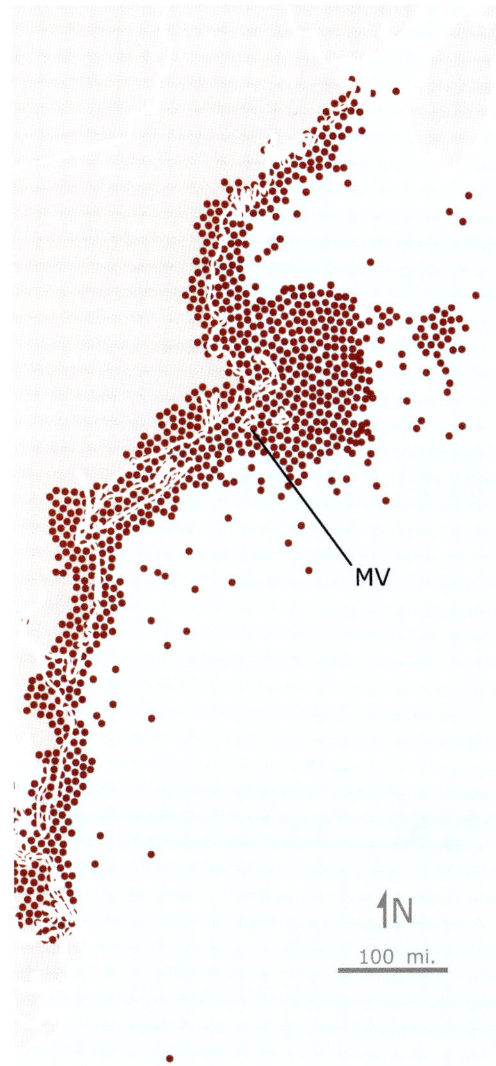

Whale-strandings (1,107 total from 1987–2011) along the northeast coast support archaeologist Betty Little's assertion that marine mammals provided an important source of food and materials. Historical evidence suggests that before their global depredation, stranded or "drift" whales were even more abundant along New England. Note: dots are slightly displaced to eliminate overlap. Cetacean data provided by NOAA Fisheries Service, Northeast Regional Office and the NER Marine Mammal Stranding Network.

hundred and eighty blackfish had been driven ashore in one school at Eastham. . . . One fisherman declared to me that they ran ashore in pursuit of squid, and that they generally came on the coast about the last of July.

About a week afterward, when I came to this shore, it was strewn, as far as I could see with a glass, with the carcasses of blackfish stripped of their blubber and their heads cut off; the latter lying higher up.

Little argued that whale beachings were common and culturally critical. Gathering records from research centers and archaeological studies, she confirmed that they were concentrated from Massachusetts to New Jersey and corresponded with the distribution of native implements fashioned of cartilage or baleen. Meanwhile, the use of drift whales is well established in the ancient Wampanoag legend of the giant Moshup who lived among the native people at Gay Head and ate whales that he caught with his bare hands and cooked over an open fire. At the end of the tale, Moshup departs mysteriously, leaving a lasting gift to the tribe—a dependable source of his favorite food.

> Moshup went away nobody knows whither. He had no conversation with the Indians, but was kind to them, sending whales, &c. ashore to them to eat.

In the early colonial period, stranded whales were considered an important property right regulated by local law and tradition. On the Vineyard in November 1652, Thomas Daggett and William Weeks were appointed "whale cutters for the year." And in 1653, a town order stated that drift whales would be "cut out freely, four men at one time, and four at another, and so every whale, beginning at the east end of town." By 1668, the English on Nantucket made a "bargain with ye Indians concerning all whales that should drift to the shores of the island and subsequently divided the shores into sections for different Indian groups."

Studies of recent strandings suggest that they result from animals that are dead, disoriented, trapped with falling tides, or fleeing predators. Since there is no prehistorical evidence in southern New England for harpoons or the ability to hunt marine mammals in deep water, it is assumed that most whales were scavenged when washed up, stranded, or herded toward the shallows. Whales must certainly be considered one of the many abundant resources on the coast.

was one of the favored species, providing meat, diverse implements, and clothing from antler, bone, hide, and sinew. This observation underscores an often overlooked ecological role of native people. Humans were a top predator that controlled the abundance of many animals. They undoubtedly played a key factor in wildlife dynamics, especially on the islands, where other large carnivores were few.

Native plants formed a major component of the Indian diet. Foremost in caloric value and abundance was a food largely ignored by modern society, the nuts of forest-grown trees—hickories, oaks, beech, chestnut, and hazel. Acorns were most abundant and readily collected. White oaks were preferred as the most palatable and easiest prepared, but all acorns contain tannins and required mashing and soaking to remove these bitter compounds before consumption. The ability to store nuts made them a critical resource for winter survival and likely a key factor determining human population levels. Coincident with the widespread use of nuts is the appearance of tools to process and store them—mortars and pestles, bowls, and storage pits.

Nut-bearing trees pose one challenge; most are "mast fruiters" that produce abundant crops every three to five years, varying by species. Hence, like the wildlife with which they competed, humans would have switched opportunistically among species. It is conceivable that extended periods of low nut production resulting from natural variation, drought, or insect defoliation may have had major consequences. On the other hand, the large number of masting species on the Vineyard generally ensured good crops in most years. Meanwhile, mast years brought phenomenal bounty. The production from seven hickory trees can support a single person, and ten square miles of oak-hickory forest can yield more than six million pounds of nuts.

Although poorly preserved in the archaeological record, berries, fruits, and starchy tubers were widely harvested, supplemented by the seeds from weedy plants. Growing on open disturbed sites, on adjacent encampments, and along the shores of streams, lakes, and coast, species like goosefoot (*Chenopodium*), sumpweed (*Iva annua*), and knotweed (*Polygonum*) produce abundant, easily dispersed seeds. The encouragement of such plants through purposeful disturbance is thought to be a precursor of horticulture.

Coastal groups also exploited a dependable bounty from the sea, mudflats, marshes, and coastal ponds. When the plant and animal resources from these productive habitats were coupled with those from streams, wetlands, and upland forests, a prosperous life could be sustained on a land base like the Vineyard. Rapid reproduction of native mollusks, fish, crustaceans, and plants and the small size of the human population made it possible to reuse sites without depleting local resources.

From an ecological perspective, there is a critical distinction between this nature-dependent subsistence and the agricultural lifestyle often ascribed to Vineyard natives. Farmers are rooted to a site and modify the land, often intensively. In contrast, the collection of wild foods is dispersed and the impacts indirect. Native people undoubtedly favored some terrestrial species and altered the local abundance of key resources. But the greatest bounty lay near the ocean margin, produced by the regular delivery of nutrients and material to estuaries and coastal waters. By tapping these resources, coastal Indians substantially broadened the food base available to inland groups. Both pursued the light-handed use of many natural products.

The Vineyard and Coast

In her synthesis of Vineyard archaeology for our collaborative study of Island history, Deena Duranleau wrote,

> [There is] evidence along almost every pond, watercourse, bay and coastline of native occupation of Martha's Vineyard for thousands of years. . . . Around Squibnocket, Menemsha, Nashaquitsa, and Lagoon Ponds, native inhabitants . . . impacted their local environment in countless ways. . . . To support year-round habitation shellfish were harvested, white-tailed deer were hunted, fresh and saltwater fish were caught, saplings were procured to construct wigwams, firewood was collected for cooking and to keep warm in the winter, wild and some domesticated plants were harvested, clay resources were identified, lithic and bone tools were made, and hundreds of other tasks which leave no trace were performed by residents on a daily basis.

As we tap the history of archaeology to examine the details of native lifestyles, there is much to explore. Island sites extend back ten thousand years and are as dense as any place in New England. Past studies give the Vineyard an enduring place in American scholarship, while new syntheses continue to refine established interpretations.

Native sites on the Vineyard have been picked over for centuries by landowners, amateur collectors, and visitors. Indeed, fascinating artifacts sit on mantles and bookshelves or lie tucked in the drawers of farmers, fishermen, caretakers, and contractors, while highly informed "amateurs" and professionals offer insights from every part of the Island's coast. Academic archaeology commenced in 1917 when Samuel Guernsey arrived from Harvard's Peabody Museum. Guernsey's work is often dismissed as descriptive, yet he was a great field scientist whose revelations span the country from the Vineyard to Mesa Verde National Park, home to the Anasazi cliff dwellers. Guernsey was also a gifted artisan who partnered with artist Theodore Pitman in a Harvard Square workshop that produced the

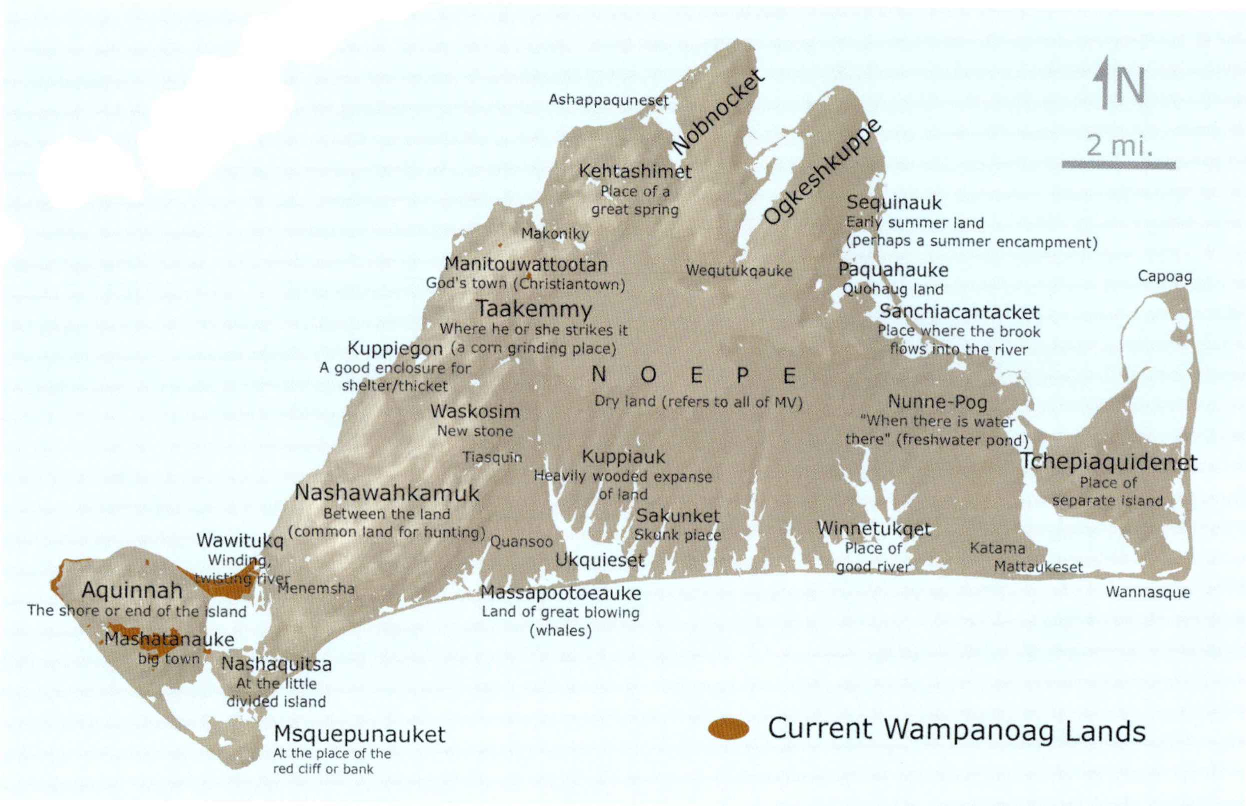

Wampanoag place names. At European settlement some three thousand Wampanoag ancestors inhabited the Island in distributed bands. Currently, the tribe owns lands in Aquinnah and at Christiantown in Chilmark. Data from the Massachusetts Cultural Council.

magnificent historical dioramas of Boston and Harvard College that graced Widener Library for decades. The teams' models remain highlights in the Peabody Museum's American Indian exhibit, but their masterpiece is the remarkable set of twenty-three dioramas at the Harvard Forest depicting the ecological history of New England.

Focusing on the Vineyard's southwestern corner, Guernsey had unique access to artifacts exposed by three centuries of sheep grazing. He began by exploring shell middens—massive accumulations of clams, scallops, and oysters discarded over the millennia at Indian encampments. Despite years of local mining by farmers and builders as a source of lime, Guernsey described mounds up to a quarter of an acre in size and four feet thick at Gay Head and along Menemsha and Quitsa Ponds. He also scoured the eroding coast for artifacts, from north of Menemsha, down around Gay Head, and on to Lucy Vincent Beach.

His discoveries were intriguing: in addition to countless artifacts he described two circular "house" sites, a small field of post-contact corn mounds above Menemsha Pond, and three cemeteries with pre-contact burials and historical gravestones. The house site soils were blackened by charcoal, heavy use, and bits of refuse—bone splinters, shell fragments, and stone flakes. Guernsey described dwellings alongside Quitsa Pond: "[They] measure approximately 19 feet in diameter, and are formed by a ridge of earth . . . two feet wide, and six to seven inches higher than the surrounding surface.

Modern shell midden on Wampanoag Tribal Lands, Aquinnah, March 2013.

The best preserved of the two rings showed a break where the door had evidently been."

To envision these classic wigwams, Guernsey quoted Thomas Mayhew Jr.'s observations from 1650: "[They were houses] made with small poles like an arbor covered with mats, and their fire in the midst, over which they leave a place for the smoke to go out."

The soils yielded artifacts from hundreds or thousands of years of occupation—diverse spear points and cutting tools, shards of Archaic soapstone bowls and Woodland period pottery, and glass beads and metal artifacts from European hands. Guernsey concluded with a statement endorsed by every generation of Vineyard archaeologist: wherever coastal ponds connect to the ocean, there is "almost uninterrupted evidence of aboriginal occupation."

By the early 1930s when Douglas Byers and Frederick Johnson arrived from the R. S. Peabody Foundation in Andover, Massachusetts, more than twenty large shell heaps had been identified in the Island's southwestern corner. These included piles of oyster, scallop, and quahog along Squibnocket Pond that helped confirm old maps showing that this pond had once connected to the sea. Bones uncovered from diverse fresh- and saltwater fish, sea mammals, and birds were added to a list of stone tools—scrapers, hammerstones, net sinkers, drills, knives, awls, gouges, choppers, pestles, and hoes, along with soapstone pipes and spear and arrow points. Their findings of many small post-holes in irregular patterns reinforced Guernsey's interpretations of wigwams erected over thousands of years. The two envisioned small encampments of people engaged in tool manufacture and woodworking, food processing and cooking, fishing, hunting, and plant collection, along with a bit of horticulture.

National recognition for Vineyard prehistory arrived with the renowned archaeologist William Ritchie from the State University of New York. Ritchie spent three summers working the southwest coast and north shore around Lagoon Pond. His book *Archaeology of Martha's Vineyard* placed the Island in a regional framework and was widely, though not uniformly, acclaimed. Ritchie's research was bolstered by students and young scientists, including three with enduring connections to the island—James Tuck, James Richardson, and Islander Dick Burt. Tuck subsequently spent his career at Memorial University in Newfoundland, where he revealed the workings of sixteenth- and seventeenth-century Basque whaling bases in southern Labrador. Based at the University of Pittsburgh's Carnegie Museum of Natural History, Richardson explored the prehistory of coastal Peru and identified interesting parallels in coastal adaptations with New England. Burt has made a lifelong passion of adding remarkable finds and insights to native and colonial archaeology across the Island.

Ritchie provided the first comprehensive summary of fish and shellfish use. A broad array of water conditions was represented by striped bass, scup, sturgeon, tautog, goosefish, cod, sea robin, and bluefish. Across six sites, fifteen species of shellfish were identified,

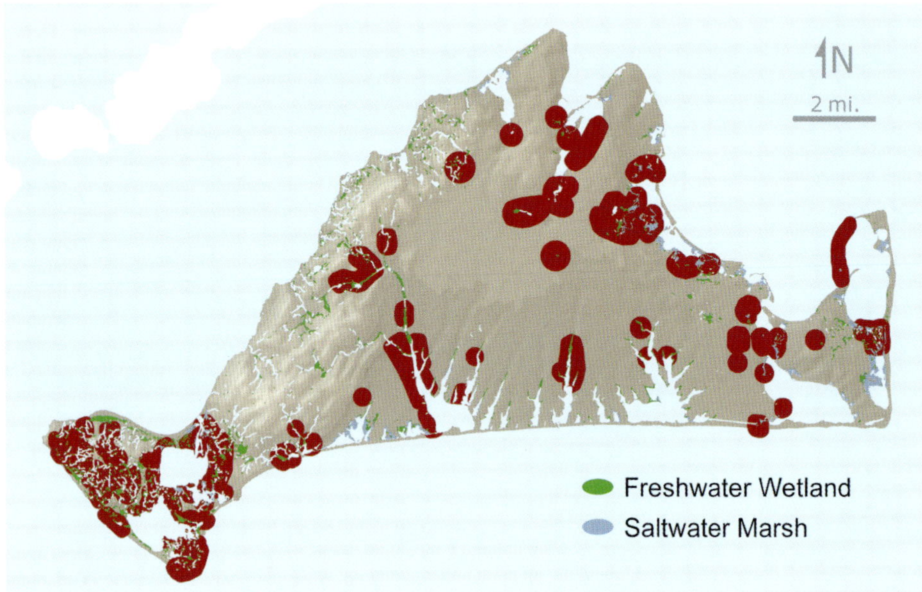

The distribution of major archaeological sites is indicative of native concentration on areas with access to shellfish and resources in estuaries and brackish water ponds. This generalized map is compiled from many published and unpublished studies but certainly underrepresents the distribution of native activity on the Island. Map redrawn based on an original developed by Dianna Doucette.

including hard- and soft-shell clams, oysters, mussels, conch, limpet, and whelks. Meanwhile, a complex of inland and marine environments was represented by other animals—otter, muskrat, gray and harbor seals, ducks (eider, brant, black, mallard, greater scaup, and merganser), common and red-throated loons, whistling swan, cormorant, and Canada goose. The diversity of terrestrial vertebrates was great: red and gray fox; raccoon; box, snapping, painted, and red-bellied turtle; mink; vole; gray squirrel; turkey; and heath hen. Surprisingly, beaver and black bear appeared, presumably as a result of trading or reuse from the era before island formation. As in most studies, the most abundant remains—15,468 bones—came from white-tailed deer.

Ritchie added detail to interpretations of repeated site use by carefully documenting wigwams supported by two- to three-inch diameter posts dug eight inches into the soil and arranged in overlapping patterns. But unlike his predecessors and successors, and based on less than a dozen grains of corn, Ritchie boldly concluded that "corn and other cultivars were certainly an important part of the food of [Late Woodland times]." He added, "It is . . . manifest even from our limited data that significant demographic changes were associated with the rise of farming."

That interpretation was soundly refuted by studies commencing in the 1990s by Elizabeth Chilton and colleagues at Harvard and the University of Massachusetts. In her paper "The Origin and Spread of Maize" Chilton wrote, "[There is] no evidence for year-round farming villages in New England . . . [and] no evidence for intensive maize horticulture until after European colonization." Chilton's work at Lucy Vincent Beach was precipitated by the discovery of a human skull below the eroding bluff. Combining emergency salvage with careful excavation by a Harvard field school, and working closely with the Wampanoag, Chilton and Tonya Largy sorted through 2,775 animal bones to add herring, shad, menhaden, porgy, and goosefish/monk to Ritchie's list. With Dianna Doucette, who has continued to work across the Island, Chilton concluded that before

European arrival, horticulture played a minor role on the Vineyard and across New England.

Professional archaeologists like Doucette working for cultural resource management (CRM) firms have added substantial information in recent decades as they help towns and landowners address the spirit and legal requirements of federal and state archaeological regulations. Due to the concentration of cultural resources, active involvement by the Wampanoag, and strong respect for human history, all Vineyard towns have commissioned archaeological and historical surveys, while town boards and the Martha's Vineyard Commission require studies on many new developments. CRM firms employ superb scientists who conduct archival reviews, site surveys, and field excavations that yield detailed reports, maps, and data. Vineyard studies have been advanced by notable archaeologists: Mitch Mulholland, Tim Binzen, and Chris Donta at the University of Massachusetts; Holly Herbster, Suzanne Cherau, and Doucette at the Public Archaeology Laboratory; and Kevin McBride from the Mashantucket Pequot Museum, among others.

These island-wide data confirm that sites occur "wherever well drained, elevated, and level soils are located in proximity to existing wetlands or poorly drained depressions." Three areas on the Vineyard display almost uninterrupted use over the past eight thousand years: the southwestern area from Mill Brook and Tisbury Great Pond through Stonewall, Menemsha, and Squibnocket Ponds; the southeast from Edgartown Great Pond and Katama Bay across to Chappaquiddick, Cape Poge, and north to Sengekontacket Pond; and the north shore from Lagoon Pond to Lake Tashmoo. In contrast, the Island's interior was the focus of hunting and foraging but displays little evidence of intensive use. People centered their lives on water, food sources, and shelter. The more diverse and productive the associated habitats, the greater the concentration of activity.

The most productive areas lay where fresh- and saltwater converged. A number of the Island's coastal ponds and streams supported annual runs of anadromous fish—alewife and herring—along with the diadromous eel, which returns annually to the sea to spawn. The large coastal ponds that connected to the sea were especially important. Breaches in barrier beaches switched the ponds from freshwater to brackish or salty and triggered huge shifts in the abundance and composition of fish and shellfish. When inlets were open, the ponds supported extensive beds of hard- and soft-shelled mollusks along with runs of migrating fish, like striped bass. When ponds were closed, these resources gradually collapsed as the waters freshened. Major shifts in human use and settlement patterns must have tracked such boom and bust cycles. The dynamism of the coast, with rising sea level, erosion and storms, wetland formation, and episodic breaching of barrier beaches, created a kaleidoscopic suite of continually varying resources.

Burial grounds in Christiantown, Chilmark, January 2015.

One recent archaeological overview is Deena Duranleau's Harvard doctoral thesis, which draws from evidence across the coastal region to paint a picture of a distinct New England lifestyle. Working with Elizabeth Chilton and our Harvard Forest group, Duranleau analyzed changes in native lifestyle, implements, subsistence, and settlement patterns in relation to records of climate and vegetation change. Her striking conclusions are drawn from more than 2,500 sites and presented in maps and graphs, some of which are included here. Through ten thousand years of pre-contact history, the word that best captures her conclusion is *continuity.*

Duranleau confirmed that the pattern of repeated site use was widespread throughout the coast. Individual localities were used over and over again. Meanwhile, the broad elements of life remained consistent for thousands of years—the dependence on plant and animal life from the woods, waters, and wetlands; the movement of people within homelands through the seasons as resources and conditions varied; and the housing, implements, and clothing that supported this flexible lifestyle. As the coastal landscape itself changed, people did adapt geographically and with modest cultural shifts. Indeed, a whole world of archaeological exploration lies in sites submerged by rising seas, while careers have been spent distinguishing among pottery and point types. These features and population sizes have varied, but the broad relationship between people and the land has remained substantially intact.

One of Duranleau's important conclusions is that the different cultural periods distinguished by archaeologists—Paleo Indian, Archaic, Woodland, and their divisions—emphasize modest differences in implements while obscuring the overall consistency of life over time. Although Late Archaic soapstone bowls were replaced by ceramics in the Woodland period, both groups employed mortar and pestles to grind nuts and seeds and bowls to consume mushes and stews. This striking continuity led Duranleau to advocate a professionally audacious (yet intuitive) proposal—that the system of cultural periods be replaced by simple dates. She argued that nature and culture change most often through gradients rather than discrete steps. While it is convenient to speak of oak forests, pine forests, and beech forests or Paleo Indian, Archaic, and Woodland periods, reality is more subtle and complex. The use of dates removes an artificial structure that encourages thinking in terms of discrete and progressive periods rather than the continuity in life displayed in the evidence.

Duranleau did note modest geographical variation. There was more plant use around Long Island Sound, greater reliance on fish across Martha's Vineyard, Nantucket, and Block Island, and more concentrated use of shellfish in Narragansett Bay and coastal New York. She also concluded that maize was incorporated in a limited fashion shortly before European settlement, and life then did become a bit more settled.

The most intriguing outgrowth of Duranleau's collaboration came from comparing her cultural data with ours on climate and vegetation conditions—information not available to earlier archaeologists. The number of sites, and presumably the size of the population, varied parallel to major changes in the environment. While many researchers had noted a peak in site numbers around five thousand years ago and again before European arrival, there now is compelling reason to link these dynamics to climate and forest change.

During the early postglacial period, site density slowly increased as the climate warmed and the land's productivity increased. The first population peak extended from about 5,000 to 3,500 years ago, coincident with the shift from maximum warmth, the collapse of hemlock regionally (and oak on the Vineyard), and the expansion of oaks, hickory, and beech. Hemlock has dense shade that inhibits undergrowth. The loss of this dominant species across many inland and some coastal areas allowed for the formation of multi-species nut-bearing forests with a diverse understory. This boon for many animals also allowed the human population to flourish. When conditions subsequently changed, with cooling temperatures, increased moisture, a resurgence of hemlock, and regional decline in oak, beech, and hickory, the human population also declined.

The archaeological record suggests that native populations began to grow again about a thousand years ago. These last five hundred

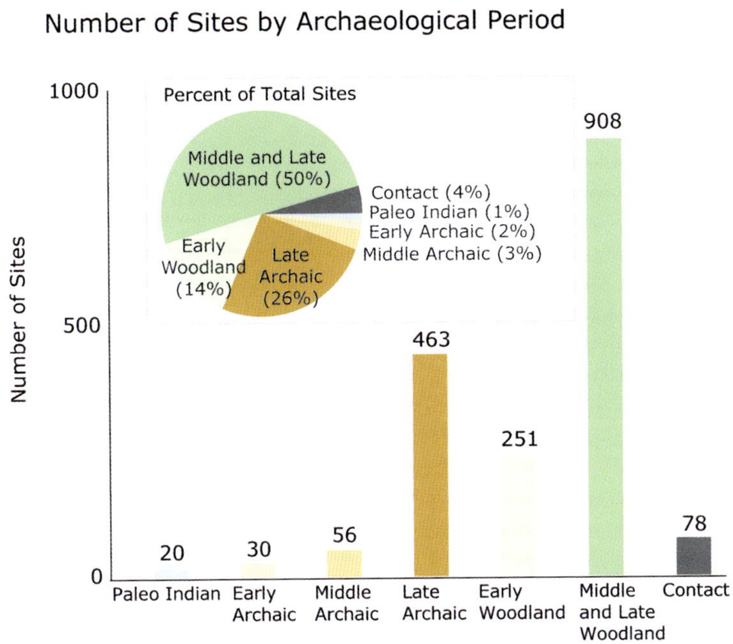

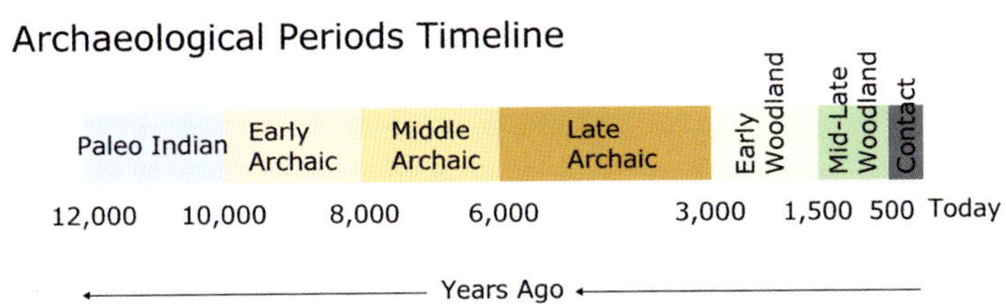

The changing number of archaeological sites over time for coastal New England and New York. From fewer than sixty sites through the Middle Archaic period, the number increased to a peak in the Late Archaic period (463), dropped through the Early Woodland period (251), and increased in the Late Woodland period (908). There are few contact period sites. Based on over 1,800 site records, the thesis by Deena Duranleau (2009), and collaborative studies by the University of Massachusetts and the Harvard Forest.

years before European contact were a fascinating period throughout eastern North America and the North Atlantic. In the far north, Viking settlements flourished for nearly four hundred years, and huge cultural transformations occurred as the Thule Inuit culture swept eastward across Hudson Bay to replace the Dorset people in northern Labrador. Closer to New England, corn horticulture became established in the northern Mississippi and Susquehanna river valleys by A.D. 900, inducing great population growth and cultural reorganization. Large villages, fortifications, and longhouses appeared among the Iroquois, who adopted agriculture and entered a period of intertribal allegiances and rivalries. Along coastal New England, the human population grew through this period of moderate climate whose end nearly coincided with the arrival of Europeans.

What About Corn?

Maize in New England dates to about A.D. 1000 or so, preceding the appearance of beans but lagging behind squash, which arrived some two thousand years earlier. Thus the "three sisters"—intertwined in lore and many minds—have quite independent histories. Corn is absent from most archaeological sites and, when present, is represented by a few kernels. While large storage facilities, including pits and cribs, occur among agriculturalists like the Iroquois, they are absent from New England. The caches discovered by the Pilgrims represent meager amounts relative to the needs of a population.

Since New England Indians traded with the Iroquois, why didn't they adopt their agricultural ways? One interpretation is that this process was inevitable but interrupted by European arrival. Farming established elsewhere, and might have spread solidly into New England with adequate time. Our colleague Elizabeth Chilton counters with another perspective: New Englanders rejected agriculture in favor of their broad-spectrum lifestyle. She points out that local wild foods met the nutritional needs of the natives and that corn, beans, and squash were used primarily as supplements

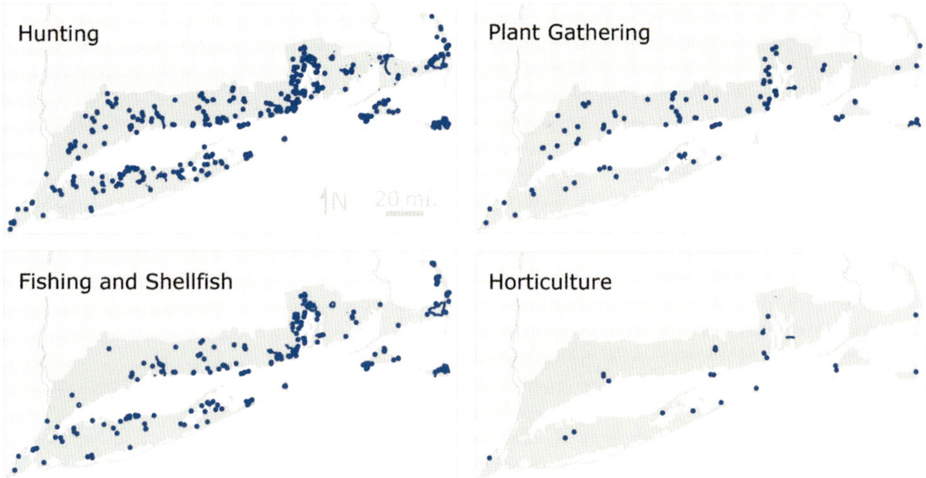

Pre-contact evidence indicates that human subsistence along coastal New England/New York relied on a broad spectrum of natural resources obtained through hunting, fishing, and collecting. Horticulture played a minor supplemental role immediately before European settlement. Based on over 1,800 site records, the thesis by Deena Duranleau (2009), and collaborative studies by the University of Massachusetts and the Harvard Forest.

and for ceremony rather than subsistence. As one of her colleagues summarized it: New England natives "chose clams over corn."

Chilton argues that agriculture brings inherent uncertainty and demands. Most important, there is great risk in relying on a tropical cultigen at its northern range limit. The timing of maize arrival made its use particularly challenging. From A.D. 1000 to 1300, eastern North America benefited from the relatively mild Medieval Warm period. But that benign period was followed by the Little Ice Age, a time of cool climate and erratic growing seasons that was especially challenging for agricultural societies. Through that period, when New England Indians were experimenting with cultigens, Norse settlements failed on Newfoundland and Greenland, and harsh weather caused failures of colonial outposts at Roanoke in Virginia and Saduhoc in Maine and misery and deaths in Jamestown and Plymouth.

Corn is challenging in other ways. In 1626 Isaac De Raiseres, an agent for the Dutch West Indies Company, described it as "a grain to which much labor must be given with weeding and earthing-up, or it does not thrive." Due to its value, the fields and grain stockpiles are vulnerable to theft or destruction by animals or desperate or antagonistic neighbors. Finally, its deficiency in key proteins and amino acids requires that corn be supplemented by other foods. These mixed blessings of the agricultural lifestyle are evident among the Iroquois, where growing hostilities and declining health paralleled the dependence on crops.

Given these downsides, Chilton argues that New Englanders chose to maintain their productive hunter-gatherer lifestyle. In this view she finds support from archaeologists in neighboring regions, including John Strong from Long Island. He wrote, "Corn, beans and squash, the three sisters so important to the Iroquois, never made much of an impact on Long Island until after European settlements destroyed the rich sources of natural foods, medicines and supplies. . . . Natives made such efficient use of the environment that there was no rush to change from the old ways when plant domestication was introduced. . . . Why besides externally imposed necessity would foragers give up a secure comfortable means of production for the uncertainty and drudgery of agriculture?"

It is clear that New England, including the Vineyard, has a distinct history. It is a tale of remarkable continuity rather than an evolutionary story that begins with stone-age man and ends

Health and Hostilities: Further Clues to Lifestyle

The study of human bones and teeth provides insights into health and lifestyle that underscore the great differences between New England natives and agriculturalists like their Iroquois neighbors. The most comprehensive evidence emerges from the Indian Neck Ossuary in Wellfleet, where the pre-contact remains of many individuals were accidentally uncovered during construction in 1979. Analyses of this cross-section of the population indicated that they shared good health with "little evidence of disease-related pathology and no unusual evidence of trauma." Especially informative was the absence of tooth decay, which accompanies a cereal diet. All evidence conforms to a diet of wild plant and animal foods and a negligible role of corn.

Equally notable is the absence of trauma or death due to combat. In fact, there is scant evidence of wounds among males from any pre-contact burials in New England. The pattern of temporary encampments across the coast reinforces the interpretation of peaceful conditions before European arrival. The distribution of sites on the Vineyard and elsewhere aligns with locations favorable for sustenance and survival in a seasonally harsh environment. Most were situated within five hundred feet of wetlands, ponds, streams, estuaries, and the coast to provide access to food resources, sheltered from prevailing and winter winds and exposed to the southern sun in late fall through spring. The remarkable finding is the complete disregard for defense in site locations or design until European arrival.

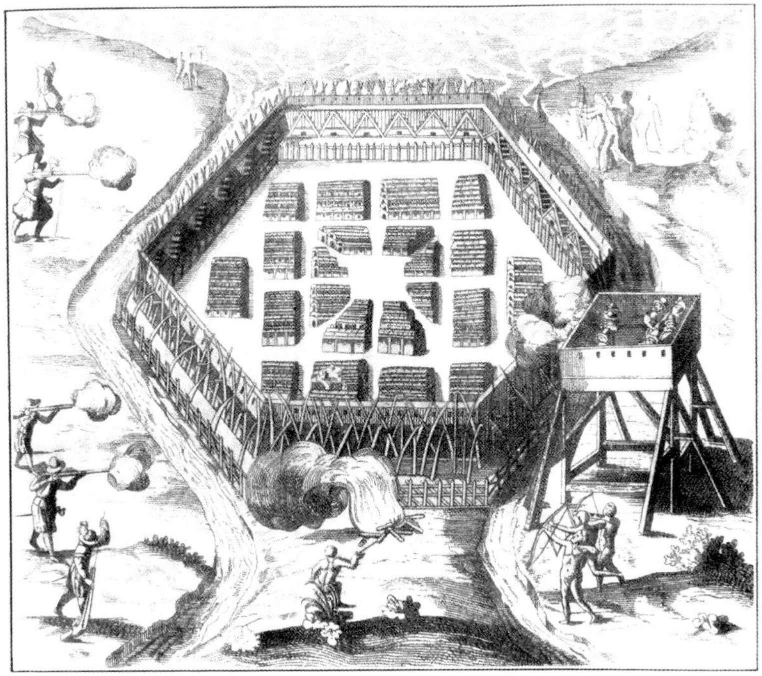

Samuel Champlain's 1615 depiction of the Huron and French attack on a large permanent Iroquois village in central New York. This well-established village consisted of more than twenty longhouses, each more than a hundred feet long and containing a dozen or more families. The village was surrounded by a multilayered palisade that afforded protection and occupied a defensible hilltop while large cornfields lay separately. Reproduced from Champlain (1632).

These patterns contrast starkly with conditions exhibited by the agricultural Iroquois and Huron. Within those groups, tooth decay and nutritional deficiencies from corn consumption were widespread, as was brutal evidence of wounds, disfigurations, and mortality from torture and warfare. Similarly, site design for defense is widespread more than five hundred years before contact. Villages were heavily fortified—surrounded by ten-foot palisades and with two or three rows of tree-size posts—and often situated on deforested hilltops that limited surprise attack but were inconveniently separated from large crop fields. Within the fortified villages, immense longhouses were often constructed with few points of egress, positioned to frustrate attack. In New England fortifications appear only after contact.

with farmers actively managing nature. For ten thousand years, Indians thrived, drawing sustenance from an extraordinarily diverse spectrum of resources. In the hunting-gathering life, people assembled in small groups and temporary structures—an arrangement that was highly adaptable. Although technology did change—throwing spears and atlatls gave way to the more efficient bow and arrow; wood and bark bowls were replaced by stoneware and then ceramics—these shifts were modest and gradual. Meanwhile, a potentially revolutionary product—maize—was incorporated erratically. Although horticulture had profound implications elsewhere, its late arrival in New England was, in the words of archaeologist Kevin McBride, a "non-event."

One could look at this lengthy history as lacking innovation. But an alternative view casts this New England lifestyle as flexible, efficient, and eminently successful. These practices endured and the populations thrived over millennia, through immense climatic and vegetation changes; major droughts, hurricanes, and the collapse of oak populations; the reconfiguration of the coastline; and changing abundances of wetlands, ponds, and streams. The native lifestyle was adaptable and attuned to nature's spatial variation and temporal rhythms. Rather than specializing on a few crops, the people deployed multiple strategies and derived calories and nutrients from many sources. This life was successful due to its simplicity and the flexible kit of tools and practices. As the environments and resources varied over time, the populations did accordingly.

The Historical-Archaeological Discrepancy

So how did historical scholarship and conservation practice come to subscribe to the interpretation of agricultural Indians shaping the New England landscape? There are a number of reasons.

- Over-reliance on historical versus archaeological and paleoecological records;
- Huge post-contact transformations in native practices to include farming, village settlement, and fortifications;
- Biases in regional scholarship that obscured a unique New England story;
- Exaggeration of the ecological paradigm of natural dynamics and fire.

These conclusions are cautionary. They emphasize the need to consider all evidence and to treat different regions with considerable independence. They also highlight the rapid way that Indian society changed in response to new cultural forces from Europe.

In large measure, the different interpretations—New England native people as hunter-gatherers versus agriculturalists—are both correct. They simply apply to different periods and result from differences between archaeology and history. While these disciplines have different strengths, perspectives, and biases, they also tend to capture very different slices of time. Archaeology affords a deep perspective on pre-contact history but an incomplete view of recent history. Its great strength is its temporal depth and ability to explore cultural and ecological changes over generations, and contrasting environments. By tapping records from many sites across a large region, studies like Duranleau's reveal a multidimensional picture of cultural and environmental dynamics. Archaeology has another major strength: its record is *created* independently of the observer. While archaeologists may bring personal perspectives to their work, the people and sites they investigate operated completely independently from the ultimate viewer's eyes.

In sharp contrast, historical observations are narrowly bounded in time and constrained to fewer and fewer sources back in time. The fact that we have observations from Sebastian Cabot in 1502 and Giovanni Verrazano in 1524 at all is truly remarkable. Some of these records, like Verrazano's travelogue from northern Florida to Maine, provide rich descriptions of the land and people. But the reality that even our very earliest documents (those from Cabot) describe natives with metal objects of European origin (presumably left by his father in 1498) is jolting; this reinforces the fact that all histories arise as European contact is already exerting major impacts

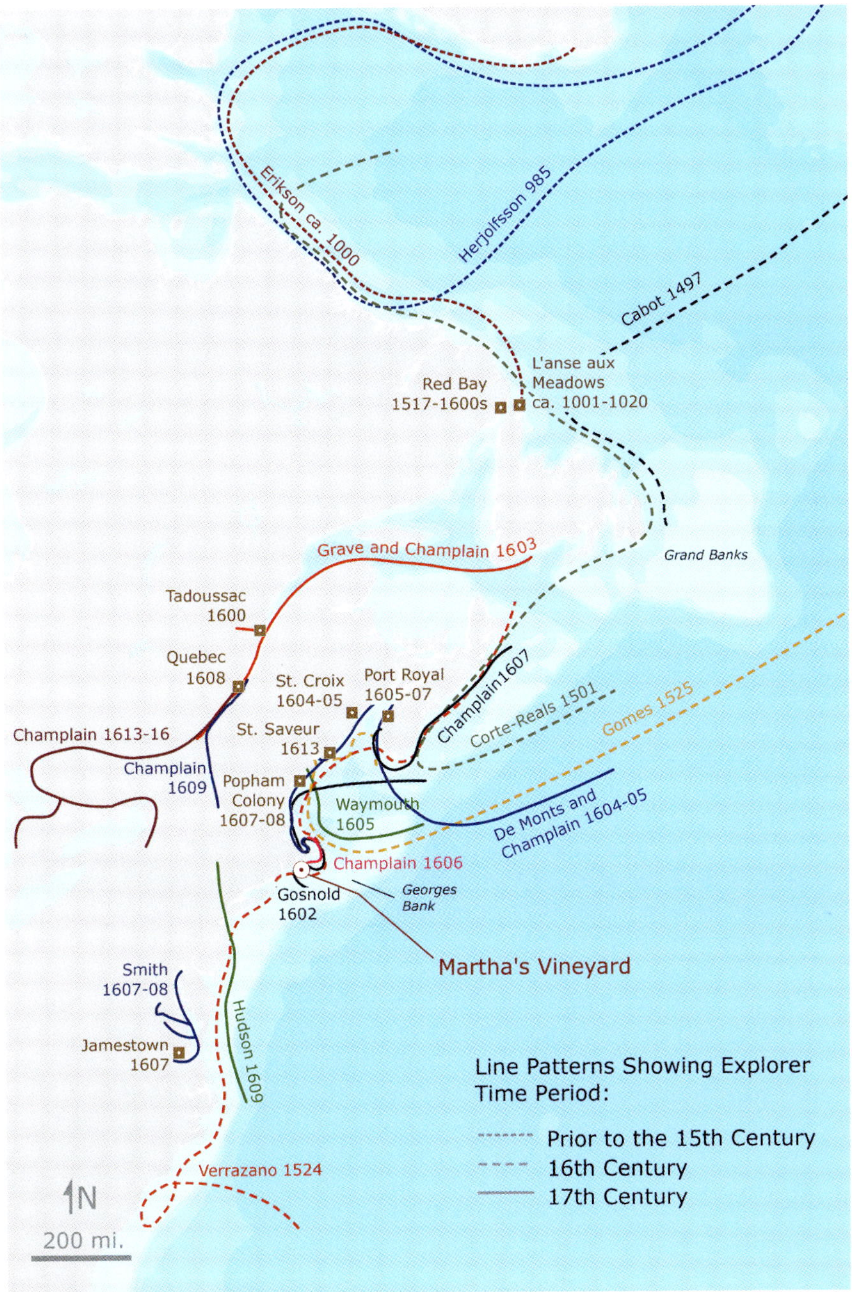

on native people. Consider these 1643 observations from Roger Williams, often cited as one authority on native practice.

> Most commonly . . . there doore is a hanging Mat . . . yet many of them get English boards and nailes, and make artificiall doores and bolts themselves. . . . Many of them begin to be furnished with English chests. . . . The Indians bring downe all their sorts of Furs, which they take in the countrey, both to the Indians and to the English for this Indian Money . . . and whatsoever they stand in need of.

The abrupt transformation of New Englanders from hunter-gatherers to farmers leads to contrasting interpretations. Both are correct, but each captures a different slice in time. Considering the span of time involved—1498 to well beyond 1620—contact was a complex process. In most cases, indirect European influences through the transfer of goods, germs, and information shaped Indian behavior long before any physical interaction. Catastrophically, the greatest impacts came from diseases that killed tens of thousands in the decade before the *Mayflower.* The vast majority of sixteenth- and seventeenth-century Indians who heard tales of visitors, traded for their goods, or were stricken ill had never set eyes on a European. This diffuse contact was facilitated by the Indian system of mobility and exchange that worked over surprising distances.

The first word uttered by Samoset when he met the Pilgrims was "Welcome." With contact, Indians began to alter their ways. Metal produced better tools, cooking utensils, and weapons. Cloth provided warmth and durability, and all items were useful barter. As exchanges proliferated, individuals and groups positioned

Due to extensive exploration, fishing, and whaling activity by Viking, Basque, French, Spanish, and English crews, it is doubtful if any of the seventeenth-century European settlers encountered Indians lacking prior exposure to European people, goods, practices, and diseases.

Open young woods with a continuous tall layer of huckleberry.

themselves geographically to interact with the powerful newcomers. The native proclivity for trade was reciprocated by explorers, fishermen, and settlers seeking food and goods. The natives responded by focusing their time at specific coastal locations and modifying their lifestyle. Corn emerged from a minor supplement to a valuable item for trade and settled life. It could be produced by women in villages as wild foods decreased and men shifted their energy to procuring furs for trade.

Increasing contact magnified the cultural changes. As populations concentrated into villages, frictions developed, and fortifications appeared. The ultimate consequences are familiar: increased conflict among natives and with Europeans; the gradual adoption of European ways; decimation by disease; and restriction of Indians into confined lands.

Throughout this tumultuous period there appeared a growing legacy of letters, journals, maps, and accounts—the historical record. These histories contradict the archaeological record in part because they describe a steep trajectory of change away from the pre-contact world. While the reports may be accurate, they represent snapshots from one vantage point in time. New England Indians did settle in villages, become farmers, and shape ancient landscapes, but not in the murk of pre-contact history. This transition largely occurred before European eyes.

Pohogonot Revisited

The important ecological lesson from this reconsideration of native lifestyle is that the predominant driver of ecological changes prior to European contact was climate and natural forces, not people. This perspective from Vineyard archaeology nicely complements that from our paleoecological study of vegetation and the environment. In New England, nature operated for ten thousand years with a cultural influence so modest that it is impossible to detect.

In that regard it is notable that all Vineyard archaeologists have interpreted the landscape as heavily wooded. Rather than describing grassland, open savannas, or large fields, they speak of old forests broken by encampments, wetlands, and the sea. In his reconstruction of the famous Hornblower II site, which was a treeless expanse alongside Squibnocket Pond when he worked there in the 1960s, William Ritchie envisioned a small opening "surrounded on north, west, and east sides by what would have been heavily forested highlands that provide shelter from the northwest winds." Similarly, on Nantucket, archaeologist Betty Little concluded that this largely treeless landscape was once thickly forested. The native people did not manage the land intensively into a mosaic of savannas, heathlands, and sandplain grasslands. They did not employ fire in the intensive manner asserted by many ecologists and historians. They resided in a largely wooded land where their influence was subordinated and complementary to that of nature. Abruptly, and only a few hundred years ago, this pattern was broken by the arrival of a new culture. The contrast, in the pace of change and face of the New England landscape, could not have been more dramatic.

III Shaping the Land

HISTORY THROUGH THE TWENTY-FIRST CENTURY

NOAA
Hurricane Bob
NOAA-10
8/19/91
8:30 A.M. EDT
NOAA/NESDIS

Winds That Roil the Island

The Indians here had a tradition that away back in the ante-historical days there was a flood which rose so high that it destroyed the entire human population of New England. . . . It is said, and with much truth, that New England has more tornadoes and cyclones [hurricanes] in proportion to its area than any other part of the United States.
— Sidney Perley, 1891, *The Historic Storms of New England*

(*opposite*) Hurricane Bob over New England, August 19, 1991, showing the broadly circular structure of the storm, the counterclockwise rotation, and the central eye surrounded by dense clouds. Courtesy of the U.S. National Oceanic and Atmospheric Administration.

In August of 1991, all eyes on the east coast were focused southward. Hurricane Bob had lumbered past the Bahamas at five miles per hour, heading north just offshore, growing in size and gaining in strength. As the week progressed, the focus of forecasters, residents, and emergency personnel gradually shifted northward to Long Island and southern New England; with the storm building and quickening in pace, newspapers, the morning television shows, and evening news elevated the drama. Tropical Storm and then Hurricane Bob, as it was called, began to be compared to the devastating storms of the recent and distant past—the Great Hurricane of 1938, the September Gale of 1815, and even the long forgotten Great Colonial Hurricane of 1635. Along the coast, evacuations commenced; windows were shuttered, taped, and covered with plywood; and water, batteries, and toilet paper disappeared off grocery shelves from New Haven to Provincetown and north into Vermont.

In the last hours before landfall, Bob's forward speed peaked at thirty-five miles per hour and the storm began to curve sharply to the east. While its clouds and broad rotating outline on satellite images engulfed much of the region, the central eye of the great storm cut an oblique path just south of Boston and then passed rapidly into the Gulf of Maine, where it dissipated and lost newscasters' attention. In the aftermath, New England shuddered and then took stock. The initial, overwhelming response was one of relief. Across south and central New England, the region that newspapers and television images had reminded us was battered and crippled in 1938, the impact was slight—leaves stripped from trees, branches broken, and a few wires on the ground, nothing that couldn't be addressed through a bit of overtime by utility crews and local police. Beating a hasty retreat from their pedestal of hyperbole, the regional media changed its tone from one of alarm to derision. "Hurricane Bobby" was the phrase employed by one newspaper from a town in central Massachusetts that had been flooded and blown apart in 1938.

But toward the region's southeastern corner, the scene and

response couldn't have been more different. The landscape was in utter havoc, as if visited by a completely different event. Wind speeds had exceeded one hundred miles per hour across eastern Long Island; storm tides rose more than ten feet in Narragansett Bay and reached twelve to fifteen feet in the Cape town of Bourne; and the devastation from wind and water spread further. Twenty-nine houses were swept away on Mattapoisett; eighty-five boats were scuttled or driven ashore in Edgartown Harbor; and a strip of beach fifty feet deep disappeared from the south shore of the Vineyard and Nantucket. Across the Cape and Islands, more trees were uprooted than in 1938. From southeastern Connecticut to Provincetown at the tip of the Cape, fallen trees crushed cars and houses and crippled local transportation. Downed power lines blackened communities for days. In this restricted part of New England, eighteen people died and damages exceeded $1.5 billion. Months later, official recognition of the truly catastrophic nature of the storm came when the name "Bob" was formally retired by the U.S. Hurricane Center.

The varied impacts of Bob and subsequent storms like Irene and Sandy and the frenzied media discussion concerning monster storms of the past generate many questions. Most broadly, what role do hurricanes play in New England? If we accumulate hundreds of years of historical information, what can we determine concerning the frequency, intensity, and size of these great cyclonic storms? What types of damage do they inflict through wind, flooding, and coastal erosion? How varied are these impacts—along the coast, inland, and regionally, as well as locally within a given landscape—and what controls this variation? In the process of developing this larger perspective, can we reconcile the two portraits that emerge from 1991: a retired name for a venerable storm on the one hand, and Bobby the wimp on the other?

Our times also make us wonder, what will the future bring? More storms have hit the region recently, displaying a wide range of characteristics that should alert us to many possibilities. In 2011, tropical storm Irene roared through the region with a media buildup reminiscent to that of Bob, followed by a similar sigh of relief due

Gay Head as tropical storm Irene passed the Island in August 2011.

to the subdued nature of the winds and coastal damage. But over the next days came shock at the unanticipated ravage from flooding that swept through parts of Vermont perceived as distant from the storm's center of activity. In 2012 it was "Superstorm Sandy" that inflicted damage from Delaware to New York, abetted by high tides that delivered devastating surges to the region.

Similar storms and even greater ones will come. What impacts can we expect? More than seventy-five years have passed since Long Island and the bulk of New England witnessed the 1938 hurricane, the single greatest natural catastrophe in the region's history. Much has happened since that ferocious storm—our forests have grown remarkably, making them much more susceptible to storm damage; our coastlines have become dense with people and vulnerable infrastructure; and our changing climate is generating more extreme weather. When we look back we recognize that Irene, Bob, and even Sandy pale in comparison to the 1938 hurricane. And 1938, as pivotal as it was in the region's psyche, still does not earn the title of "most powerful hurricane in New England history." That claim falls to the storm of 1635, just fifteen years after the Pilgrims landed, and a time when few people witnessed the consequences. Looking to the future, we should contemplate what will happen when a cyclone like that suffered by William Bradford, Miles Standish, and other colonists hits our region.

To address these questions, we need a basic understanding of hurricanes and their meteorology and some means of reconstructing their history regionally and back to 1620, if possible. Given the low frequency and extreme variation in these storms, it is clear that only with a regional perspective and the longest lens of history can we understand their natural role and consider their future consequences.

Hurricanes are distinctive tropical storms, each a great revolving vortex five or more miles in height and up to one thousand miles across. They are distinguished by a central area of low pressure and a pattern of counterclockwise rotation in the northern hemisphere that continually draws in warm and moist ocean-fed air. This saturated air provides the energy to drive the storm, for it condenses as it rises and releases heat. This thermal energy, in turn, increases the updraft and inflow of air and generates mechanical energy, more commonly known as "wind." At the eye, air descends and clears of moisture, often yielding nearly cloudless skies with little wind. From that center, wind speeds rise rapidly to a peak in the eye wall some ten to fifty miles out, where the strongest winds, heaviest precipitation, and greatest cloud cover often coincide with severe thunderstorms that may spawn local tornadoes. Beyond the eye wall, the winds attenuate progressively, but gradually. In these immense storms, winds exceeding sixty miles per hour may easily extend out one hundred miles.

Given their warm-water engine, these storms typically originate in the tropics and are confined to the world's great oceans and their immediate coasts. In the western Pacific they are called typhoons, in the eastern Pacific and the Indian Ocean they are known as

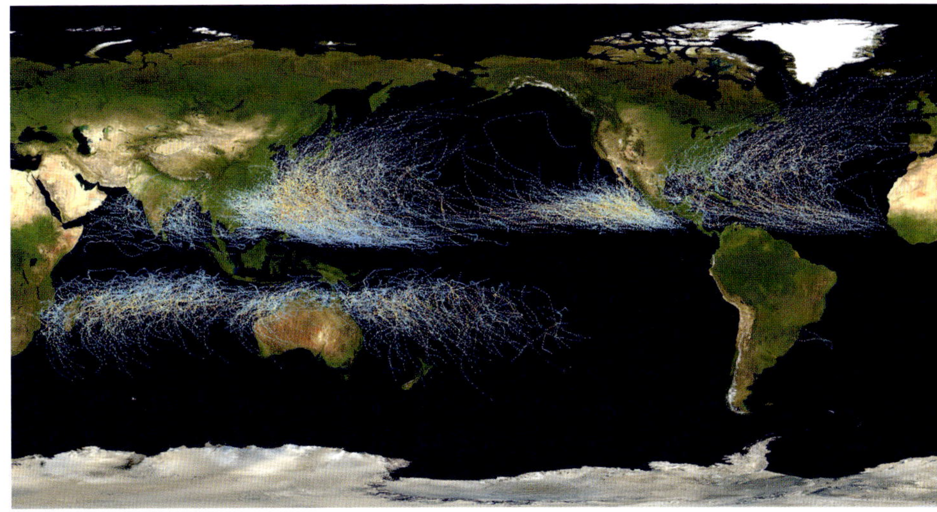

Cumulative tracks of all tropical cyclones, 1985–2005. The western Pacific Ocean experiences more storms than any other region. Courtesy of the U.S. National Oceanic and Atmospheric Administration.

cyclones, and in the Atlantic the word hurricane has been in common use since the mid-nineteenth century.

From their tropical birth, hurricanes are drawn slowly poleward by Coriolis forces that increase with latitude and enhance the pace of movement as storms travel along their path. In the Atlantic, hurricanes typically carve a long, arcing path, steered first in a westerly direction by the tropical trade winds toward the Caribbean and around the immense Bermuda high, a large, persistent, high-pressure system that lies over the north Atlantic. Approaching North America, they tend to curve back to the east with the prevailing westerlies. This recurvature out from the tropics is often accompanied by a notable shift in forward velocity. Nearly stagnant in their early days, hurricanes can roar across northern regions at speeds of thirty to fifty miles per hour, a characteristic that gave the 1938 storm two nicknames, "Long Island Express" and "Yankee Clipper."

The fact that warm water provides the energy for these storms

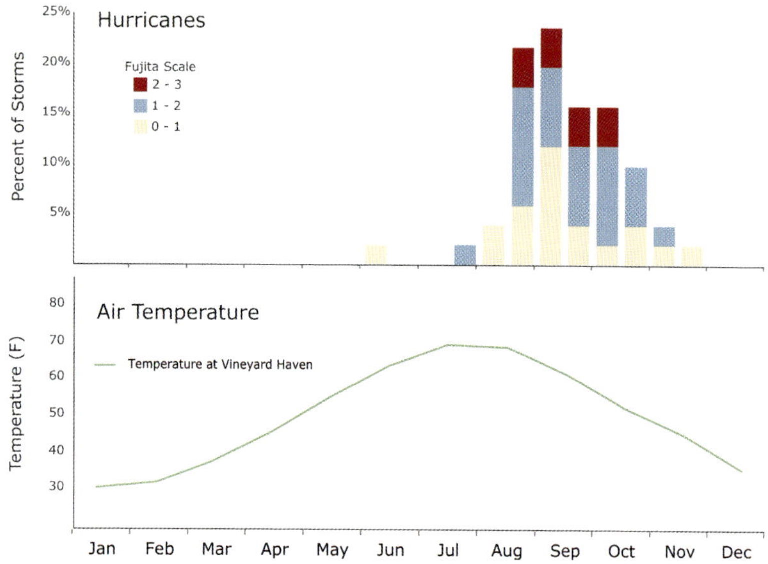

Asynchrony of New England hurricanes and air temperature. Atlantic hurricanes depend on warm ocean temperatures. Because the sea warms and cools more slowly than the air and it takes days to weeks for hurricanes to travel to New England, the peak of hurricane activity in the Vineyard occurs as air temperatures are falling, tourism is declining, but ocean temperatures remain high. The Fujita scale of wind velocity and damage: (1) light: less than 73 miles per hour (mph); branches broken; (2) moderate: 73–112 mph; trees heavily damaged, shingles peeled; (3) considerable: 137 mph or greater; trees uprooted, roofs torn off. Hurricane data from Emery Boose and the Harvard Forest; ocean temperature data from U.S. National Oceanic and Atmospheric Administration.

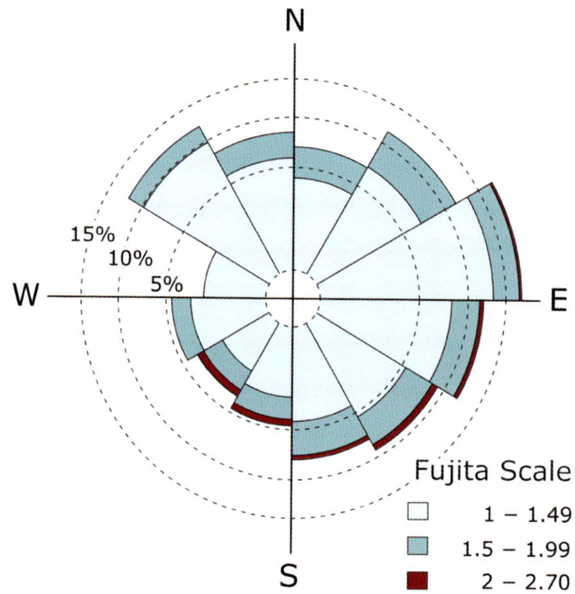

Wind direction and damage level for major hurricanes. This composite wind profile for all major storms that have affected the Vineyard shows that peak winds come from the east and south, but intense winds range from north of east to southwest. Each "petal" in the wind rose shows the percentage of hurricane winds from that compass direction, categorized by Fujita damage scale. Hurricane data from Emery Boose and the Harvard Forest.

explains much about their distribution in space and time. Originating over tropical waters, hurricanes often dissipate rather quickly when they encounter land or cold ocean regions. Hurricanes interact strongly with the surfaces they traverse. Over the ocean their strong winds churn the water deeply, bringing cold water to the surface and restricting the passage of subsequent storms until the upper waters warm again. Solid ground offers more resistance to the wind than water, and so following landfall the resulting friction and declining supply of warm, wet air reduce a storm's energy. Meanwhile, other storm systems may interact with hurricanes, disrupting them with turbulence or feeding into them in ways that sap or, in some cases, even increase their energy.

The seasonal distribution of hurricanes is likewise controlled by energy supply. Water has a much greater heat capacity than land and thereby warms more slowly in the spring and cools more gradually in the fall and early winter. As a consequence of this lag, the hurricane season is offset significantly from the summer months. In July and August, the waters off the east coast are still warming; only as the vacation season winds down toward Labor Day does water temperature peak and the hurricane season become most active. In New England the frequency of storms peaks around September 10. The warm coastal waters and hurricane season then persist well into October.

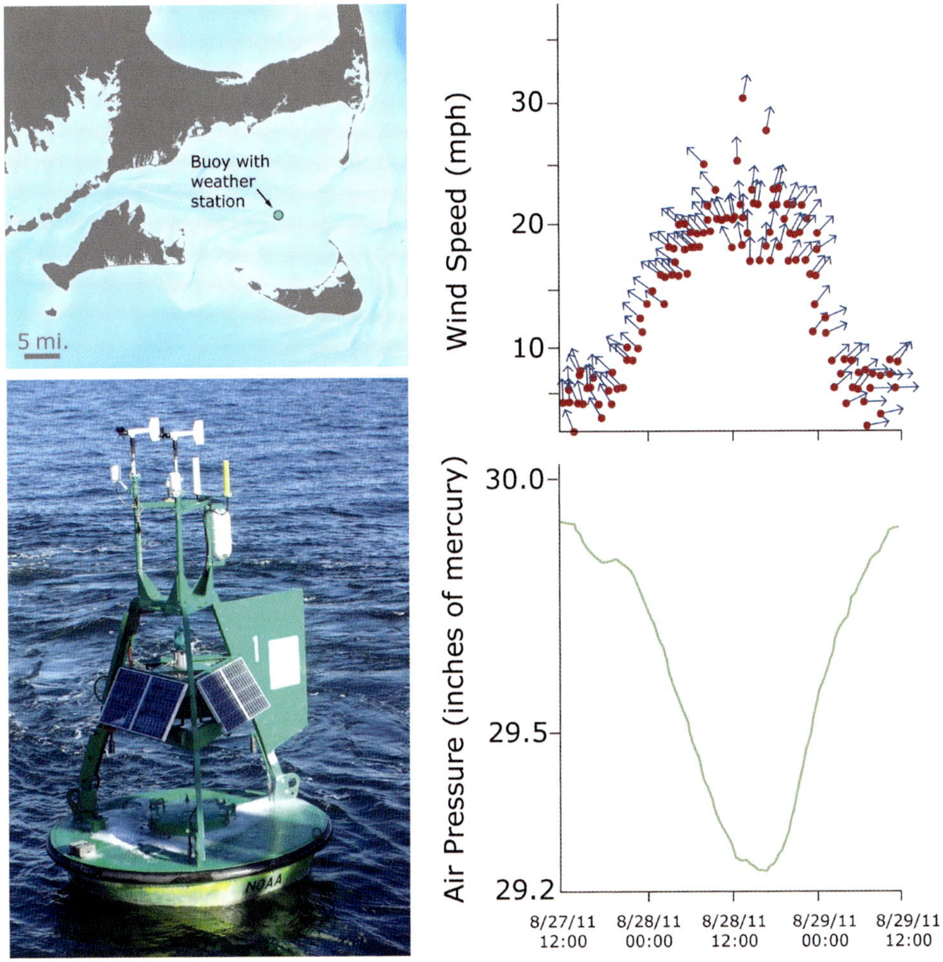

Meteorological characteristics of Hurricane/Tropical Storm Irene as it passed the Vineyard. Buoy data from Nantucket Sound captures changes in wind speed (*top graph, red dots*), wind direction (*blue arrows*), and pressure (*bottom graph*) as the storm passes west of the island. Photo and data from the U.S. National Oceanic and Atmospheric Administration.

A final factor to consider is the structure and movement of individual storms. When a great tropical storm is stationary or moving slowly, the winds circulate symmetrically around the eye, spiraling inward at the surface because of the effects of friction. The wind direction at any given location is determined by its position relative to the storm's center. For simplicity's sake, let's assume the storm is heading due north. As a storm passes to the west of a location, there is a single peak of wind primarily from a southeasterly direction. If the storm tracks to the east, northwesterly winds prevail. If the storm's center passes overhead, there is a double peak of wind from opposing directions, separated by an intervening lull representing the eye. In the tropics, where forward movement is slow, such symmetrical behavior prevails and the wind's intensity is largely governed by the strength of the storm and distance from the storm center. Equally intense winds can come from any direction and to positions on both sides of the track of the storm.

But as the storm's forward movement increases, the wind velocity and the distribution of damage become increasingly asymmetrical; the forward speed of the storm adds to the strength of winds moving in that direction and reduces those in the opposite direction an equivalent amount. Thus in the northern hemisphere, as storms gather speed as they move northward, peak winds begin to concentrate on the right side of the track and the difference in damage on either side of the storm increases. In a rapidly moving storm such as the 1938 hurricane, the difference in wind speed on the right and left sides of the track can exceed fifty miles per hour. Thus, as demonstrated so well in Hurricane Bob, location makes a huge difference. In that 1991 event, areas on the left side of the track in central New England experienced modest winds. In contrast, the Cape and Islands, located to the right of the track and close to the eye wall, received the combined fury of the storm's rotational velocity and its forward momentum.

Synthesizing this information, we recognize that the hurricanes with the greatest role in shaping the New England landscape occur in late summer to early fall, arrive by skirting the eastern coast, and make landfall across Long Island or just to the east. Storms that swing inland south of New England dissipate when they interact with land, whereas those that trend east of Cape Cod concentrate their fury out to sea even as the body of the storm envelops a considerable part of our region.

Given this information, how do we gauge the frequency and impacts of historical storms and interpret their ecological role and future consequences? Anecdotes, statistics, and even native legends hint to the long-standing importance of hurricanes. In recent decades, their global average annual damage to human infrastructure has approached five billion dollars. Over the past two hundred years, nearly two million people have died in hurricanes worldwide. In New England the greatest toll came in 1938, when 580 people died. That storm remains one of the nation's costliest natural disasters—more expensive than the 1906 San Francisco quake, the 1871 Chicago fire, or even the Galveston hurricane of 1900 when more than 6,000 people died. In today's currency, the 1938 storm caused over $45 billion of damage in a landscape with considerably less development than today.

Ecologically, September 21, 1938, remains a defining date for New England. Forest damage covered 35 percent of the region's 42 million acres. With some 275 million trees broken and uprooted, the storm's ecological legacies persist across our landscape, and its memory continues to shape the thinking of ecologists, meteorologists, and coastal planners. From completely reconfigured coastlines to the bent shapes of trees generated more than seventy-five years ago, it is clear that big events can shape the land nearly indefinitely. The age and nature of the word itself suggests the long-standing importance of these storms, as it is variously ascribed to Huracan, the Mayan storm god, or to the Norse giantess Hyrrokkin. But how do we develop an ecological perspective of the role of hurricanes?

It is often said that hurricane impacts are so chaotic as to be indecipherable. This is certainly one's first impression. However, as just described, hurricanes are highly structured by interpretable physical forces. So, while the interplay of many factors makes precise forecasting highly challenging, the reconstruction of their characteristics after they have passed is surprisingly straightforward when the appropriate tools are applied. As a consequence, it turns out that with historical information, some basic meteorological knowledge, and an ingenious model developed by my close Harvard colleague Emery Boose, we can do a good job of reconstructing most storms in recorded history. From that compilation we can interpret the long-term pattern and role of hurricanes for New England and the Vineyard.

We can illustrate this point using a pair of observations that are notable by the fact that they come from the first decades of American history, and the authors—William Bradford and John Winthrop—were leaders of the first New England colonies in

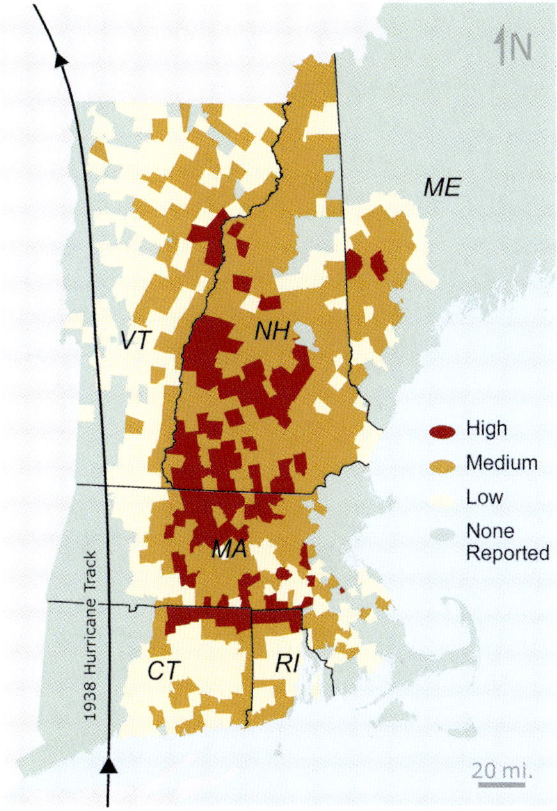

Forest damage from the 1938 hurricane. Greater wind speeds and forest damage occurred on the right side of the track due to its rapid forward movement and counter-clockwise rotation. Connecticut and Rhode Island display less damage than northern states due to less forest cover and less white pine, which is highly susceptible to blow-down. Damage is represented by the volume of timber windblown by town: (none), no damage or data; (low), <1 million board feet; (medium), 1–10 million board feet; and (high), >10 million board feet. One board foot equals a board one foot square and one inch thick. Hurricane track data from Emery Boose and the Harvard Forest. Forest damage data from the New England Timber Salvage Administration (1943).

Postcards showing Menemsha Harbor immediately before and following the 1938 hurricane. Courtesy American Art Post Card Co., Boston, Mass.

Plymouth and Massachusetts Bay (Boston). The storm captured in their notes of mid-August 1635 remains the single greatest natural event in the region's recorded history. These were the right men, in the right place, at exactly the right time to provide unique insight into American ecological history. Using their accounts and the characteristics of this great storm, we can also begin to envision possibilities for New England's future.

From Governor William Bradford in Plymouth comes a vivid depiction of the storm's wrath on land, coast, and man.

> This year, the 14th or 15th of August (being Saturday), was such a mighty storm of wind and rain as none living in these parts, either English or Indians, ever saw. Being like . . . those hurricanes and typhoons that writers make mention of in the Indies. It began in the morning a little before day, and grew not by degrees but came with violence in the beginning, to the great amazement of many. It blew down sundry houses and uncovered others. Divers vessels were lost at sea and many more in extreme danger. It caused the sea to swell to the south of this place above 20 foot right up and down, and made many of the Indians to climb into trees for their safety. . . . And if it had continued long without the shifting of the wind, it is like it would have drowned some part of the country. It blew down many hundred thousands of trees, turning up the stronger by the roots and breaking the higher pine trees off in the middle. And the tall young oaks and walnut trees of good bigness were wound like a withe, very strange and fearful to behold. It began in the southeast and parted toward the south and east, and veered sundry ways, but the greatest force of it here was from the former quarters. It continued not (in the extremity) above five or six hours but the violence began to abate. The signs and marks of it will remain this hundred years in these parts where it was sorest.

And from Boston and the Massachusetts Bay Colony we receive a complementary account from Governor Winthrop.

> The wind having blown hard at S and SW a week before, about midnight it came up at NE and blew with such violence, with abundance of rain, that it blew down many hundreds of trees, near the towns, overthrew some houses, and drove the ships from their anchors. The *Great Hope,* of Ipswich, being about four hundred tons, was driven aground at Mr. Hoffes's Point, and brought back again presently by a NW wind, and ran on shore at Charlestown. About eight of the clock the wind came about to NW very strong, and, it being then about high water, by nine the tide had fallen three feet. Then it began to flow again about one hour, and rose about two or three feet, which was conceived to be, that the sea was grown so high abroad with a NE wind, that, meeting with the ebb, it forced it back again.

Although vocabulary challenges easy comprehension, these two accounts offer strikingly informative descriptions of the storm. Each provides specific details that lend insights into the hurricane's character. Though these are historic men, the fact that so much information can be deciphered at such an early date through rudimentary observations and qualitative descriptions lends optimism to our efforts to interpret storms over the subsequent four centuries as the landscape fills with more and more observers and measurements.

The two colonists write discerningly concerning three types of impacts: wind damage, river flooding, and coastal tidal surges. Regarding forest damage, William Bradford discriminates among processes with distinct ecological consequences and legacies that would collectively "remain this hundred years." "Many hundred thousands of trees" were uprooted, disrupting the earth and forest canopy, or snapped or broken when the roots proved to be stronger than the trunk of the tree, and yet others were "wound like a withe," forming tangles of prostrate stems.

And then there was the sea. While it may seem counterintuitive to climb trees as the wind was felling entire forests, Bradford's description of Indian behavior underscores the rate at which the

The Great Colonial Hurricane, August 26, 1635. The storm supported the greatest winds and surge of any hurricane in New England history and was perfectly positioned to inflict maximum impact on Narragansett Bay, Buzzards Bay, and the southeast coast. *Left:* Reconstructed track, with hourly positions in local time, pressure in millibars (mb), and maximum wind speed in miles per hour (mph). Circles represent the approximate location of maximum wind velocity, thirty-five miles from the center. Red arrows show the reconstructed wind speed directed into the major bays. *Center:* Model results of the storm tides (storm surge plus normal tides) in feet above mean sea level (two-foot contours). Circles are estimated maximum heights of the storm tides for Boston, Plymouth, Wellfleet, and the heads of Buzzards Bay and Narragansett Bay. Heights are extreme due to the coincidence of high tide and maximum wind speed. *Right:* Modeled time courses for storm tides for the heads of Buzzards, Narragansett (Providence, Rhode Island), and Boston Bays. The rapidity with which the tide rose and fell matches Governor William Bradford's description. Data from Brian Jarvinen at the Hurricane Research Division of U.S. National Oceanic and Atmospheric Administration.

rising sea swamped the low-lying coastline. Indeed, recent work by the National Oceanic and Atmospheric Administration (NOAA) corroborates Bradford's phenomenal statement concerning the twenty-foot storm surge to his south in Buzzards Bay. It is thought that a dozen Indians were killed by these waves.

It is also clear that although Bradford and Winthrop witnessed the same storm separated by only forty miles, their experiences were distinctly different. From Plymouth, Bradford saw the cyclone peak with winds from the south and east. In contrast, Winthrop in Boston witnessed the storm's fury pound the coast from the northeast and then northwest. Winthrop also speaks much less about wind damage and more about the impact of driving waves and pounding seas. From these subtle but meaningful differences we can discern that the hurricane passed right between the two men. Their views came from opposite sides of the storm track.

Other eyewitnesses add critical details to our understanding of the storm. Among the most poignant is the astonishing tale of Richard Mather, grandfather of Cotton Mather, the towering religious force who set the tone for colonial Christianity. On the last leg of a transatlantic voyage, Mather and passengers on the ship *James* were engulfed by the hurricane off the Isles of Shoals along the coast of southern New Hampshire. He recounts the fate of their companion ship, the *Angel Gabriel,* a 240-ton vessel that had also departed Bristol, England, on May 25: "Ye Lord sent Forth a most

Nor'easter in March 2014 driving waves onto North Neck on Chappaquiddick.

terrible Storme of rain, and ye *Angel Gabriel* lying in at anchor at Pemaquid, was burst in pieces, and cast away in ye Storme and most of ye cattle and other goodes with one seaman and three or four passengers did also perish therein."

According to Mather's son Increase, the lives of his father's family and their one hundred shipmates on the *James* were saved by divine intervention.

> And I must confess, I have peculiar reason to commemorate that solemn providence, inasmuch as my Father and Mother, and four of my Brethren were then in a Vessel upon the Coast of New England, being at Anchor amongst the Rocks at the Isles of Shoals, when the Storm began; but their Cables broke, and the Ship was driving directly upon a mighty Rock, so that all their lives were given up for lost, but then in an instant of time, God turned the Wind about, which carried them from the Rock of Death before their eyes.

God's hand may have shepherded the group to safety, but the abrupt reversal in the wind resulted from the hurricane's meteorology. As the *James* was dashed toward the rocks, the center of the storm passed over them, bringing a change in wind direction that delivered them from destruction on the rocks.

These early narratives yield crucial lessons for our efforts to reconstruct the hurricane regime. With a small number of careful observations, it is possible to interpret many essential characteristics of a hurricane as it passes through a region. By assembling the commonplace observations of farmers, ship captains, newspapers, and diarists—on the time of day, direction and intensities of wind, rainfall, and damage—we can develop a composite picture of the storm track, size, movement, and likely magnitude over time. When the ability to interpret this data is coupled with the passionate dedication of a historian willing to scour every newspaper, county courthouse, and town hall for such information, it is possible to methodically reconstruct each significant hurricane for a region. This was the goal of historians Sidney Perley, who published *Historic Storms of New England* in 1891, and David Ludlum, who penned the authoritative *Early American Hurricanes, 1492–1870*.

David M. Smith, a legendary forestry professor at Yale, was the first to interpret these histories ecologically. The result was an insightful 1940s master's thesis that unfortunately went largely unnoticed. The approach was greatly improved half a century later by Emery Boose at the Harvard Forest, who developed a computer-based meteorological model to take historical observations and quantitatively reconstruct each storm.

From our consideration of 1635, we can see that even a rudimentary understanding of hurricane meteorology helps to interpret the apparent chaos of these storms. In a ship at sea, a sudden wind reversal is no easier to interpret than the jack-strawed confusion of uprooted trees on land. Likewise, the variation in damage across a landscape or the difference in winds noted from Plymouth to Boston may appear baffling. But, when these patterns are assessed with background knowledge and geographical perspective, they readily fit an interpretable picture. By placing an individual site in a larger context and employing our understanding of these storms' movement and rotation, apparently uninterpretable details may be fit into a larger pattern of order.

One key is understanding the interactions between storms and the landscape, something that became abundantly apparent after the 1938 hurricane. The storm traversed New England at upwards of fifty miles per hour, generating peak winds from a southerly and easterly direction on the right of the storm. As a consequence, as one travels today from New Haven, Connecticut, to St. Johnsbury, Vermont, the predominant orientation of ancient bent and tilted trees, windthrow mounds, and decaying logs is to the north and west. Everywhere this evidence of ancient damage lies on sites exposed to southerly and easterly winds; conversely, the oldest and largest trees occupy slopes and valleys protected from these. Regionally, the greatest damage was concentrated east of the track, where the rotational and forward velocities coincide. There is a sharp break in damage to the west, where the two forces were in opposition. At the coast the barrier beaches of Long Island were overwashed and breached because they lie perpendicular to the strongest winds. At the northern head of Narragansett Bay, Providence, Rhode Island, received the brunt of the ocean surge that was driven up the long, shallow bay, and the city center was flooded with over fifteen feet of water. Meanwhile, the Vineyard and Nantucket lie far from the track of the 1938 storm, which ran just west of the Connecticut Valley. Though few who witnessed the event would necessarily describe it this way, the Islands were spared from the worst impacts. After all, the waters surged through Nashaquitsa and Menemsha Ponds and destroyed the harbor at Menemsha. But that particular storm was only moderate on the scale of great hurricanes to hit the Vineyard. To consider truly catastrophic damage on the southeastern coast, we need to return to 1635.

Using input parameters derived from the observations of Bradford, Winthrop, and others, we have used Emery Boose's HURRECON model to reconstruct the storm. Bradford's observation was dead correct in one enduring way—1635 remains the greatest storm in recorded New England history. Thus its reconstruction helps us understand the general approach used in all storms, envision the seventeenth-century aftermath on the Vineyard, and anticipate a future extreme storm. Intriguingly, the reconstructed path for 1635 lies directly beneath that for Hurricane Bob.

1635 Revisited

There is no record of the origins of this great storm or its path across the Atlantic, although it may have terrorized any number of ships along its course. But its appellation as the "Great Colonial Hurricane" is well earned by its interactions with three New World centers of the British Empire. Just days before it hit the two New England outposts, the storm swept past the tidewater settlement at Jamestown, bringing "pounding waves and shrieking winds" but only modest damage.

The glancing blow to the Virginia coast helps explain the storm's intensity farther north. Sea temperatures were approaching their annual maximum in August and were ideal to feed the hurricane's growth. By skirting the southern shores, where waters reach eighty degrees Fahrenheit, the storm built to a huge and powerful condition, likely the highest intensity—Category 5 (winds exceeding 155 miles per hour)—as it headed north unimpeded by any landmass. On landfall at Long Island, the storm's southeast winds topped 130 miles per hour. The hurricane coursed diagonally to the northeast across the corner of southern New England, leaving Cohasset just a couple of hours later. Wind speeds peaked some thirty-five miles from the storm center, and hurricane-strength winds extended across more than one hundred miles. With a forward movement of thirty to forty miles per hour, winds were highly asymmetrical. And with its compact structure and rapid passage across the low, flat landscape, the hurricane sustained its intensity through its brief transit.

The region was only sparsely populated, and the storm's human consequences were slight relative to subsequent hurricanes. But the storm and landscape were optimally positioned for maximum ecological impacts on the southeast coast. Old forests of large

Recently uprooted beech in the woods on the Polly Hill Arboretum with large soil-covered root mound.

trees were prone to massive damage, and so Bradford's scene of "many hundred thousands of trees" must have played out across sandplains and morainal hills from the Vineyard to Provincetown and the entire south shore below Boston. Evidence of moldering trunks and immense pits and mounds from uprooted trees would have frustrated colonial land clearance. And, as suggested by Bradford and confirmed by our studies of old-growth forests uprooted in 1938, these storm legacies would have persisted more than a century.

Coastal damage would have been even more staggering and begs for translation into our modern landscape of low-lying cities and beachfront developments. With onshore winds the immense storm surge would have overtopped barrier beaches and sandbars, flooded coastal lowlands, eroded headlands and bluffs, and covered vast areas of marsh and shoreline with sand and debris. The strong directional winds of a hurricane generate subsurface currents in addition to immense waves. When these currents reach the shallow waters of the continental shelf and approach shore, the water builds into a storm surge. The rising seas accentuate the tides and push into constricted bays, where they pile up. The greatest surges occur at high tide and at the heads of bays oriented to the south and east, exposed to the full force of hurricane winds. As the water rises to flood beaches, marshes, and shores, the associated waves break, batter, and erode the land and any structures encountered. The magnitude of the impact is partly determined by chance—when the strongest winds hit relative to the tidal cycle.

An ingenious team of scientists at the National Hurricane Center and NOAA have employed a model called SLOSH (Sea, Lake, and Overland Surge from Hurricanes) to reconstruct the storm tides for major hurricanes. Their model results confirm Bradford's narrative—the rise and fall of the surging water can be extremely rapid, with peak swells persisting a mere hour or so. The shape and topography of the coast in relation to the prevailing wind direction also would have generated the great local variation he reported. For 1635 the model estimates that seas rose three to six feet across the Cape, five to ten feet around the Vineyard, and, due to the funneling effect of their long inlets, up to fourteen feet in Providence at the head of Narragansett Bay and the astonishing twenty feet Bradford reports for Buzzards Bay. Based on these results, it appears that some three centuries before construction of the Cape Cod Canal, the Cape became an island for about an hour.

Reconstructions of 1635 set the standard for coastal hurricane impacts. But, the huge difference from today is that this great storm concentrated its damaging forces on nature's infrastructure, which could readily repair itself. The next great storm will encounter a landscape of coastlines and forests occupied by us.

New England's Hurricane Disturbance Regime

Led by Emery Boose, our group has reconstructed the meteorology of each major tropical storm from 1635 to Sandy. The resulting

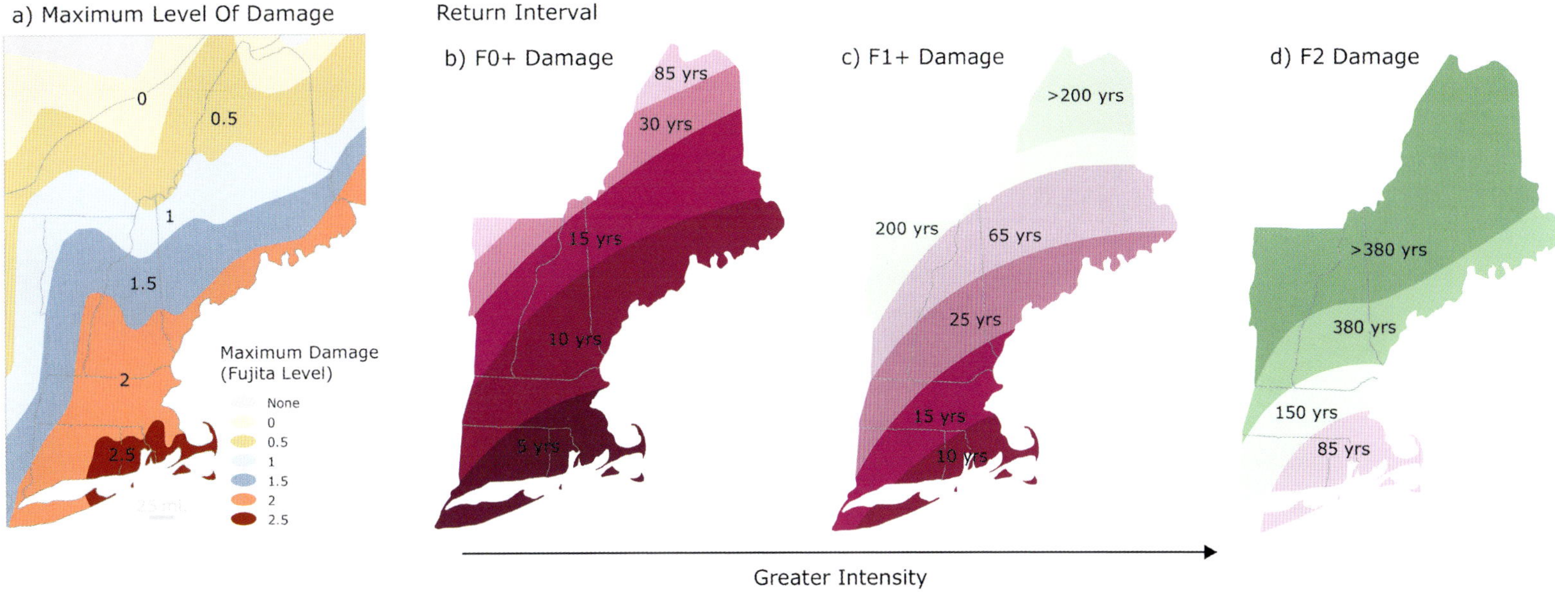

New England's hurricane disturbance regime, showing a pronounced regional gradient of hurricane frequency and intensity from southeast to the northwest. Based on storm reconstruction (1620–1991), the maps depict (a) the distribution of maximum damage inflicted and (b–d) the mean return intervals for the Fujita categories of increasing hurricane damage. Data from Emery Boose and the Harvard Forest.

understanding of the hurricane disturbance regime for New England places the Vineyard in a regional perspective.

Though unsurprising, our first conclusion is worth considering—every storm is unique. Differences in seasonal timing, sea temperatures, and weather patterns generate distinct paths, intensities, and impacts for each hurricane. Even storms sharing similar paths up the eastern seaboard diverge in many characteristics once over land. Extreme variation also occurs within a single event, as seen recently in Irene when the Cape and Islands received strong winds and local coastal erosion, but little rain, while western New England experienced a deluge, powerful flooding, and modest wind. When coupled with the changes in natural conditions and human infrastructure that transpire over time, each storm brings a unique story of ecological and human consequences.

But when we synthesize the information on the nearly one hundred storms that have crossed New England in recorded history, general patterns emerge that allow us to speak of mean conditions, frequencies, and return intervals. Historically, less intense storms are most common, with nearly seventy in the lowest two categories and only ten reaching the most intense category of impact. Over time we see substantial variation in the intervals between storms, but no clear patterns or trends. We see lengthy periods with few storms, some periods of storm clusters, and a few notable years with more than one storm. Long periods with only one or two storms, like 1870–99 and 1900–1925, lie alongside multi-decadal stretches in 1840–70 and 1926–60, when New England was hit by nine or more storms.

Across the four centuries, the hurricane season is well defined. Hurricanes hit New England from July through November, but the peak of numbers and intensity fall from August 15 through

September 30. The truly powerful storms on the Vineyard occurred on August 26* (1635), August 31 (Carol 1954), September 15 (1944), September 21 (1938), and September 23–24 (1815). This seasonal timing has important implications, beyond whether hurricanes land in the tourist season. Many historical reports, including those from 1635, focus on impacts to crops, both from a financial perspective and with an eye toward the ability of local communities to recover and survive the following winter. Whether crops had been harvested or remained in the field made a huge difference for local lives. Meanwhile, late-season storms can occur after trees have lost their leaves; this reduces the surface area intercepted by the wind and the likelihood of major blowdown.

Geographically, hurricanes exhibit consistent southeast to northwest gradients for all important characteristics—frequency of occurrence, return interval, and maximum damage. The region from eastern Long Island and Connecticut through southeastern Massachusetts has received the most damaging storms and is exposed to weaker events every five to ten years. Providence, Rhode Island, lies at the epicenter of historic storms. For most of the region, hurricanes figure regularly into small-scale forest dynamics—causing scattered blowdowns numerous times within the lifespan of a typical dominant tree in the forest—and descend only once every century or two to leave enduring legacies. Meanwhile, much of western New England is largely free from hurricane winds, but not floods.

For forests, the composition and height of the dominant trees, which in turn are largely determined by site conditions and history, control the ecological impact of major storms. Unfortunately, scant data on this subject come from the coast. Surprisingly, the best insights are still drawn from 1938 and the thesis of a Harvard graduate student who assiduously picked his way through the devastated Harvard Forest to measure tree sizes, age, and damage.

* A calendar adjustment explains the difference between Bradford's date (August 15) and that used here.

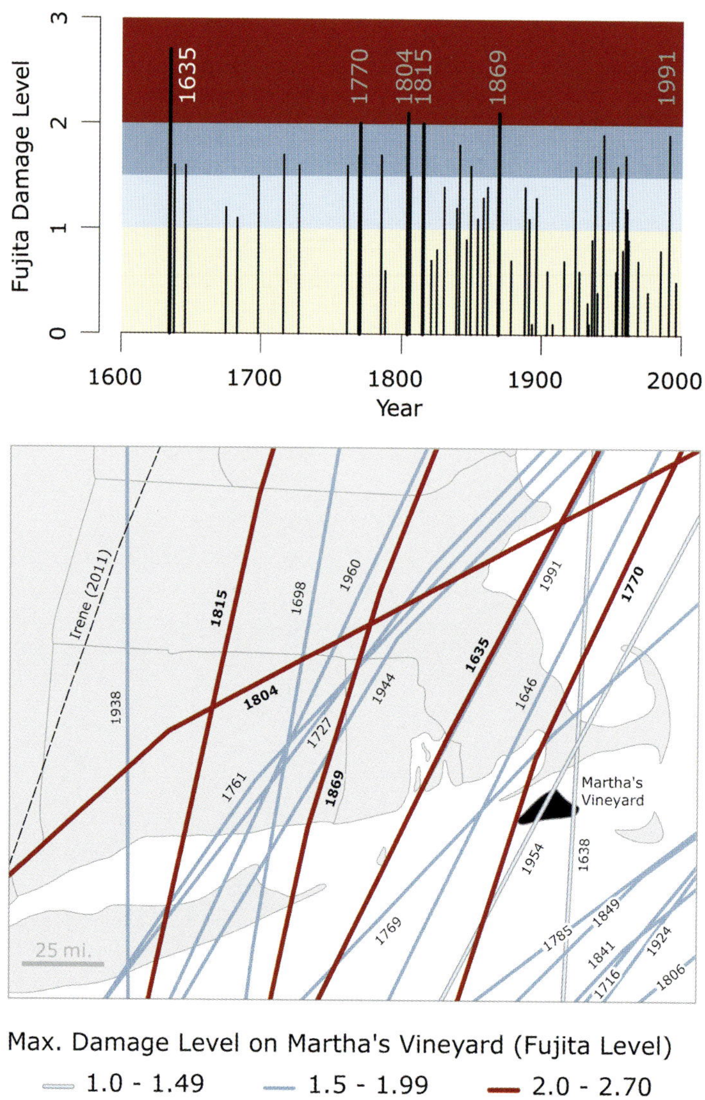

Hurricane history on Martha's Vineyard. Timeline (*top*), storm tracks (*bottom*), and Fujita damage levels (*both*) for major hurricanes (1620–2000). Storms with stronger winds and greater damage pass to the west of the Island. Some storms in eastern New England that had minor importance for most of the region, such as 1635, 1770, and 1991 (Bob), exerted substantial impacts on the Vineyard and much of the southeast coast. Tropical storm Irene (2011) is included in the storm tracks map due to its discussion in the text. Data from Emery Boose and the Harvard Forest.

Oaks above Nashaquitsa Pond browned by salt spray from tropical storm Irene, September 2011.

Willett Rowland's data depict a regular increase in damage with forest age and height along with striking differences among species. Pines concentrate their growth high in the forest canopy and are relatively shallow rooted; consequently, they are much more susceptible to blowdown than hardwoods like oaks, hickories, beech, and maples. This pattern was seen on the Vineyard in Hurricane Bob, when many of the State Forest's pines were broken or uprooted, but surrounding oak forests suffered only slight damage. Other tree-related patterns emerged on the Island in 1991. Native pitch pine proved to be much more resilient than non-native Scots, white, and red pine. Another readily damaged tree was black locust, a southern species that has spread after being planted historically for fence posts.

Rooting conditions also influence wind firmness. In swamps and on shallow soils, roots spread superficially and trees are easily

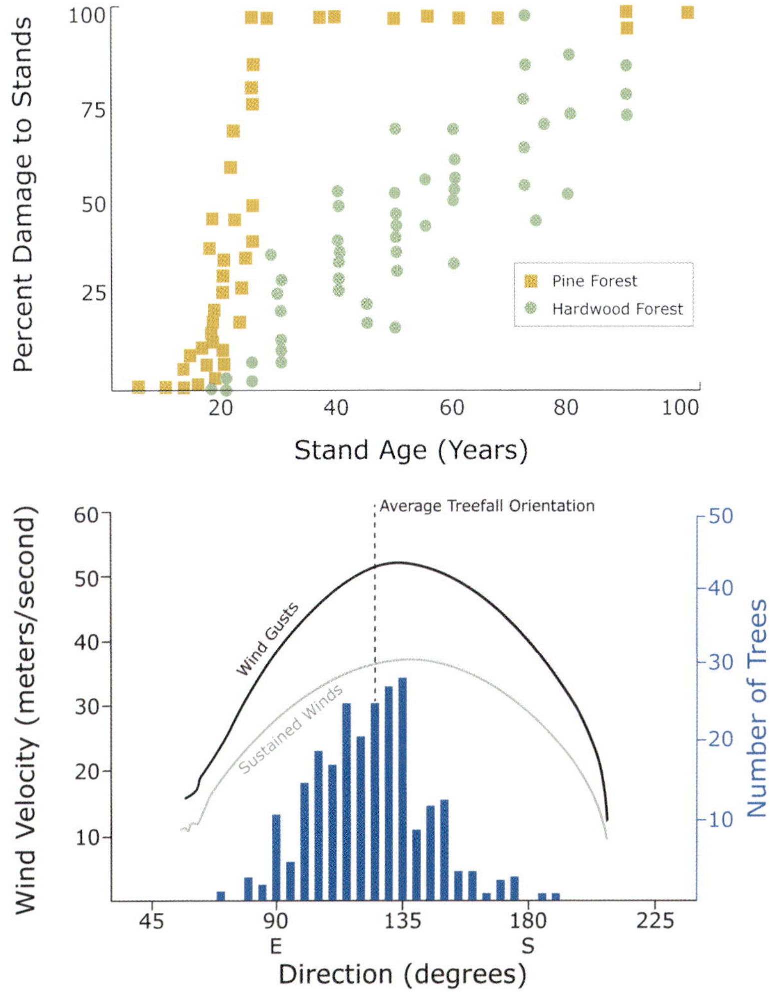

Hurricane damage and forests. *Top:* Following the 1938 hurricane, Harvard Forest researchers determined that forest damage (percentage of uprooted trees) increased with forest age and height but varied greatly between white pines and hardwood species. *Bottom:* As the 1938 hurricane passed the Harvard Forest, the winds picked up in velocity from the east (90 degrees), peaked from the southeast (135), and then declined to the south (180). The orientation of fallen trees indicates that most trees blew down during the first half of the storm's passing as the wind picked up intensity. Note that treefall direction is represented as the direction that the wind came from. Data from Emery Boose and the Harvard Forest.

uprooted. In contrast, in rocky terrain or on bedrock the trees may secure a tenacious grip and fail only if snapped. The type of damage is determined by the relative strength and ability of the many different parts of a tree to withstand the growing horizontal force applied as the wind increases. Young trees are supple, receive less force than taller trees, and generally survive by bending. As their height increases, trees become effective rigid levers that transfer the wind energy to the roots, which can then fail.

Although branch breakage and snapped boles are a familiar image after storms, the majority of trees are uprooted. Crushing or damaging smaller individuals as they fall, windthrown trees generate large mounds of upturned soil and broad depressions or pits whose physical and chemical imprints can persist for centuries as fading legacies of ancient tempests. In general, the signature of hurricane blowdowns is a distinctive pattern of prostrate trees oriented in a northwestern direction, quite distinct from damage caused by prevailing winds (trees oriented to the east) or nor'easters (southerly orientations). In coastal forests on the Cape, Vineyard, and Elizabeth Islands, uprooted trees and remnant pits and mounds primarily date to 1944 or 1991.

A careful analysis of Harvard Forest trees blown down in 1938 reveals other distinctive patterns and relationships. While the peak orientation is to the northwest, coinciding with the direction of the maximum wind velocity, the distribution is skewed toward winds in the earlier stages of the storm's passage. Apparently, the wind reached a threshold when trees began to fall, and this peaked with maximum wind speeds. As winds subsided and shifted to the southwest, few susceptible trees remained to fall. In the aftermath, only trees that could withstand the peak velocities, or were sheltered, remained standing. Many of these were in a battered condition. Within this pattern is another signal, that of tree height. The tallest and presumably most susceptible trees exhibit a more westerly orientation, suggesting they succumbed to lower wind speeds during the early stages of the storm.

The general wind environment for a tree and forest strongly

Foliage of pitch pine seedlings browned and dying due to salt spray from tropical storm Irene, September 2011. The resilient pines readily regrew.

conditions susceptibility to hurricanes. Trees sitting on the coast, a ridge, or singly in a field that routinely experience windy conditions are more wind-firm than those in most forest stands. Meanwhile, a second model developed by Emery Boose called EXPOS reveals enormous variation in wind exposure across a varied landscape like the Vineyard. Locally, opposing hill slopes are exposed to winds from different directions, hilltops are fully exposed, and narrow valley bottoms offer protection from almost any wind. At broader scales, level terrain like that of the Great Plain is broadly exposed, whereas the morainal hills and valleys provide various degrees of shelter. At the regional scale, the flat to gentle topography of the southeast coast yields modest variation in exposure, in contrast to the tall bedrock hills and mountains in central and northern New England.

Hurricanes exert other subtle influences worth considering. For example, ocean salt has an important influence on soil chemistry and plant growth and can damage, kill, and even limit the

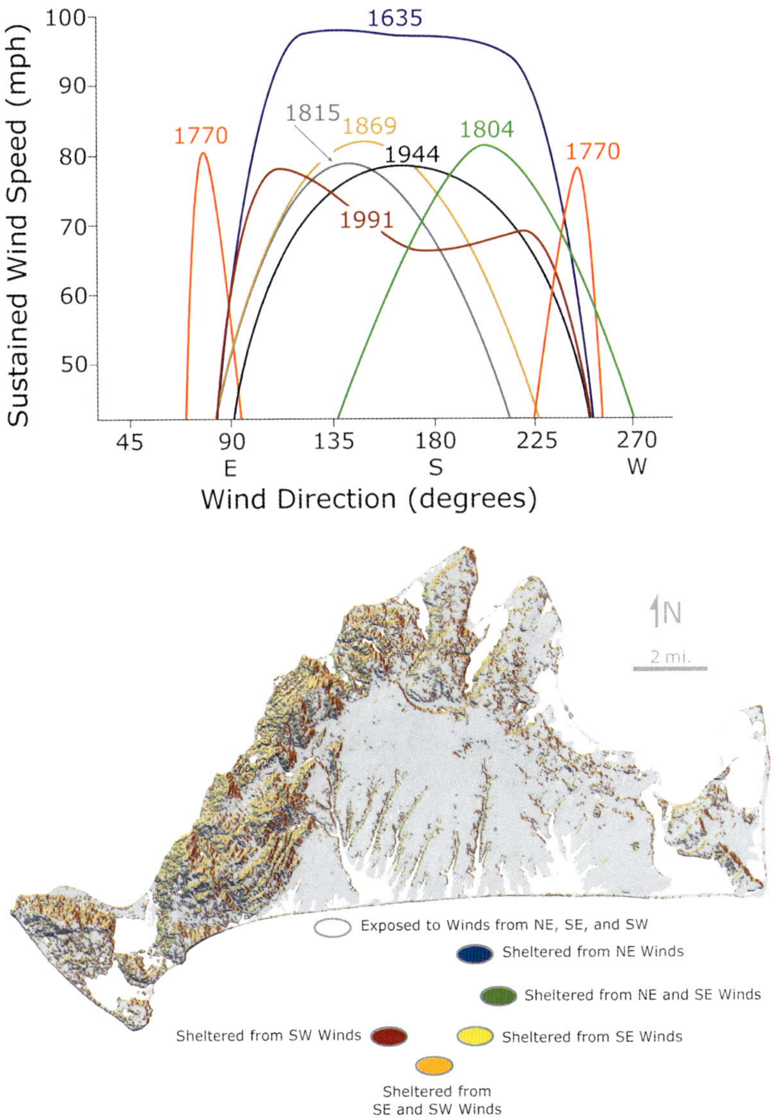

Wind direction and the orientation and exposure of specific sites in the landscape exert a major effect on the resulting patterns of forest damage. *Top:* Historically, the strongest hurricane winds ranged from the southwest to southeast. *Bottom:* Slight differences in wind direction interact with the Island's topography to produce complex patterns of sites protected from or exposed to damaging winds. Produced using the EXPOS model developed by Emery Boose, Harvard Forest.

occurrence of some species. Strong gradients of salt distribution occur with distance from the coast and are controlled by prevailing wind directions. But hurricane winds can bring pulses of salt far inland, where they can leave a coat of brine, turn waters brackish, or brown the needles on salt-intolerant species like white pine. Salt damage occurs routinely on the Island after strong winds. Indeed, it has been long held that salt spray was a major factor limiting the growth of white pine on Nantucket and the Vineyard. While this may have been true when the islands were largely open grasslands interspersed with scruffy woodlands, the role of salt and wind has progressively changed as the islands have reforested. Expansive forest cover now offers some shelter from many effects of wind and weather.

The Human Element

Dealing with hurricanes would be much simpler in the absence of people. Nature has coped with great storms quite effectively forever. It is only when we insert human property, life, and values into the equation that normal processes of nature are cast in strongly negative terms as "disturbances" or "catastrophic." Humans also add complexity to the patterns of damage and consequences of hurricane impacts through their activity before and following storms. With our impact on the earth's climate, we also are altering the very nature of the storms themselves.

We have conditioned the landscape in many ways that can accentuate or reduce the damage from storms and many natural processes. By altering the structure and composition of the vegetation, we create a continually changing mosaic that varies the potential for wind impacts across the landscape. Through the placement and design of our structures, we determine the likelihood that property and life will be in jeopardy from wind or rising seas. On the coast, the construction of structures that protect the shore in one location deflect wave energy and currents to deliver greater impacts elsewhere. In even our most hurricane-prone areas, the return interval between storms is long enough that the memory of the last

Beech uprooted by the 1944 hurricane on the Chilmark Town Forest currently supporting stems that were produced from branches along its trunk.

great impact has often faded before the next great storm arrives. Consequently, our planning efforts seldom adequately prioritize storms. Our zeal to promote other values—a good view or the near-term economy—have promoted many shortsighted, dangerous, and harmful practices.

On the other hand, when it comes to forests, the central concern regarding human activity comes after rather than in advance of major storms. In our determination to fix or remediate what we interpret as negative, we often respond in ways that exacerbate the impacts on nature.

We typically cast the aftermath of major natural processes in strongly pejorative terms—damage, destruction, devastation, deterioration, and catastrophe. Yet these human value judgments have little relevance to nature's condition or ongoing functions. Plants and animals have evolved with great physical and biological forces and have a wide capacity to persist with and recover from them. Like glaciers, erosion, fire, and the browsing of animals, hurricanes are yet another process that modifies the earth. Inherently these processes are neither good nor bad. Though they may kill and maim individual plants and animals or eliminate entire populations, they simultaneously create new habitat and opportunities for others. All studies addressing the impact of storms on natural ecosystems show a remarkable capacity for plants and animals to survive, for species to recover, and for ecosystems to sustain themselves in the face of what we view as calamity. Although the physical appearance may be one of devastation, the many processes (such as the cycling of nutrients and water) that are inherent in an ecosystem continue, and new opportunities often emerge for species that previously had little space. Hurricanes foster a mosaic of habitats—intact mature woods, jumbled areas of piled trees and uproot mounds, and exuberant growth of herbs, shrubs, and young trees. This diversifying action of natural disturbances is one key to the abundance of life. And yet, as we apply our values to the resulting disorder, we often decide unnecessarily to attempt to "repair" and "restore" situations in which nature would thrive if just left alone.

The single greatest example of this impulse comes, again, from the 1938 hurricane. Having blown down three billion board feet of forest across 15 million acres, the storm triggered the single greatest salvage timber operation in U.S. history. One motivator was an irrational fear of fire, but financial interest in the timber played a role, as did the desire to restore order to "a frustratingly chaotic landscape." The consequences can be seen today in nearly every forest affected by that storm. Instead of moldering dead wood and bizarrely shaped and leaning trees, we find rotting stumps and multi-stemmed trees. These are the enduring monuments to a staggering achievement—a nearly comprehensive logging of downed trees in the New England landscape by landowners and timber crews brought in to assist the federal New England Timber Salvage Administration. Every sawmill in the area was enlisted in the effort, and hundreds of ponds and fields were commandeered to hold the logs.

Nearly two-thirds of the damaged trees were extricated from

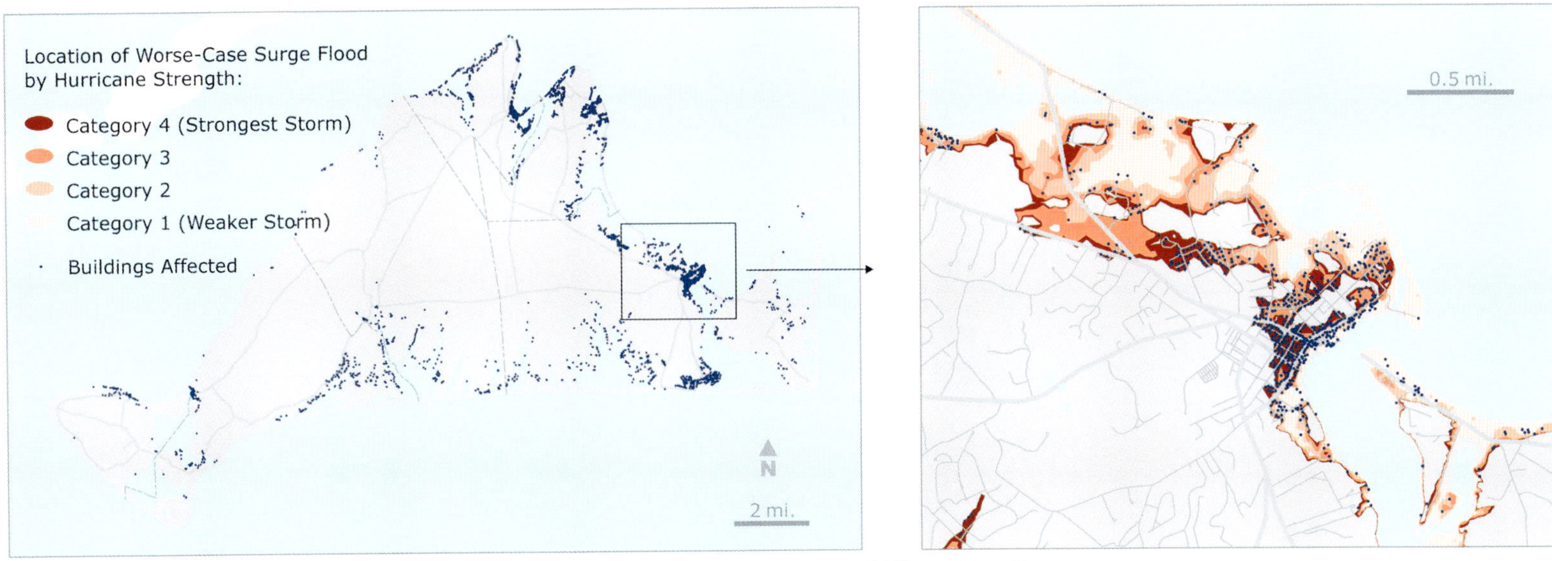

Predicted worse-case flooding for (*left*) Martha's Vineyard and (*right*) Edgartown by hurricanes of different intensity. The worse-case flooding is determined by tidal stage and wind direction. Map generated from data of the Martha's Vineyard Commission.

the woods. As a consequence, our forests may be viewed as having survived a regional salvage operation as much as experiencing a natural event. As loggers fanned across the landscape, they opened new roads, extracted any downed, leaning, or damaged trees, and burned the resulting slash. As trees were cut and hauled away, the land was scoured and then burned, leaving much bare soil. A similar but less intensive effort at cleanup and timber salvage occurred on Naushon after the 1944 hurricane and in the Correllus State Forest following Hurricane Bob.

Salvage logging never helps the forest, but it does interrupt many ecological processes. Despite proclamations that forests are "destroyed" by storms or fires, the ecosystems remain intact. Leaning and broken trees are alive; indeed, even uprooted trees may survive a year or two, and many will live for decades as branches become new upright stems on a horizontal trunk. Surviving plants and even dead trees provide valuable functions—shading and binding the soil, diverting moisture into the atmosphere, and providing habitat. The ecosystem thrives though structurally transformed. Removing trees and wood alters the environment, many processes, and forest composition, as different species are favored. Important forest structures are removed or damaged—downed and decaying logs, leaning trees, root mounds and pits, and the myriad of growth forms resulting from the unique attempt of each tree to deal with its specific damage and compete with others for light. These diverse structures are replaced by features already common in our forests due to their history of logging: multiple stemmed sprouts. Efforts at cleanup simplify and homogenize our woods.

The Vineyard—Hurricanes Past and Future

Due to its location at the region's most exposed periphery, the Vineyard has a distinctive disturbance regime. Hurricanes are frequent and often intense; storm surges are a fundamental process in the landscape's evolution; and a different array of storms figures into the Island's history than inland. Though the 1635 storm was defining on the Vineyard, it influenced less than 10 percent of New England. The 1938 storm had big impacts through waves and the surge but was not accompanied by powerful winds. The four other most powerful storms on the Vineyard were 1770, 1804, 1815, and 1869, with return intervals ranging from 11 to 142 years. Although the storms of 1938, 1944, and 1991 were strong, it is a long time since the Vineyard has been visited by winds as powerful as these earlier ones. Because of the lengthy history of land use and the absence of large trees over past centuries, the Vineyard bears little evidence of these great storms. But due to broad-scale forest recovery and a huge upsurge in human population and building, the landscape stands poised and vulnerable to an immense blow from the next major storm. The consequences will be unprecedented and will leave an enduring impact on the Island and the human psyche.

Harvard University's Pisgah old-growth forest following the 1938 hurricane. Respecting the pristine history of this forest and the scientific value in understanding the natural dynamics of New England forests, the director of the Harvard Forest chose not to salvage the downed trees on this landscape. Today, immense moldering trunks lie in a westerly direction in a thriving forest of hardwoods and hemlock. Reprinted with permission of the Harvard Forest Archives, Harvard University, Petersham, Massachusetts.

Waves That Waste the Coast

Everyone who hears these words of Mine and does not act on them, will be like a foolish man who built his house on the sand. . . . The floods came, and the winds blew and slammed against that house; and it fell—and great was its fall.

— *Sermon on the Mount,* Matthew 7:24–29

In April of 2012, the Chappaquiddick world of Rick and Jennifer Schifter began to change dramatically. Just five years earlier, they had finished their dream summer retreat—a twenty-something-million-dollar compound prominently set on the spectacular bluff at Wasque, the extreme southeastern corner of Martha's Vineyard. From the main house came commanding views south and east across Muskeget Channel toward Nantucket, north over Poucha Pond toward the famed Dike Bridge, and west down the beach that offers some of the best fishing and most exhilarating bathing in New England. But that was then.

By spring of 2012, the football-field-sized lawn that had hosted rock bands and news-catching weddings was largely gone. First, the magnificent beaches of fine fluffy sand were stripped away by waves and the currents. Then, the bluff and lawn began falling into the sea. Along with the land fell hundreds of pitch pines and wind-sculpted oaks that had stretched across the three-hundred-acre conservation reserve owned by The Trustees of Reservations that abuts the Schifter compound. Waves lapped through a tangle of trunks below a thirty-foot cliff of the most delightfully pure but highly erodible sand. At the top, the cliff terminated just two dozen feet from the family's patio and 8,300-square-foot residence.

Eye-catching scenes of the immense compound filled local papers and the *Boston Globe* as contractors applied to the Edgartown Conservation Commission for emergency permits to armor the bluff temporarily while they negotiated the equipment, permits, and land purchases required to move the structures to a neighboring lot. A coastal geologist working with the owners summarized the situation for the *Martha's Vineyard Times.* The rate of erosion—more than two hundred feet of upland was lost in seven months—was "unprecedented," he said, and "an unforeseen emergency."

Yet, in the online version of that article, Tom Dunlop, seasoned writer for the *Vineyard Gazette,* immediately posted an intriguing rebuttal to this expert opinion. Dunlop's note caught my attention, for we had exchanged perspectives on Wasque frequently, and a year earlier he had penned a superb overview on the area's history

Fallen oaks along the eroding bluff at Wasque in April 2012.

of erosion. That piece had documented a close relationship between erosion of the bluff and breaches in the barrier beach at Norton Point that serves as the tenuous connection between Chappaquiddick and the Vineyard. Dunlop's comment was concise. "Far from being 'unprecedented' and 'unanticipated,' the rates of erosion at Wasque Point after an opening through Norton Point are well documented in a report by J. Gordon (Pete) Ogden III . . . in 1974."

How could two informed perspectives be so diametrically opposed? Recent storms have generated enormous erosion on the Vineyard and down the eastern seaboard. What's the cause of these damaging events? How do they compare with past episodes, and what does this portend for the future? Finally, what triggered the apparently radical shift in the rate of erosion and pace of life for the Schifter family at Wasque? As in most discussions of current conditions and future expectations, looking as far into the past as possible is good strategy. Fortuitously, when it comes to coastline dynamics, the Vineyard and Wasque, in particular, offer as comprehensive a record as any place in the United States.

In 1886, Henry L. Whiting, assistant to the superintendent of the U.S. Coast and Geodetic Survey and owner of a picturesque farm in West Tisbury that remains in his family today, summarized his four decades of detailed surveys along the Vineyard coastline. Whiting's crew had first mapped the Island in 1845–46. With leadership from men with elite military training like Whiting, the Coastal Survey was an ambitious federal effort commissioned to document the nation's shoreline and harbors in order to advance the country's strategic military, nautical, and commercial enterprises. Noting changes in just a few years after his first surveying campaign, and spurred by his love for the Island, Whiting updated the original cartographic sheets nearly every decade thereafter. When the field season closed in 1886, he retreated to West Tisbury to compile a set of maps and reports. He was particularly interested in what his Island friend and colleague on the Massachusetts Cartographical Commission, Professor Nathaniel Shaler at Harvard, called "rolling beaches." These long and narrow barrier beaches separate the Island's great ponds and saltwater marshes from the ocean and are capable of migrating or "rolling" inland over time as the coastline retreats, all the while retaining their general form.

In a precise and matter-of-fact style, Whiting's report to the superintendent of the Coastal Survey lays out an astonishing tale of change in the Island of geological proportions. In forty years, the entire south shore of Martha's Vineyard had receded a staggering 175 to 200 feet. Isolated areas like the crest of Chilmark's Nashaquitsa Cliffs had lost more than 220 feet. In the process, the barrier beaches had rolled into the great ponds, filling their southern ends, pinching arms into isolated ponds, and completely erasing smaller bodies of water. Whiting noted evidence of waves overshooting the barrier beaches, carrying sand more than 500 feet into Chilmark Pond and nearly 700 feet into Great Herring Pond (Edgartown Great Pond). But none of this rivaled what he reported for Wasque and the three-mile-long sinuous beach that we call Norton Point.

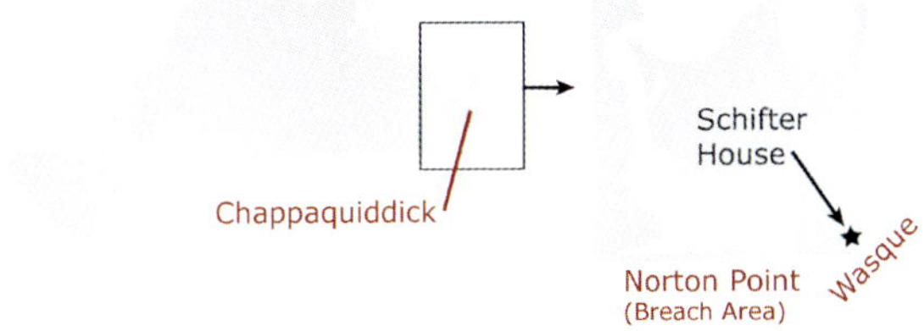

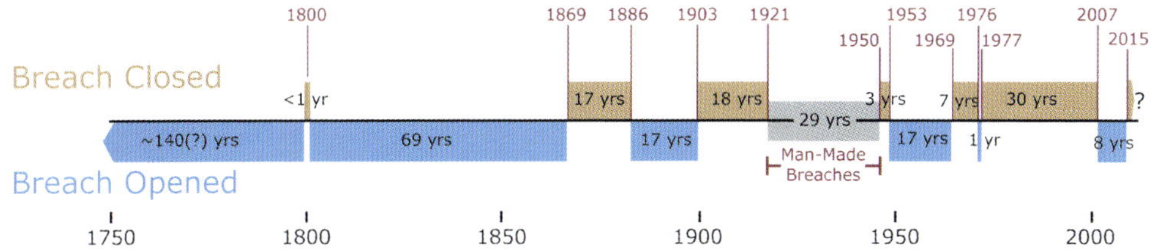

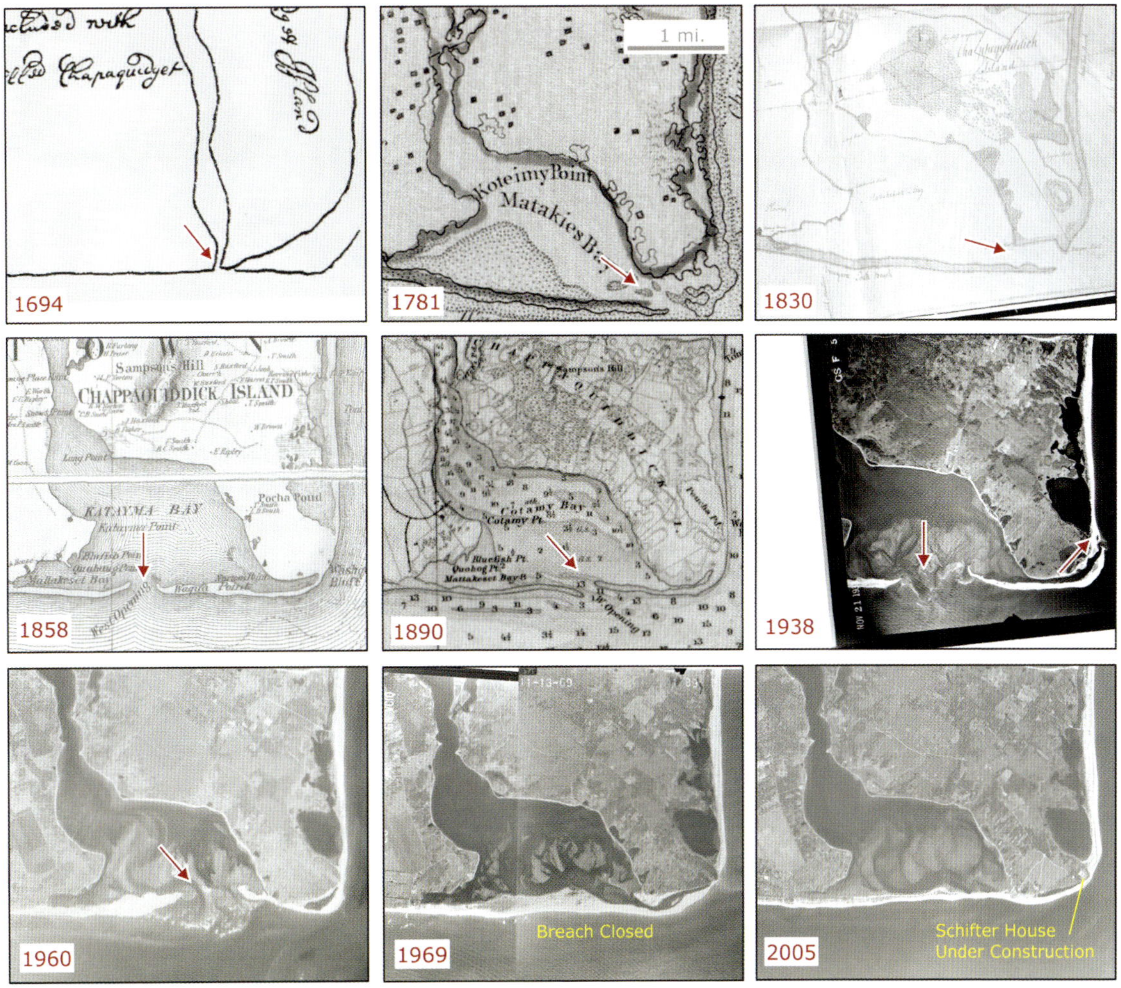

The map (*top left*) shows the location of Chappaquiddick, Wasque, and the Schifter house in 2007 before the Norton Point breach. The graph (*above*) shows the history of opening and closures of the Norton Point breach over time. Freeman (1807) and other sources indicate that before 1799, Chappaquiddick was a separate island back to the beginning of records. In contrast, since the late nineteenth century, the barrier beach has been closed more often than not. The period preceding the Schifter house construction was the longest known duration for the intact barrier beach. Data gathered by David Foster and Tom Dunlop from many sources.

Historical maps (*left*) depicting the Norton Point breach (arrow). When the breach closed in 1969, with the exception of a brief period between 1976 and 1977, it remained that way until 2007. In 2005 the intact barrier beach enclosed the narrow Swan Pond south of Wasque. 1694 map by Simon Athearn (from Banks, 1911); 1781 map by Joseph F. W. Des Barres, Courtesy of the Norman B. Leventhal Map Center at the Boston Public Library; 1830 map by Henry Crapo, Courtesy Massachusetts Archives; 1858 map by H. F. Walling, Courtesy of Harvard Map Collection, Pusey Library, Harvard University; 1890 map by U.S. Coastal Survey; 1938, 1960, and 1969 aerial photos from U.S.G.S. Earth Explorer; 2005 aerial photo MassGIS, Commonwealth of Massachusetts EOEA.

Henry Laurens Whiting, 1821–1897

He could make a better map of the far side of a hill without seeing it than others could while standing before it.

A recognized leader among the elite topographers associated with the Coastal Survey, Henry Whiting was much more than a brilliant scientist and part-time professor at MIT. He also served in military ventures, assisted Massachusetts and the nation in developing their most economically important and strategically positioned harbors, promoted Vineyard agriculture and infrastructural development, and helped interpret one of coast's most important natural processes—erosion. A son of an Albany-based judge and descendant of Governor William Bradford, Whiting was eulogized in *Science*, the world's leading scientific journal, in endearing terms that match the visage in his portrait: "[having] a dignity of manners which is usually associated with 'a gentleman of the old school' along with a simplicity of character and openness of heart that made him beloved by all who came in contact with him."

As related in an 1891 letter, Whiting's decision to settle around the old parsonage and farm in West Tisbury was prompted by his first encounter with the Island: "In my survey of this coast in my younger days, I was much taken with this Island of M's V, and it has become the only home I have had since." Here he was joined by Nathaniel Shaler, fellow scientist and member of the state topographical survey commission. The two forged an enduring friendship grounded in their studies, love for the Vineyard, and promotion of local agriculture.

Henry Laurens Whiting (*center, sitting*), chairman, with other topographers at the convention of the U.S. Coast and Geodetic Survey, 1892. Photograph from the U.S. National Oceanic and Atmospheric Administration.

Whiting maintained an active though economically marginal farm and joined Levitt Thaxter and Ichabod Luce as founding members of the Martha's Vineyard Agricultural Society in 1858. That agricultural connection continues as his descendants remain farm landowners and in 1994 provided the land for the Agricultural Society and Fair to move from the center of West Tisbury to its present location on Panhandle Road. Whiting's grandson Edwin Newhall Woods placed a conservation easement in memory of his mother (Henry's daughter), Virginia Whiting Newhall, on 532 acres of family land (the Woods Preserve) to prevent future development and the family ultimately donated the property to The Nature Conservancy.

Through four decades, this narrow barrier of sand had breached five times, in each case producing an inlet that allowed the free exchange of water between the ocean, Katama Bay, Edgartown Harbor, and Nantucket Sound to the north. The breach severed Chappaquiddick from the Vineyard, forming two distinct but nestled land masses. Each episode of breach formation followed a similar storyline of repair and closure. Each began with a narrow opening that first expanded westward toward the Vineyard, often wildly and reaching a mile or more in breadth. At that point, the opening began to close eastward and narrow as the barrier beach gradually extended toward the southern base of Chappaquiddick. Once the opening was positioned adjacent to the smaller island, it directed the strong tidal currents that rushed into and out of Katama Bay against the southwestern corner and south shore of Wasque. As these waters roared through the breach twice daily with the alternating tides, they combined with coastal currents and storm waves to tear at Wasque's corner and exposed shoreline. The beaches were variously stripped and rebuilt in a baffling confusion of dynamics that could change daily as the current's flow changed velocity and configuration. But the general dynamic was one of terrific erosion of the landmass and the sweeping of sand, vegetation, and chunks of peat out to sea.

With time amid stretches of calm weather, and following the deposition of great quantities of sand in the waters adjacent Wasque, the barrier beach extended parallel to the base of Chappaquiddick and eventually fused to its southeastern end. The south beach at Wasque then often reestablished a pond called Swan Pond that formed in the old channel, and residents and visitors could once again travel over solid land between the connected islands. Once the breach was closed, the extreme erosion ceased and the southern end of Chappaquiddick settled down to the much slower recessional pace of the rest of the south shore. Through Whiting's four decades of observations and five openings of the breach, the upland eroded back over five hundred feet, more than twice the rate of the south shore. From Whiting's knowledgeable perspective, Wasque was the single most erosive spot on Martha's Vineyard and one of the most rapidly disappearing landscapes on the northeastern coast.

The barrier beach at Wasque encroaching across the wetland and pond following the breach at Norton Point, August 2009.

Thus, while the Schifters had set their house back some three hundred feet from the cliff edge with little thought of confronting the sea, nature took its own, hardly unprecedented course. The year they completed their home, a breach was triggered by an April storm. Whiting's script then played out in a spectacular, and highly predictable, fashion. As the breach migrated toward Chappaquiddick, the southwestern corner of the island took the first hits, and the land there was stripped away at an astonishing pace. Then the currents and waves attacked the southern end and eastern corner. Storms took the entire beach out, and waves began eating into the bluff in daily chunks. Over a six-month period in 2012, the currents swept away more than two hundred feet of upland.

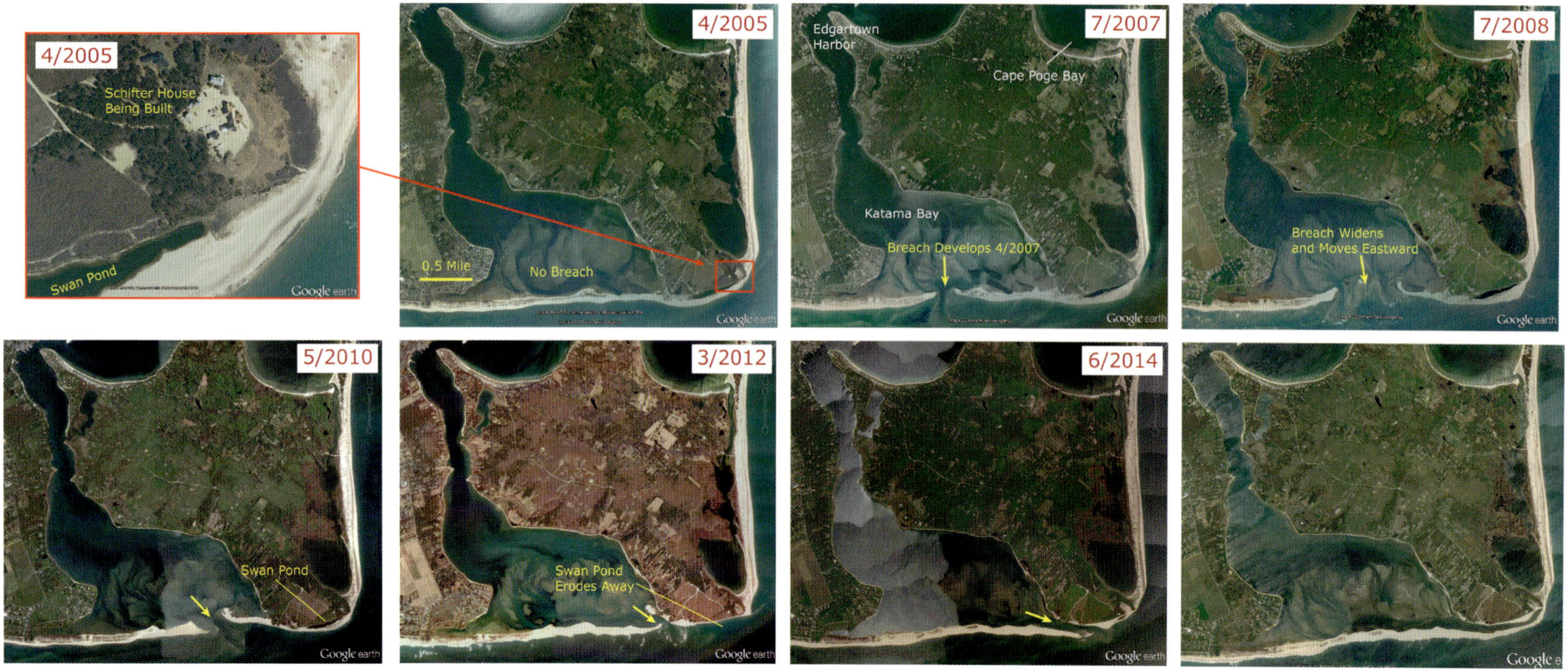

Eastward movement of the 2007 breach in the barrier beach that connects Chappaquiddick to the Vineyard. When the breach reached the southeastern part of Katama Bay, it directed strongly erosive currents on Wasque's south shore. By 2014 the barrier beach began extending past the western edge of Wasque. 2005 photo: Google, MassGIS, Commonwealth of Massachusetts Executive Office of Environmental Affairs; 2007 and 2008 photos: Google, NASA, USDA Farm Service Agency; 2010, 2012, and 2014 photos: Google, NASA; 2015 photo: Google, SIO, NOAA, U.S. Navy, NGA, GEBCO, TerraMetrics.

The problem underlying the decision to build the house is clear. The Schifters, their planners and engineers, the town boards that approved their building plans, and even the coastal geologist who opined on the situation after the fact were deceived by taking too narrow a view of history. When the house was first proposed, the barrier beach connecting the two islands had been intact for more than thirty years. During this unusually lengthy quiescent period, residents and visitors became accustomed to driving from the Vineyard to Chappaquiddick and fishermen and bathers enjoyed routine access to a broad and benign expanse of beach. This long stretch of time erased the tumultuous past from most minds. Above all, it allowed everyone to overlook everything that Henry Whiting had carefully documented and published. But this lengthy quiet period was historically anomalous. It turns out that the Schifters chose to construct their house toward the end of the single longest period without a breach in recorded history.

A historical perspective yields an even greater surprise. It turns out that in Henry Whiting's day and previous centuries the existence of separate islands—Martha's Vineyard and Chappaquiddick—was the typical condition, and periods of intense erosion were the norm. Although the beach occasionally connected during the period of his study, Whiting states that a connection was so unusual that he had to reach far back in time to identify a previous period when it was possible to walk between the islands. He wrote, "There is the tradition of a continuous beach . . . as at present, in the early part of this [nineteenth] century; and that people passed with teams from the main island to Chappaquiddick; but that the link was only temporary, a new opening breaking through the beach again in a few months after it was closed. *With this exception, there is no record or tradition of the non-existence of an inlet through Cotamy Beach since this section of the coast has been known.*" Early maps and reports confirm Whiting's conclusion; there was a continuous breach going back well before the American Revolution.

From the perspective of Whiting's records and Pete Ogden's and Tom Dunlop's research, the past century is distinct from the long-term record. The lengthy periods with intact beaches and moderate pace of change that twentieth-century residents and tourists became accustomed to, and which the Edgartown boards relied on in permitting construction, is not the historical norm. The dynamic condition of breach and erosion that confronted the Schifters fits much more comfortably with the deeper view of time. That condition also better fits the Wampanoag name for Chappaquiddick—"separated island."

The larger importance of Whiting's work goes far beyond Wasque. More broadly, his meticulous measurements along the entire coastline confirm that erosion is a central element of the Island's geological story, one that remains active today and will figure strongly into its future. For twenty thousand years, the Island has been shrinking. Although a few limited stretches of the coast, such as Dogfish Bar north of Gay Head, are building through accretion, the continual loss of land is the predominant process. The coastal currents that receive material stripped from the bluffs and beaches do not give it back. In the 1970s and 1980s, Whiting's assessment was updated Island-wide by geologist Clifford Kaye, who confirmed that the south shore continues to recede nearly five feet annually. When we update these measurements using Whiting's map, we find that since 1850 the south shore has receded anywhere from 250 to 1,000

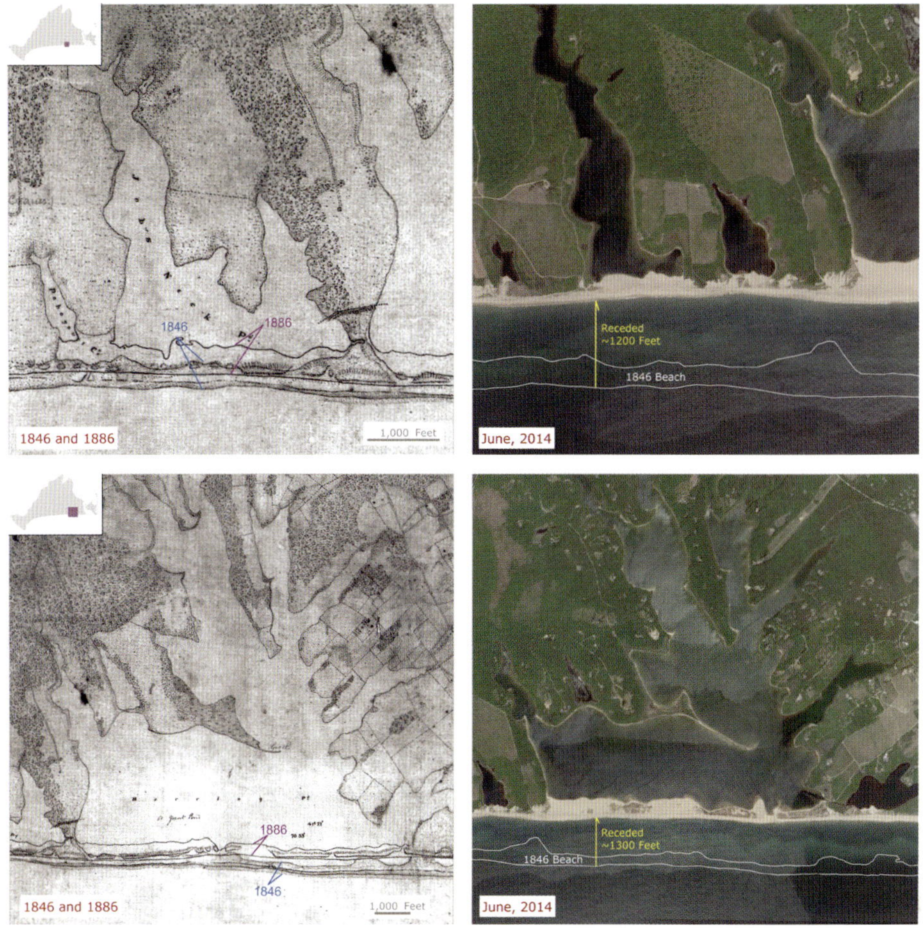

Martha's Vineyard is shrinking visibly. Historical and modern depictions of the Jobs Neck region (*top pair*) and Edgartown Great Pond (*bottom pair*) highlight changes mapped by Henry Whiting in the nineteenth century and those to the present. As the shore erodes northward, coastal ponds shrink and separate into a series of smaller water bodies. Black and white maps provided by U.S. National Oceanic and Atmospheric Administration. Aerial photos: Google, NASA, 2014.

Erosion and deposition around Martha's Vineyard based on data from Clifford Kaye's 1973 study. The size of the blue bubbles indicates erosion rates (in feet per year) whereas the positive brown numbers indicate areas of deposition. Dogfish Bar, on Aquinnah's north shore, is one area with active accretion of sand.

Scattered hay bales from a failed effort to restrict erosion and overwash into Chilmark Pond, April 2011.

feet and the Island has lost 1,200 acres of prime beachfront, some 1.7 percent of the total land area of the Vineyard. The northwest and the northeast coasts angle toward the gentler waters of Vineyard and Nantucket sounds and experience an erosional loss only 10 percent of this rate. Still, the Island is shrinking on all sides.

Mechanisms of Coastal Erosion

In addition to their detailed maps, Whiting and his colleagues left a clear explanation for the dynamics they recorded and the landforms that these coastal processes produce. The coastline evolves through the interactions between persistent daily processes and episodic storm events. The dominant chronic process is "longshore drift," the incessant action of waves that arrive obliquely to the shore, gently erode sands, and then progressively sweep these into the stream of coastal currents along the shore. This continuous coastal conveyor

Coastal armament of imported quarried rock along the east side of Squibnocket Ridge. While protecting the land and house immediately behind them, these barriers divert energy to adjoining shores, where they enhance erosion.

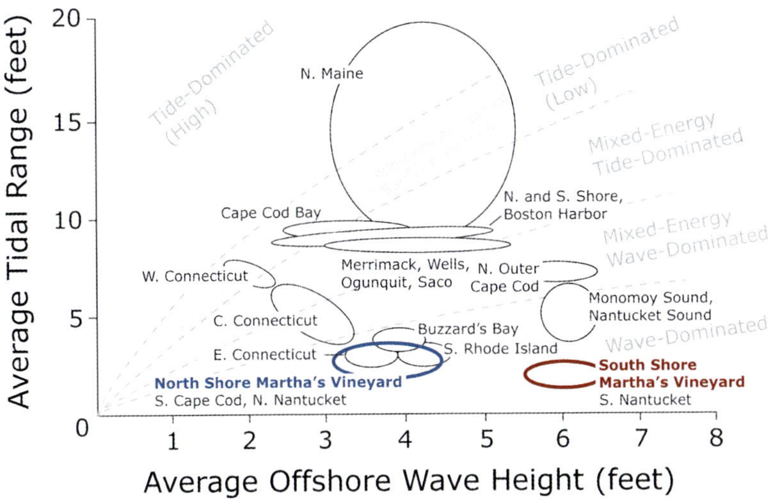

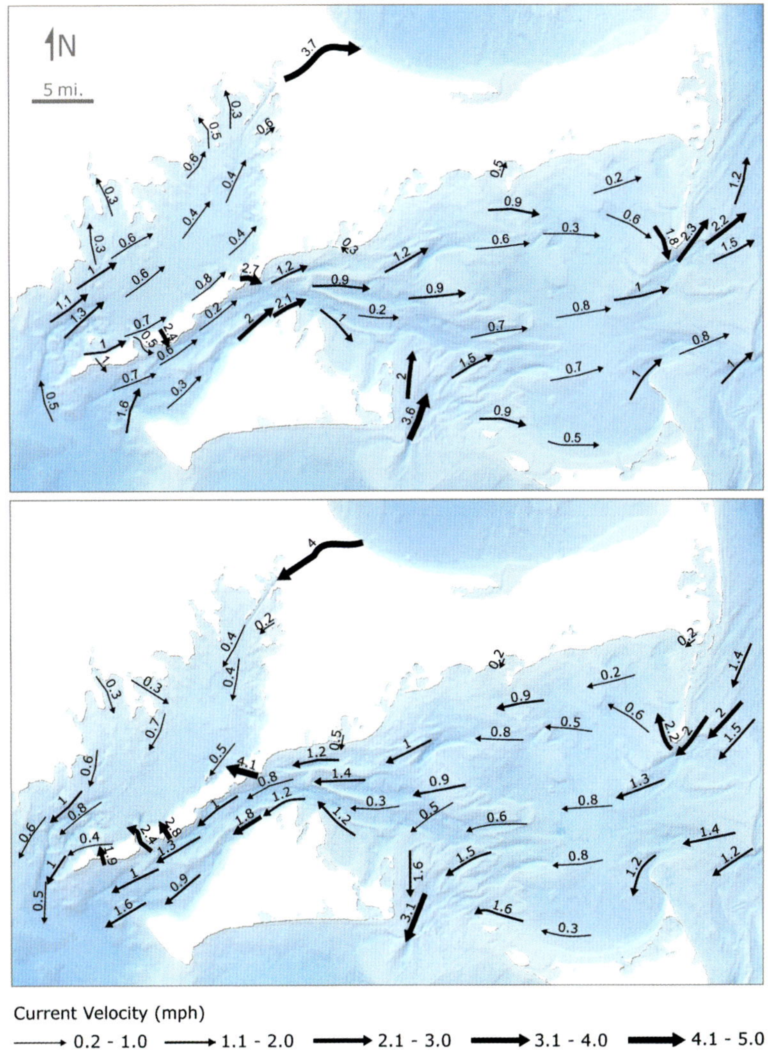

Above: The tidal range and offshore wave height for northeastern sites. While the Vineyard has a small tidal range, the south shore receives high-energy ocean waves. The north shore faces Vineyard and Nantucket Sounds and experiences less wave action. Figure redrawn from Fitzgerald and Van Heteren (1999). *Right:* Twice a day, strong currents course in opposite directions through Vineyard and Nantucket Sounds shaping the shoreline, shoals, and ocean bottom. The current direction and velocity displayed here correspond to three hours after flood (*top*) or ebb tide (*bottom*) at Pollock Rip Channel southeast of Monomoy Point. Redrawn from White (2002).

replenishes eroding beaches, bolsters baymouth bars, spits, and barrier islands, and eventually builds ocean shoals in locations where the currents sweep off the Island toward deeper waters. Through this ongoing process, the "destructive" erosion of any given area yields supplies of sand that nourish downshore beaches and landforms. While the entire Island continues to shrink, this regular resupply of material guarantees the dynamic stability of features like barrier beaches. While appearing stable, these beaches are actually changing incessantly as they continually erode and migrate inland. This conveyor of sand also allows these features—spits, barrier beaches, and shoals—to adjust and persist in the face of changing sea level and to re-form after major coastal storms. But this process of landform maintenance is contingent on continual erosion and resupply. The process cannot be altered in any given place without changing the status of sites further downshore. By preventing erosion or retaining sand at any one place, adjoining areas will be starved of essential new material and will erode away.

The cumulative power of these coastal processes is astonishing, as summarized by Whiting in his mapping of the three-and-a-half-

Beach and adjoining dunes on Red Gate Farm, Aquinnah.

mile barrier beach between Katama and Wasque: "In other words . . . 116,500,000 cubic feet [of sand], has been beaten in upon the bay and shore, by the steady encroaching action of the ocean-waves, in a period of twenty-five years. . . . This is not a case of any great convulsion or powerful action. It is but a fair illustration of the gradual but unceasing *waste* which is going on upon such shores."

A comprehensive description of this longshore process was provided in a companion report in 1869 by Henry Mitchell, Whiting's colleague on the Coastal Survey, who went on to become a professor at MIT and member of the National Academy of Sciences. Mitchell had an even longer connection with the islands than Whiting, having been raised in a Nantucket home with his sister Maria, who would become a renowned astronomer. In his analysis, Mitchell begins by distinguishing among three types of materials that are eroded and shaped by coastal processes: *upland bluffs* that were originally laid down by the glacier; *beaches* that had been deposited more recently by the waves and can reach a maximum height of eight feet or so; and *dunes* that are whipped by winds oftentimes to great heights and distances from the shore. He then proceeds to outline the dynamics of these landforms that can help serve as a basis for their conservation.

> Professor Whiting's repeated topographical surveys along the south side of Martha's Vineyard show that the bluffs are being continually undermined by the sea, and since the waves, running before the prevailing westerly winds, strike the shore diagonally, there is a tendency of all materials to work down to the eastward, a tendency which seems to have resulted in the choking up of all the indentations of the original shore-line, and finally in the formation of shoals. . . . I have no doubt that Wasque Shoal has received supplies from the entire length of the south shore of the island. . . .
>
> The material which falls from the headlands into the seas is not borne far off, but continually crowded toward the shore. As the waves drive up on the beach, great masses of sand and stones are forced on shore, but the reacting "under-tow" carries back always a portion of the finer material. By the grinding action of the breaker, and the selecting operation of the under-tow, the strand becomes largely composed of fine sands, easily caught up by the dry winds, and formed into dunes. . . .
>
> The most casual observer may distinguish in Cotamy [Norton] Beach a contrast of appearance between the wind-worn and the water-worn districts; the former characterized by irregular mounds, the latter by smooth surfaces, declining in either direction from a crest-line to the sea. . . .
>
> The wind plays its part upon a stage previously raised by the water, and builds higher still. The coarse grasses which take root in the dunes arrest the flying sands still more effectively, and with them the fertilizing dust of the shore. In this way the levee thrown up by the sea is gradually strengthened, till it becomes a permanent barrier against the storm.
>
> Where exposed to the open ocean, the levee maintained by the sea is seven to eight feet above ordinary high water; in bays and

The opening in the barrier beach at Norton Point in October 2011 when the channel leaving Katama Bay was positioned against the southwestern corner of Wasque. The Wack house, still well protected by the intervening beach and uplands, is at the far right.

sounds, where somewhat sheltered, it is less. Of course, storms easily pour over it, and these, by sweeping sands from front to rear, cause the levee to move back without changing its form usually. . . .

Whiting added to this understanding by documenting a second set of processes—the convulsive geomorphological changes induced by hurricanes and other major storms. These episodic events ratchet forward the loss of beaches and uplands followed by a return to the longshore processes described by Mitchell that smooth and sculpt the shoreline. Extreme events are especially important in breaching barrier beaches.

In Whiting's day, Vineyarders were accustomed to and preferred the breached condition. Inlets to the sea allowed ponds to flush and circulate with clean salt water, providing entry to ocean fish and increasing the productivity of many types of shellfish. The opening of Katama Bay was especially desirable as it provided a southern access to Edgartown Harbor for pilot boats, fishermen, and pleasure craft, and opened the brackish waters to herring and eels. These desirable qualities led to a long history of pond openings, especially at Chilmark, Tisbury, Edgartown, and other great ponds and an immediate effort to breach the Cotamy (Norton) Beach whenever it connected to Chappaquiddick.

In his account of the natural opening created in January 1886, Whiting described a two-step process required for the successful breach of the barrier beach guarding Katama Bay. First, the beach is weakened and lowered by overwash from waves largely coming from the ocean. This action is generally facilitated by powerful onshore winds that coincide with high tide. The next step is then initiated by winds from the north that push the waters of a high tide on the bay side over and through the beach. As the powerful outgoing tidal current sweeps through the opening, it cuts down and widens the outlet. In short order the breach exceeds fifty yards in width. Once created, the persistence of the breach is dependent on what Whiting describes as a battle between the tidal currents that scour the opening and the "sea dash" of waves and longshore currents that work to seal it by rebuilding the barrier beach.

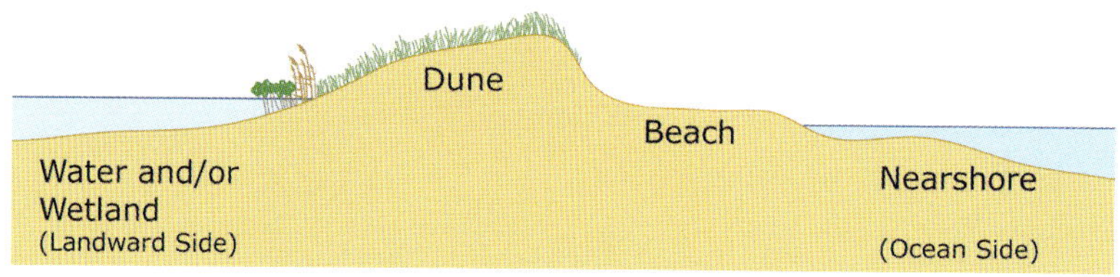

Barrier beach cross-section (*below*) and major processes structuring the Vineyard's south shore through supply, erosion, and transport of sand and materials (*above*). Waves strike the shore obliquely and then retreat perpendicularly (*yellow arrows*); longshore currents transport sand primarily in an easterly direction (*light blue arrows*); under higher tides and storm conditions, waves overwash the beach transporting sand into coastal ponds and migrating the beach inland (*red arrow*); following a breach tidal currents flow into and out of coastal ponds. Photo: MassGIS, Commonwealth of Massachusetts EOEA, 2009.

Whiting keenly noted a critical and readily interpretable difference in the behavior of breaches in coastal ponds and in Katama Bay. In ponds, the power eroding and maintaining a breach comes from the height difference between the pond and low tide in the ocean. Once the pond is opened to the sea, the water flows

Erosive power of the water flowing through an opening purposefully cut in the barrier beach of Tisbury Great Pond. With the rising and falling of the tides, the water runs freely into and out of the pond, flushing nutrients, creating brackish conditions, and allowing ocean fish and organisms to enter, July 2014.

out until the pond and ocean are of equivalent height. Thereafter, sea water flows in at high tide and out as the tide reverses. These relatively modest flows are weak in relationship to the constant building action of the waves and longshore processes. Consequently, breaches in ponds are generally short-lived, closing in a matter of days, weeks, or occasionally months. In contrast, a breach in the barrier beach stretching across Katama Bay does more than simply open a closed body of water to the sea. It creates an open passageway between the north and south shores of Martha's Vineyard, which differ in tidal rhythms, as revealed by Henry Mitchell's detailed studies. This difference in the timing of the two tides generates extremely strong currents and great erosive forces through the harbor, bay, and channel that alternate twice daily in direction and are capable of maintaining the openings for years or decades.

It is the extreme power of this tidal current, in association with the coastal erosion, that leads most people today to dislike the breach at Norton Point. At its most powerful, the current creates havoc in Edgartown Harbor, leading boats to swing around and pull wildly at their moorings, challenging boats navigating the channel, and greatly increasing fuel usage for boats in regular transit like the Chappaquiddick ferry *On Time.*

The Future of Coastal Erosion

With up to 70 percent of the Massachusetts coastline eroding, more than 10 percent suffering from severe damage, and over one billion dollars needed to simply restore existing seawalls (where desirable), there has been much recent effort to anticipate future dynamics. While the pounding surf, storm waves, and the sweep of currents shape the shoreline, it is the relentless increase in sea level that is driving much of the ongoing loss of land. Although the land has given way for twenty thousand years, the process of sea level rise has changed significantly over time. The rate was most rapid when the great glaciers melted in the Northern Hemisphere, after which it slowed substantially but never ceased. For the past few thousand years, the ocean has crept upwards at less than one millimeter per year. Since the mid-nineteenth century, that rate has increased significantly, but in patterns that are uneven around the globe. In the recent past the Northeast has experienced one of the fastest rates in the United States—nearly six inches (fifteen centimeters) in the past half-century, or three times the long-standing rate.

The recent rise in sea level is driven by numerous interacting factors: expansion of the ocean's volume as it warms; ongoing melting of glaciers; land subsidence in the region just beyond the furthest extent of the last continental glacier; and tectonic shifts within the earth. New evidence also suggests that climate change is slowing the circulation of the Gulf Stream in the north Atlantic, causing water to pile up and deepen in our region as strong currents from the south continue to push water into the Northeast.

Among these factors, climate warming and glacial ice melt represent the largest uncertainties for the future. Current projections of three to six degrees of warming through the end of the century lead to large differences in the rate of conversion of ice into seawater. Based on this range of projections and associated increases in sea level of from 1.5 to more than 4.5 feet, estimates

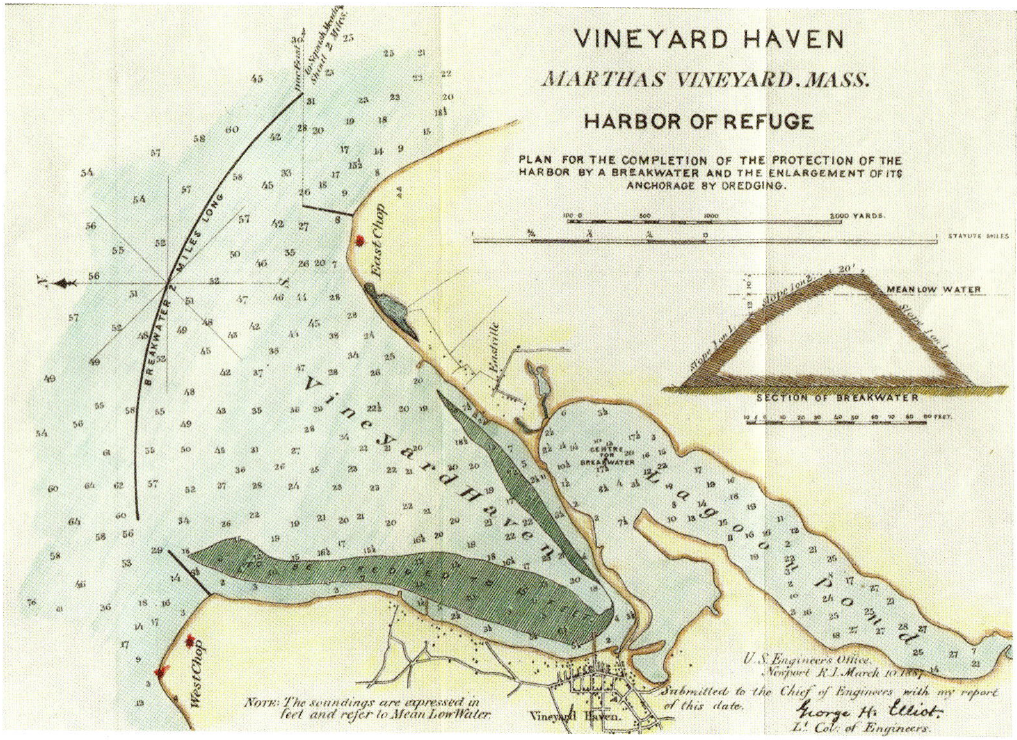

An 1887 plan for a massive breakwater (seen at far left) and dredging of the anchorage at Vineyard Haven harbor was rejected by Congress due to the great expense and declining need for a coastal layover. Map from U.S. National Oceanic and Atmospheric Administration and produced originally by the U.S. Corps of Engineers.

suggest that the south shore will erode an additional five hundred to two thousand feet by 2100.

As we look to this future, three factors may lead to much greater rates of erosion and damage than in recent memory. The first is simply that sea level is rising at an accelerating rate. Another is that the most extreme damage is associated with intense coastal storms, which should increase as a result of both natural cycles and climate change. For hurricanes, the period from 1970 through 2010 was relatively quiescent, with only half as many powerful storms as the previous four decades. As we enter another period of more frequent extreme storms, we can expect that these will be magnified by additional energy from global warming. Finally, the quiescent period coincided with a building boom that situated considerable new infrastructure—especially residences and commercial buildings—in harm's way. On the Vineyard, we are fortunate because the oversight of a coastal District of Critical Planning Concern by the Martha's Vineyard Commission has greatly constrained construction in the most vulnerable areas. Nonetheless, there is increasing need to apply a historical understanding of coastal erosion with anticipated changes in sea level and storm intensity in the permitting of new construction and removal or retrofitting of existing infrastructure.

Back to Wasque

As we anticipate the future, we can revisit Wasque and other rapidly changing parts of the Vineyard for insights. We have much to learn from the past. Remarkable sources of information and perspective emerge with just a little digging into archives and published articles. As a simple case in point, all the information needed to predict the

Precariously positioned houses above the eroding Wequobsque cliffs, October 2011.

Chappaquiddick by and in the sea. A 1913 development proposal for Wasque that was never built included hundreds of house lots bounded by avenues to be called Watcha, Wasque, and Poucha Drives. Today approximately one-third of the proposed development would be in the Atlantic Ocean. Courtesy of the Martha's Vineyard Museum. Original blueprint map by M. A. Millhouse, Civil Engineer 1912. May 2015 base aerial photo: Google, SIO, NOAA, U.S. Navy, NGA, GEBCO, TerraMetrics.

erosional catastrophes experienced by families on Chappaquiddick and the Vineyard's south shore is clearly laid out in nineteenth-century reports from the Coastal Survey, which are all available online. Whiting, Mitchell, and colleagues do not simply outline the dynamics of these coastal landscapes; they provide comprehensive and robust explanations of the mechanisms that drive these changes along with estimates of rates that can be projected into the future.

The fact that the latest projections of coastal vulnerability by the Massachusetts Coastal Erosion Commission employ Henry Whiting's maps from 1850 along with the sophisticated SLOSH model (described in the previous chapter) provides a hopeful sign of new reliance on historical data. Efforts by the Martha's Vineyard Commission and the Vineyard Conservation Society are making these projections and other sound information widely available to the Island population and municipal boards.

Regrettably, we have often failed to invest the modest effort required to uncover the historical information needed to save us from ill-advised conclusions and costly mistakes. Instead, we frame our expectations and base our actions on the narrow and insufficient record derived from recent memory. As the bluff was collapsing at Wasque, one news article stated, "Historically, openings have occurred infrequently in the two-mile barrier beach." The recent and anomalous stability of the beach at Wasque became the baseline condition for planning. Indeed, from 2003 to 2006, when the original Schifter house proposal came before town boards in Edgartown,

Groins on Sylvia Beach adjacent Sengekontacket Pond create a scalloping pattern of sand retention, October 2011.

One fascinating lesson from Wasque's history is that the management impulses of society shift wildly with changes in the physical and social context. In Whiting's era, when an open Katama Bay was familiar and much of the population looked to the sea for livelihood, a wide-open inlet was valued. The open bay provided easy transit for fishermen seeking Atlantic waters, for pilots heading to the assistance of the large ships navigating Vineyard waters, and for residents seeking a day trip to the south shore for pleasure. The opening was also critical to the Mattakesett herring run and to good fishing in the bay and harbor. In his review, Whiting completely ignored the erosional downsides of the opening. After all, in his day no one would think of living perched on an exposed bluff high above the sea. Therefore, when a barrier beach connection was formed between the Vineyard and Chappaquiddick in 1869, his admonition

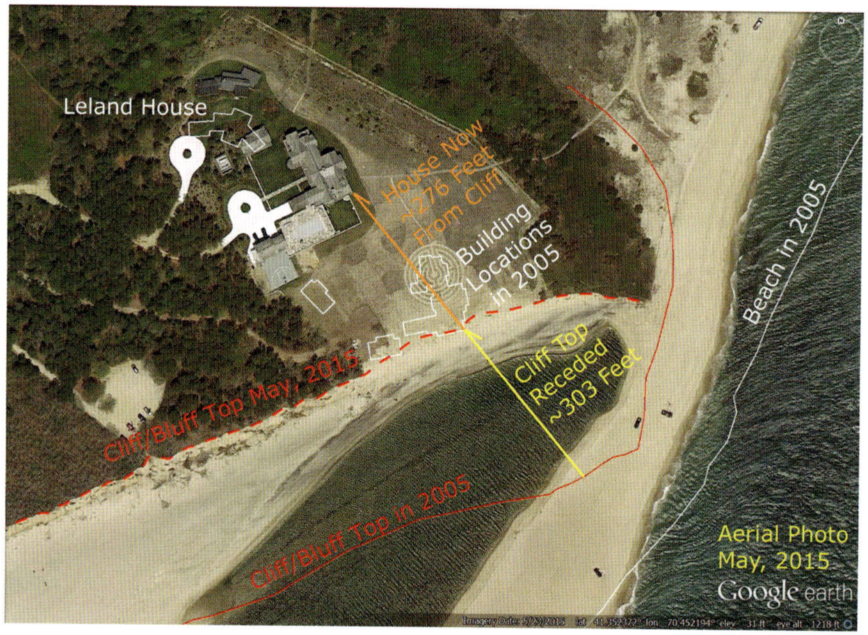

By June 2014 the Schifter buildings were relocated and the bluff had eroded up to their original location. In 2015, following the move, the main house stood 276 feet from the cliff's edge, slightly less than it had originally. Base aerial photo: Google, SIO, NOAA, U.S. Navy, NGA, GEBCO, TerraMetrics, May 2015.

public opposition and official scrutiny were largely focused on the massive size of the structure and its lack of fit to the site and Chappaquiddick. There was almost no discussion of the location's vulnerability to the sea.

In sharp contrast, in Henry Whiting's day, the baseline condition was an open inlet. Consequently, as we review this history of a dynamic process that is contingent on many other processes—sea level rise, coastal storms, and tidal surges and their serendipitous interactions—we are challenged as we seek to make detailed long-term projections. The record from the 1600s to the present displays two completely different modes of operation for this section of the coast, against a backdrop of incessant erosion of the entire south coast. To add complexity, the landscape and ocean context continue to change, as does the global environment. Therefore, while we can expect the general processes of coastal erosion to continue and accelerate over time, many details may vary century by century, making it impossible to predict the future of any given place.

Groins near Cow Bay in Edgartown absorb the impact of the waves from a December 2012 nor'easter. Houses on Chappaquiddick appear in the background.

was immediate and diametrically opposed to the modern inclination: "With regards to the question of opening Cotamy Beach by artificial means, and thus restoring the former regimen of the bay and harbor, there seems to be but one answer to be given—it should be done at once." The early, historical era of management of the inlet at Katama Bay was focused on keeping it wide open.

That inclination shifted with time and context. A successful breach in 1937 was followed by the 1938 hurricane and major damage to Edgartown's waterfront. That catastrophe brought allegations that the absence of a solid barrier beach to the south enhanced the damage. In the context of this debate and historical closures in 1903–21, 1950–53, 1969–76, and 1977–2007, sentiment shifted to supporting closure. In the intensively built environment of the twentieth century when recreationists were equipped with four-wheel-drive vehicles, many advantages of the beach were heralded: protection from a storm surge, gentler tidal flows, greater mooring capacity in the harbor, more consistent shell-fishing conditions, and access to spectacular beaches and sights on Chappaquiddick.

Two notable efforts aimed at closing the breach in the 1970s. The first was an Army Corps of Engineers proposal to erect a ten-foot-high berm south of South Beach. Environmentalists, including *Vineyard Gazette* editor Henry Beetle Hough, vehemently opposed this plan, which was later deemed impractical given the extreme currents and storms. The second was a futile effort by Edgartown resident George Silva to choke the breach with hundreds of loads of sand. Such sentiments continue today. In 2008, after the most recent breach, Edgartown's town manager filed a $511,000 proposal with the Federal Emergency Management Agency (FEMA) to reestablish the barrier beach. The stated rationale was to provide an emergency escape route back to the Vineyard for Chappaquiddick's 375 households. While local resident and environmentalist Edo Potter quickly characterized this effort as ridiculous ("What would we escape from?" queried Mrs. Potter from her delightful setting at Pimpneymouse Farm), the proposal made clear that our approach to managing nature is highly contingent on recent experience and social and economic context. With regard to coastal dynamics, the priorities of the nineteenth and twenty-first centuries stand widely apart.

When we do seek to manage nature, the best methods are generally gleaned from nature's history. In his 1872 article, Henry Mitchell provided a recipe for opening the bay based on his many records of natural breaches. His solution: prepare a trench in the beach on the ocean side; await a high spring tide on the Katama Bay side of the beach when the moon and sun are aligned; and then, with a strong wind blowing from the north, open the breach and allow the erosive forces of the elevated bay water to do the rest. "In the natural formation of an opening through the littoral cordon [barrier beach] of the coast, the waves play only a preliminary part; *it is the current which digs the channel,*" he wrote. "The overleaping waves weaken the dike, but it is by filling the basin behind to overflowing that the breach is opened wide and deep."

Finally, Wasque tells us much about the subtle factors that

The Wack house on the southwest corner of Wasque in December 2012 as the waves beat against a single line of hay bales and debris deposited by the current.

influence our perceptions of nature and how these can be altered by modest changes in circumstances. As I have watched the bluff and south shore erode, I often wondered about our many different relationships with the sea—source of food and resources, focal setting for recreation and relaxation, but also a terrifying force that takes lives, land, and possessions. For many of us, the sounds of the sea, along with the foghorns, bells, and breakers, are a soothing backdrop to lives near the ocean and a relaxing presence drifting in our bedroom windows at night. How quickly that changes when the sea begins to approach too closely.

For three decades Jerry Wacks and his family enjoyed life tucked away in a modest house at the southwestern corner of Wasque, on the point opposite the Schifters but originally much farther from the ocean. Following the breach in 2007, it was their home that drew first attention, for the tidal flows in and out of Katama Bay cut the beach and upland on both the bay and ocean sides toward them. As the shore eroded, they lost most of their land, then their well and fresh water, and finally their septic system. Then, for two years, backed into the far corner of their property, they and their mostly packed-up house waited either for the day when it needed to be disassembled or for the breach to close and stop the erosion. According to Jerry it was during those days when the sounds that used to soothe began to change. "Listening at night, listening to a pounding storm surf, and you know the sands are shifting, you know the sands are coming at you, but you don't know how big the waves are, and you don't know where the wrack line is going to be this time," he said.

As for the Schifters, a new, but in some ways familiar episode has been initiated. With the house, pool, and guesthouse lying perilously close to the thirty-foot cliff of sand, they launched one of the biggest

The colossal effort to protect the Schifter house before it was moved included building tiers of coir logs, sand in burlap tubes three feet in diameter and fifty feet long.

The Schifter house in April 2015 following the move and immediately after the breach had closed. The highly erodible nature of the sandy bluff is apparent.

home relocation projects on any U.S. coast. A terraced system of immense burlap-wrapped and sand-filled tubes, each three feet in diameter and fifty feet long, provided temporary protection against the waves. Teams of engineers, contractors, machine operators, and laborers worked methodically to excavate the compound's structures and then dig an eighteen-foot-deep trench (to accommodate the oversized basement and bowling alley) and new foundation onto the neighbor's property that was purchased and whose house was moved for the relocation. Despite repeated recitations of the newly rediscovered erosional history of the Wasque landscape, the local conservation commission granted permission for the house to be moved back just a few hundred feet. All that was accomplished with great media fanfare. The house now sits a shorter distance from the bluff than it did in 2007 when the entire episode began.

While this may appear reckless or extravagant—the scene darkened when deep erosional channels appeared in the bluff during the winter of 2014, and then brightened as the barrier beach extended past Wasque and the breach finally closed in April 2015—it is not possible to predict which of the patterns from the past will transpire in the near future. The closure followed exactly the pattern documented by Whiting and Ogden and anticipated by Tom Dunlop. But which future will Wasque follow? If Henry Whiting's era repeats itself, the buildings may need to be moved again or torn down in a few years. But, if the twentieth-century story plays out, the family could have a decade or more of enjoyment before the sounds of the sea begin to gnaw at them in bed.

Woods Preserve with massive stonework adjoining the barn with a view of the house and Fisher Pond, one of the sources of water power for Dr. Daniel Fisher's gristmill downstream on Mill Brook at Crocker Pond.

1850: The Landscape That Became Today

Tisbury. Three miles westward of Holmesville [Vineyard Haven] . . . the traveller is brought to the top of Merry's Hill, so called. . . . Taking your position there, and looking off in the direction of South and East, a level plain is seen spreading out its even surface . . . till it touches the line of the impinging ocean. Twelve miles distant, a continuous forest intervening, is Edgartown with its towers and spires. Beyond are the heights of Chappaquiddic; and the tongue of land, whose termination is named Cape Poge, is clearly visible. Holmesville, with its neat white houses, is seen; and far off the South is the old ocean—dim, mystic and vast.
— *Vineyard Gazette*, June 15, 1855

Chilmark is a very superior farming town.—All the south shore below the central range of hills . . . is of a superior quality—excellent for both grass and grain—easily tilled, and exceedingly productive. . . . We will hazard the assertion that there is not in the State, with the exception of some portions of the banks of the Connecticut river, a tract of as many acres lying together, with less poor land.
— *Vineyard Gazette*, July 9, 1846

Edgartown. Of the numerous villages in New England, which I have visited there are few that will compare in beauty of external appearance with this. . . . With its position upon one of the best and most picturesque harbors in the world, with its view of the sound with its busy fleets and the far reaching Cape [Poge] on the East, the Highlands in the West, and the broad-spread, leaping, foaming, dashing and thundering Ocean on the South, it may bid defiance to all the New England coast to present a more inviting retreat to the lovers of pleasure and the admirers of the beautiful. I have ever liked to ride across your South plain, enlivened by its flocks of bleating sheep and skipping lambs, and stand alone upon the ocean's shore.
— *Vineyard Gazette*, May 28, 1846

The graph of forest decline and recovery below is deceptively simple given the magnitude of cultural and environmental shifts it portrays and their profound implications for the Vineyard and region. The lines trace the history of the New England landscape and reveal much about our world and why it is changing so rapidly today. Focus on the mid-nineteenth century. Following two hundred years of slow rise, the line depicting the region's population inflects upward, assuming a steep trajectory to the present. About that time, the lines for forest cover, which had been on a steady decline since the first colonists began clearing forests to farm and feed themselves, reverse and begin to climb. This upward trend for forest cover by every state signals a remarkable watershed—around 1850 New England commenced the transition from a society that obtains its food and resources from its own land to one that imports most materials and sustenance while living increasingly separated from the land. Due to this change in food base, a region-wide recovery of forest began.

The mid-nineteenth century marked a fundamental shift in the region's and nation's economy and a change in the relationship of people to the land that extended from Massachusetts to California. At this time New Englanders uprooted geographically from rural villages where they had settled for centuries. Many moved only a few miles to new industrial "mill towns" in adjacent river valleys or to rapidly growing cities, joined by a steady stream of immigrants. Others headed west, drawn by inexpensive land in Pennsylvania, Ohio, the Lake States, or Oregon. And quite a few from the Vineyard boarded ships for exotic seas or California, following the lure of employment, adventure, or riches.

As farmers sold and slaughtered their livestock and abandoned less productive pastures, a wave of tree seedlings swept across the fields of New England. The consequences could not have been more profound or unexpected. In the 1840s, Henry Thoreau despaired for the fate of the remaining forests at the hand of farmers and woodchoppers, exclaiming, "Thank god they cannot cut down the clouds." Yet, by the early 1860s he was engrossed documenting the spread of pitch pines and oaks across scrubby pasturelands and the scraggly emergence

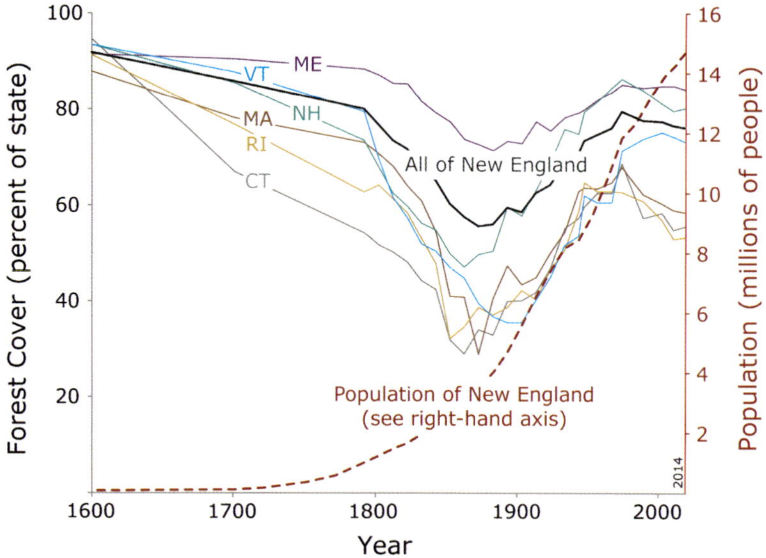

New England's singular environmental story is that of forest clearance for agriculture through the mid-nineteenth century, followed by forest regrowth as the region industrialized, the population continued to increase, and farm production shifted westward. Regional changes in forest cover were paralleled on Martha's Vineyard with profound ecological consequences. Data from the Harvard Forest.

of new forests whose future he would not live to see. He called this process "succession," but little recognized that it would make New England what it had been two hundred years earlier—the most heavily wooded region in the country. From mountain pastures in Vermont to the once sheep-filled hills of Chilmark and West Tisbury, forests and natural processes began to dominate the land.

The mid-nineteenth century, when Thoreau lived, represents the peak of human impacts on the land. Of course, more intensive insults would come to many places. The region would eventually witness sprawling factories, housing developments, mega-mansions, and big-box stores; miles of asphalt and looming cities; and the demolition of entire trap rock mountains and excavation of gaping gravel pits to support all manner of development. But these are localized impacts. In 1850 the entire landscape was heavily shaped by the incessantly busy hands of people and grazing of domestic animals. Most of the region was farmed—in expansive pastures, hay and crop fields—while the remaining woods were cut heavily and repeatedly for timber, fencing, and especially fuel to heat homes. Swamps were drained and wet meadows and salt marshes yielded hay for the sheep, cattle, horses, and hogs that vastly outnumbered humans. The landscape became open and controlled by cultural rather than natural processes.

This land use erased many features in the landscape and left indelible legacies. Native Americans had drawn resources from nature for thousands of years. But by the nineteenth century any evidence of this earlier activity was subsumed in an agrarian landscape of fields and woodlots, a spiderweb of narrow roadways connected most of the countryside, and fencelines formed a lacework pattern across the land. On the Vineyard such activities played out in patterns as varied as the glacially derived landscape. When the forests returned, many marks of that agrarian past remained imbedded in the land.

That mid-nineteenth century period of intensive human activity strongly shaped the land we have inherited. Thus, if we are going to interpret our modern landscape, if we hope to explain why one woods is different from another, why a field sits here, a heathland there, and forest beyond, or why stonewalls, cellar holes, and the old road beds that we call *ancient ways* fill our countryside, we need to envision and understand the land and people back then. In concert with the geologically defined lay of the land and the past century and a half of forest growth and development, the activities from 1850 have made the landscape that we have today.

If we could visit that nineteenth-century landscape, we'd frame questions to explain the origins of our modern woods, fields, ponds, and shorelines. What was the look of the land and how was it used? Where were the fields, farms, villages, houses, and mills? Where did forests survive through time, and what practices shaped

Deeply eroded ancient way winding through the woods near Ripleys Field in West Tisbury.

these ancient woodlots? And how varied were these patterns across a given landscape in Chilmark versus Edgartown, or over in Chappaquiddick or Gay Head? And why?

It turns out that we can immerse ourselves in the mid-nineteenth-century landscape and understand it deeply, because good fortune and unrivaled skill have given us an absolutely extraordinary map. While many sources illuminate New England's past—diaries, town histories, census records, farm journals, and newspapers—nothing matches a finely detailed map when we seek to visualize history and relate it ecologically to the present. For the most critical time in Martha's Vineyard history—1850—there just happens to be an unsurpassed map, one that provides a comprehensive and thoroughly accurate depiction of nearly every woodlot, fenceline, building, road, and wharf.

The larger story behind this map is the early tale of the emergence of the United States as a global leader in science, engineering, and maritime activity. But the poignant fact for the Vineyard is that the central actor in that tale became a resident and leading figure in Island society—Henry Whiting—the man who lent so many insights into coastal dynamics. It was Whiting, in his role as chief cartographer for the U.S. Coastal Survey, who led the nation's field effort to map its coastline for navigation, to chart important harbors, and to improve military security. In one of his very first efforts Whiting led a company of surveyors and hydrographers in mapping the Vineyard in the late 1840s. Based on the stroke of historical fortune that brought one of the world's preeminent cartographic efforts to focus on New England shores and the nation's foremost cartographer to reside in West Tisbury, we have an incomparable ability to interpret how the past controls the present on Martha's Vineyard.

Two opposing but equally accurate impressions jump out from a first glance at Henry Whiting's 1850 Vineyard map.* On the one hand, the Island is remarkably open and farmed; on the other, it supports a surprising expanse of uninhabited woodland. Indeed, the disparity in these observations emphasizes a larger conclusion—the Vineyard and its towns were as highly varied back in time as they are today. Across the great central sandplain sat a vast woodland of scrub oak, low-sprouting oak, and pines that extended in broad arms toward the south shore at Quansoo and Pohogonot and northward to the top of West Chop and lower half of East Chop. In contrast, the morainal hills of Chilmark and West Tisbury were substantially treeless. There, rolling pastures were dotted with massive boulders, sheep, cows, and a few barns and houses and contained by a network of winding stonewalls, all interrupted by small woodlots and low-lying swampland with sluggish streams. The broad necks between the coastal ponds and across the upper end of East Chop were also open and grazed, as was the vast, level plain stretching from Edgartown down through Katama.

The Island's two large and topographically varied appendages—Gay Head and Chappaquiddick—supported a mixture of pasture and scrubby woodland. Gay Head's predominantly Indian lands formed a mosaic of irregular fields, brushy swamps, and grassy

* We call it 1850 for simplicity's sake, though the fieldwork was undertaken in 1844–45 and the engraving we use dates to 1860.

"With No Compromise for Accuracy or Precision"

In February 10, 1807, the cartographically enthralled president Thomas Jefferson signed legislation authorizing a national survey with the bold goal of "completing an accurate chart of every part of the coasts." Ferdinand Rudolph Hassler, a brilliant Swiss scientist and cartographer, was appointed to lead the nascent U.S. Coastal Survey in collecting "all the data . . . to enable the government to judge . . . the plan of any [coastal] public undertaking" from navigation and military defense to commercial expansion.

Applying exacting standards for precision, accuracy, and quality in map production and aided by funding that reached a staggering 1 percent of the federal budget, Hassler spent decades assembling the expertise, tools, and methods for his assignment. In the process he developed one of the world's great cartographic teams, hiring an elite corps with military training, acquiring the finest European field and map-making equipment, and refining exacting standards for field measurements, chart production, and engraving. In 1846, the ambitious effort was launched as teams directed by Henry Whiting headed north to survey New York Harbor, Long Island, the Vineyard, and the Massachusetts coast. The resulting maps eclipsed contemporary rivals and transformed coastal navigation, harbor construction, and military planning.

Hassler's corps comprised three divisions: a geodetic group that established fixed

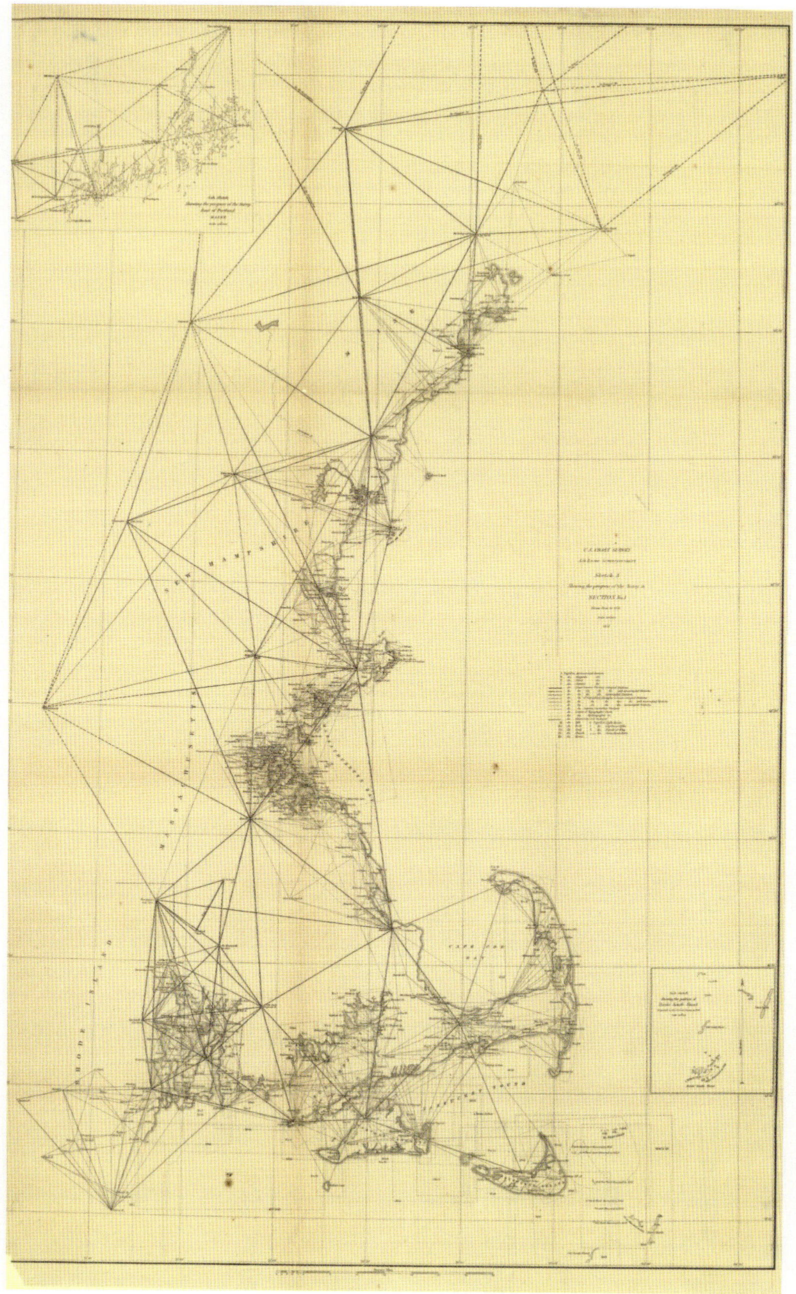

An 1860 map of the surveying triangles and points established by U.S. Coastal Survey to map the coast. Major Vineyard benchmarks (1845–46) include Indian Hill, Prospect Hill, and Sampson's Hill. Minor points include lighthouses, Holmes Hole and Edgartown harbors, Watcha and Herring ponds, and Nashaquitsa Cliff. Map from the U.S. National Oceanic and Atmospheric Administration.

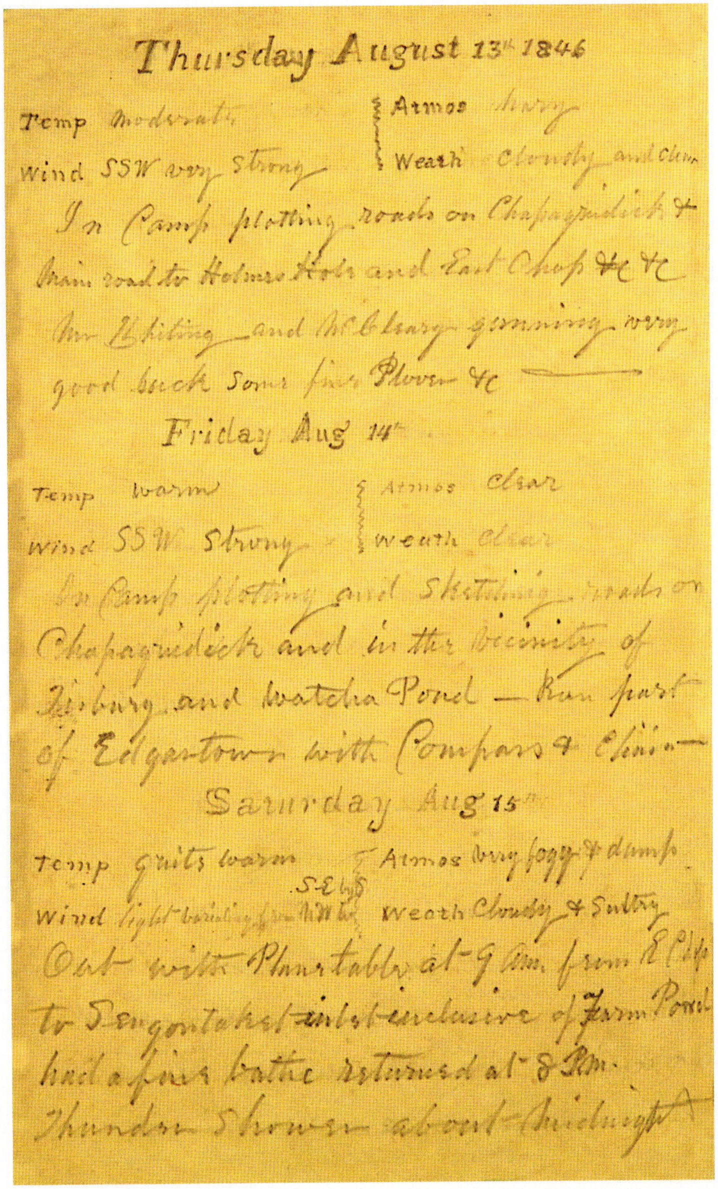

Thursday August 13th 1846

Temp moderate	Atmos hazy
Wind SSW very strong	Weath Cloudy and clear

In Camp plotting roads on Chapaquidick & main road to Holmes Hole and East Chop &c &c Mr Whiting and McCleary gunning very good luck Some fine Plover &c

Friday Aug 14th

Temp warm	Atmos clear
Wind SSW Strong	Weath clear

In Camp plotting and sketching roads on Chapaquidick and in the vicinity of Tisbury and Watcha Pond — Run part of Edgartown with Compass & Chain

Saturday Aug 15th

Temp quite warm	Atmos very foggy & damp
Wind light varying from NW to SE	Weath Cloudy & Sultry

Out with Plane table at 9 Am from E Chop to Sengontaket inlet inclusive of Farm Pond had a fine bathe returned at 8 P.M. Thunder Shower about midnight

Page from the 1846 journal of Samuel Augustus Gilbert, a twenty-year-old member of Henry Whiting's Coastal Survey crew who went on to a distinguished career as a civil engineer in the Survey and Brevet Brigadier General with service in the Civil War. From the Collections of Martha's Vineyard Museum.

Thursday August 13—"In Camp plotting roads on Chapaquidick + main road to Holmes Hole and East Chop etc. etc. Mr. Whiting and McCleary gunning very good luck some fine Plover etc."

Friday August 14—"In Camp plotting and sketching roads on Chapaquidick and in the vicinity of Tisbury and Watcha Ponds—ran part of Edgartown with Compass + Chain"

Saturday August 15—"Out with Plane table at—9 am from E Chop to Sengegontaket inlet inclusive of Farm Pond had a fine bathe returned at 8 Pm. Thunder shower about midnight."

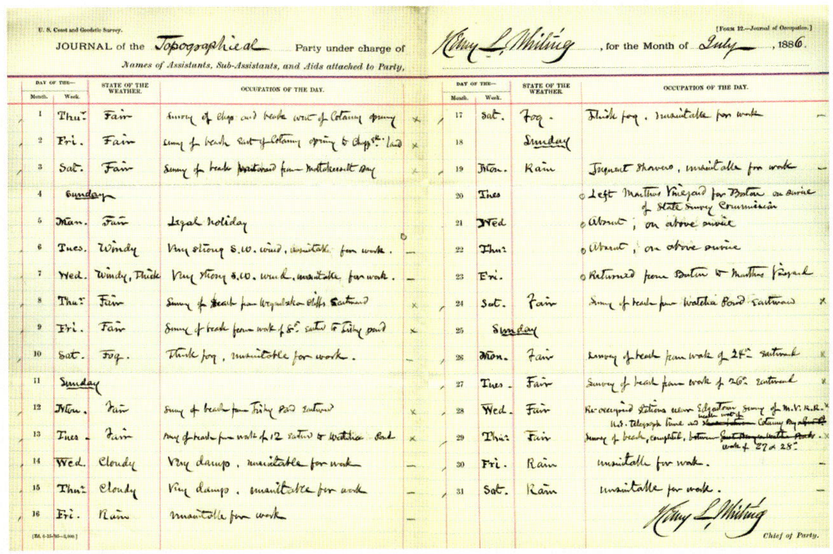

U. S. Coast and Geodetic Survey. [Form 12.—Journal of Occupation.]

JOURNAL of the Topographical Party under charge of Henry L. Whiting, for the Month of July, 1886.

Names of Assistants, Sub-Assistants, and Aids attached to Party,

Day of the— Month.	Week.	State of the Weather.	Occupation of the Day.		Day of the— Month.	Week.	State of the Weather.	Occupation of the Day.	
1	Thu.	Fair	Survey of Chop and beach west of Cotamy opening	x	17	Sat.	Fog.	Thick fog, unsuitable for work	—
2	Fri.	Fair	Survey of beach east of Cotamy opening & Chappa. land	x	18		Sunday		
3	Sat.	Fair	Survey of beach [illegible] from Mattakeesett Bay	x	19	Mon.	Rain	Frequent showers, unsuitable for work	—
4	Sunday				20	Tues		Left Marthas Vineyard for Boston on service of State Survey Commission	
5	Mon.	Fair	Legal holiday		21	Wed		Absent; on above service	
6	Tues.	Windy	Very strong S.W. wind, unsuitable for work.	—	22	Thu.		Absent; on above service	
7	Wed.	Windy, Thick	Very strong S.W. wind, unsuitable for work.	—	23	Fri.		Returned from Boston to Marthas Vineyard	
8	Thu.	Fair	Survey of beach from [illegible] cliffs eastward	x	24	Sat.	Fair	Survey of beach from Watcha Pond eastward	x
9	Fri.	Fair	Survey of beach from work of 8th eastward to Tisbury pond	x	25		Sunday		
10	Sat.	Fog.	Thick fog, unsuitable for work.	—	26	Mon.	Fair	Survey of beach from work of 24th eastward	x
11	Sunday				27	Tues.	Fair	Survey of beach from work of 26th eastward	x
12	Mon.	Fair	Survey of beach from Tisbury Pond eastward	x	28	Wed.	Fair	Re-occupied stations near Edgartown, survey of M.V. R.R. [illegible]	x
13	Tues.	Fair	Survey of beach from work of 12 eastward to Watcha Pond	x	29	Thu.	Fair	Survey of beach, completed, between [illegible] work of 27 & 28.	x
14	Wed.	Cloudy	Very damp, unsuitable for work	—	30	Fri.	Rain	unsuitable for work.	—
15	Thu.	Cloudy	Very damp, unsuitable for work	—	31	Sat.	Rain	unsuitable for work.	—
16	Fri.	Rain	unsuitable for work	—					

Henry L. Whiting
Chief of Party.

Journal entries by Henry L. Whiting during his Vineyard re-survey in July 1886. Entries include descriptions of sites (Cotamy opening, Chappaquiddick, Mattakeesett Bay, etc.) and weather conditions ("thick fog, miserable for work"). From the Collections of Martha's Vineyard Museum.

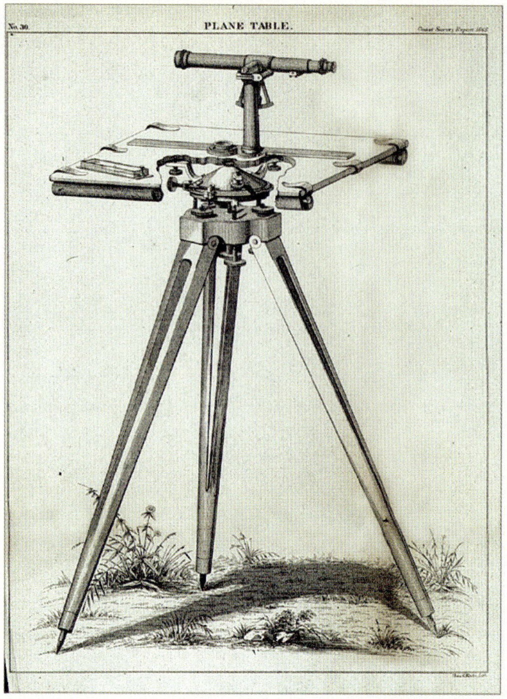

Plane table. This portable tool used extensively by the Coastal Survey consisted of a tripod, table, and alidade with telescopic sight and allowed crews to make precise surveys while taking notes, sketches, and observations on rough but accurate field maps. Illustration from the U.S. National Oceanic and Atmospheric Administration.

benchmarks including church spires, hilltop boulders, and chimneys; the rugged topographers who conducted rapid but detailed field surveys; and a hydrographic group that plumbed coastal bathymetry. Field mapping relied on the simple but flawless approach of triangulation. From a master baseline personally established by Hassler, progressively smaller triangles were surveyed on more local features. The crews employed the plane table, a square board affixed on a tripod that allowed a field map to be created containing the surveyed lines, points, notations, and "all the features of the county and details of the topography." A year or more after the field measurements yielded this first draft, a division of elite engravers produced published maps.

The initial 1845 Vineyard survey resulted in numerous products, including the remarkably comprehensive map *From Muskeget Channel to Buzzards Bay and entrance to Vineyard Sound, Massachusetts* that is utilized extensively in this volume. Despite its regional breadth, the map is exquisitely detailed, even distinguishing cultivated fields and orchards from hayfields. But how much faith should we put into these details? We can let the *Vineyard Gazette* from January 28, 1848, answer that question.

> By the report of the general superintendent of the Coast Survey . . . we learn [of] a . . . map of the island . . . were our people aware of the manner in which these charts are engraved and published, there is not a family in town but would procure one as soon as possible.
>
> What surprizes us most is, that the roads running to and on the island, are laid down so correctly. A description of the country through which they pass is seen at a glance; and with one of these charts, a perfect stranger on the island, can find his way to any part of the country. . . . A correct description of every pond, with its outlet, or path leading thereto, is laid down, besides a scale of miles, by which anyone can ascertain the exact distance (within a foot or less,) from one point to another. But this is not all. . . . The outlines of the coast, the position of the light houses, buoys and rocks, with shoals and soundings, are laid down without any errors. It is perfect, or as near perfection, as science, at the present day will admit. It is, in fact, beyond improvement. . . . Perhaps some will ridicule the idea of placing one of them in their parlor; but we are satisfied, upon inspection, they will agree with us, that they are an ornament to the room of any gentleman.

The preparation by Hassler and the exacting effort by Henry Whiting are reflected in this praise. One may also read Whiting's passion for the Island in the history of his local surveys. Though most of the coast was covered in a narrow strip extending just a mile or so inland, the Vineyard was mapped in its entirety, the largest landmass to receive such treatment. Moreover, he resurveyed the Island numerous times until late in the century. The result is a legacy of maps that allow us to explore the nineteenth-century Vineyard in its full glory as a vibrant agricultural landscape and maritime center.

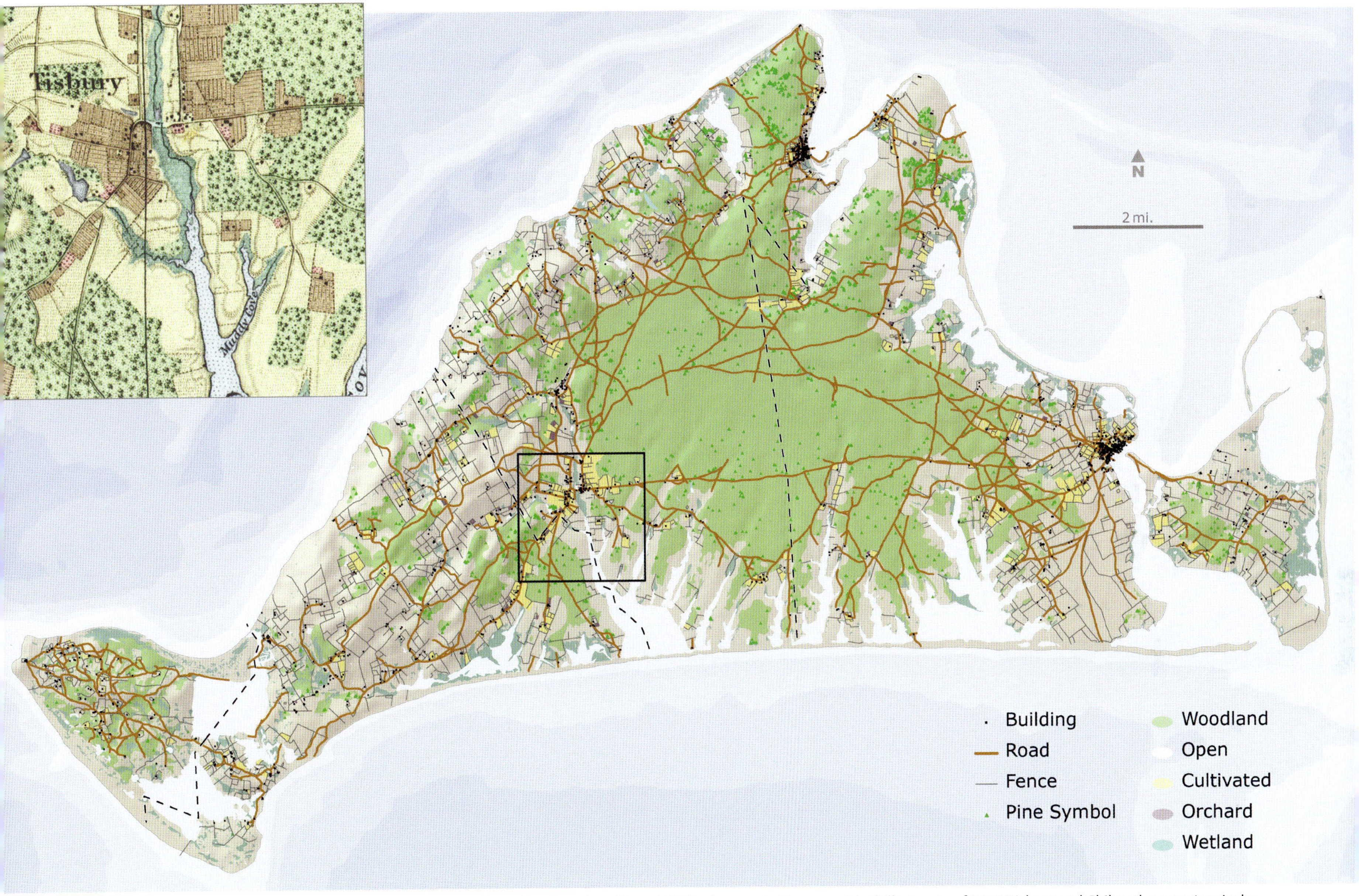

The landscape that became today. Detailed map of the Island, ca. 1850, based on Henry Whiting's remarkable survey. The hilly terrain of West Tisbury and Chilmark was extensively cleared in fields, scattered woodlands, and wetlands, whereas much of the Great Plain and large portions of West and East Chop remained wooded. *Inset:* Detail from the original survey showing West Tisbury's pattern of woodlands, fences, croplands, houses, and roads. Map inset from the U.S. National Oceanic and Atmospheric Administration, 1860, Chart No. 12, *From Muskeget Channel to Buzzard's Bay and Entrance to Vineyard Sound Mass.* Main map created by the Harvard Forest from the original U.S. Coastal Survey map sheets.

The Old Whaling Church lit above the elegant Dr. Daniel Fisher house on Main Street, Edgartown.

marshes with few sizable trees. Chappaquiddick was more regularly separated into open fields and cutover woodlots. A long, straight fence and parallel lane (Chappaquiddick Road) cut broadly eastward across the northern half of the small island, separating Indian lands toward North Neck from English ownerships extending to the south shore and Katama Bay. Elsewhere, the Indian presence had faded substantially. Scattered ownerships remained near the south shore and in the set-aside at Christiantown with its community of houses, the meetinghouse, and hill-slope cemetery.

In 1850 the Island comprised just three incorporated towns. Edgartown and Tisbury accounted for the bulk of the land, essentially all of the maritime activity, most of the woodland, and almost all of the Great Plain. Chilmark still included the Elizabeth Islands along with Noman's, and was centered on farming and shore-based fishing. Gay Head remained decades away from incorporation and, along with Chappaquiddick, was soon to enter a complex federal process of surveys and oversight of "the condition of all Indians and the descendants of Indians" that sought to resolve major questions concerning ownership and control but left legal issues that would entangle courts to the present.

Both Edgartown and Tisbury were bifurcated by miles of scraggly plain. On the southwest of East Chop, Eastville was a tiny village of houses, taverns, shops, fields, and a wharf facing out on Holmes Hole Harbor. Just southeast of East Chop was the Wesleyan Grove, an emerging summer tent community, then Farm Pond, Sengekontacket, and the Great Woods that extended south to the county seat and village of Edgartown.

Old Town, as Edgartown was also known, faced a true harbor that was "safe and excellent," according to Reverend Freeman's 1807 assessment, and "esteemed one of the best in the United States." Having grown from settlement in the 1640s through fishing, whaling, and trade, Edgartown embraced a set of rugged wharves, many shops, a few stately homes, thriving workshops, and varied housing for workers, tradesmen, town leaders, visitors, and, of course, sea captains, all within a compact center framed by farmland and water.

Holmes Hole, Tisbury's commercial center, shared Edgartown's pace but differed in character and activity. This "deeply embayed haven" was "one of the famous refuge harbors of the coast," as Shaler would later put it—part true port for commercial activity and part layover for ships seeking refuge from rough conditions before rounding Cape Cod for Boston and other destinations. The immense harbor could accommodate up to two hundred 200-ton ships at anchor, according to hydrographer Henry Mitchell. But, as Mitchell states in his 1872 report *Vineyard Haven: Its Character as a Port of Refuge,* it was "distinguished . . . pre-eminently for its convenience." For as a port the harbor had one fatal flaw—it lay fully exposed to the north. As Mitchell characterized it, "The ship that runs into this anchorage before a northeast gale leaves an open door behind her."

Despite this fundamental failing, Holmes Hole's location along major routes gave this harbor and village great importance. Vivid

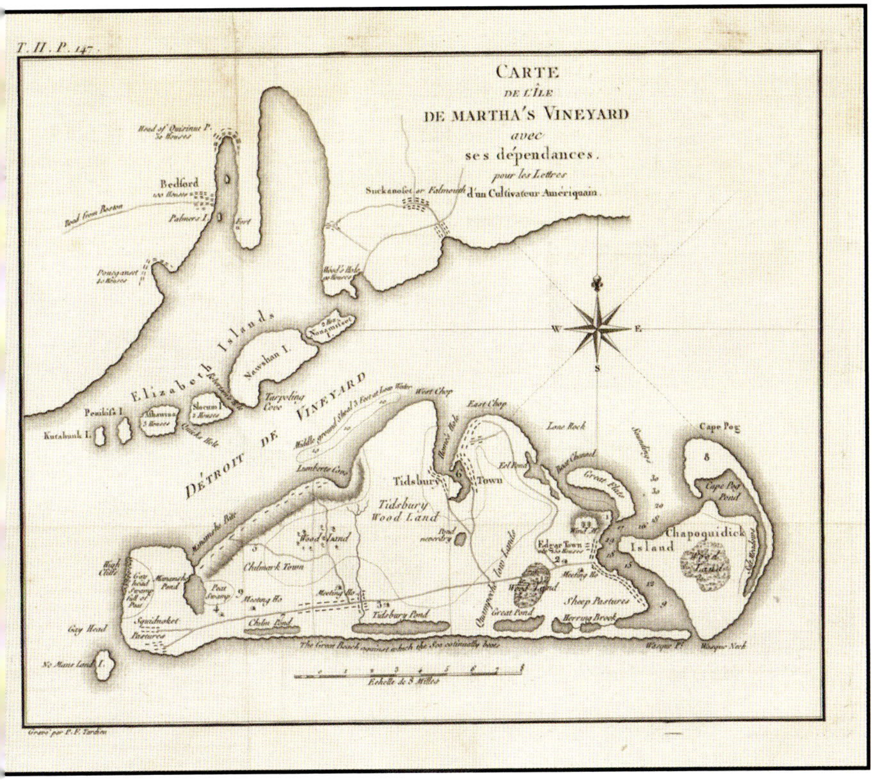

Persistent land-use patterns. The stylistic 1782 map by Hector St. John de Crevecoeur. Woodlands feature prominently across the center of the Island, broadly in the Pohogonot area, and on Chappaquiddick, whereas Katama and Squibnocket are expansive pastures, and the south shore is described as "The Great Beach against which the Sea continually beats."

In contrast to the bustling harbors with their kaleidoscope of commerce, vessels, grime, and foreign faces, the villages of West Tisbury and Chilmark were a picture of New England agrarian tranquility, as depicted by painter David Strother in 1859: "[West Tisbury] is a quiet rural village . . . in the midst of . . . the best agricultural portion of the island . . . looking . . . to the west is a tumult of hills, grass covered, specked with flocks of sheep, and broken with . . . massive granite rocks."

Beyond the fertile soils framing Mill Brook, Tiasquam River, and Tisbury Great Pond, and the productive hillside pastures, this up-Island region was prized for its water power—a critical resource almost entirely absent from the eastern two-thirds of the Island. Rains falling across the morainal hills yielded ponds for grazing animals, shallow wells for human use, moist soils that supported a diversity of trees, and extensive swamps that sourced a half-dozen streams to power many small mills. Behind rock dams lay waters that served as the nuclei for the villages of North and West Tisbury—Athearn's or Priester's, Look's, and Mill Ponds.

One striking quality of the nineteenth-century landscape is the maritime descriptions remind us that in contrast to most of rural New England, farming was hardly the Island's central economic enterprise. Surrounding waters were vibrant with craft of all sorts, as captured by Daniel Webster on an 1849 visit: "The number of vessels which pass up and down this Sound is prodigious. A hundred of them sometimes put into Holmes Hole in a day, if a head wind arises. Nearly all the coasting trade between the East and South, goes through this passage, as do often ships from South America, the West Indies, and India."

Headstone in the West Tisbury cemetery for Caleb Rotch, who left with hundreds of Vineyarders for California in the mid-nineteenth century:

No mother soothed his dying hour,
No father watched with anxious care,
No brother took the farewell hand,
No loving sister knelt in prayer;
But, in a distant, stranger land,
He yielded up his feeble breath;
And there within a lonely grave,
He sleepeth in the arms of death,
But, as the tears do freely flow,
While stricken 'neath the chastening rod,
Oh! May we in submission bow,
And cling the closer to our God.

great physical separation of these Island villages. Indeed, travel from one to another meant crossing somewhat desolate stretches of scrub and woods on frequently miserable roadways that were regularly disparaged in contemporary accounts. Like Daniel Webster, some visitors underscore the sense of isolation and inconvenience. Quite a number remark on features that surprise a modern reader, such as Devens's 1838 description of opening and closing some thirty fence gates as he traveled from West Tisbury to Gay Head. Given the seafaring orientation of much of the population, it is small wonder that many residents of Holmes Hole and Eastville felt better connected to Woods Hole than Edgartown, that residents of Edgartown were often more familiar with other parts of the globe than remote places up-Island, and that some farmers from Chilmark never set foot on East Chop, let alone visited Chappaquiddick.

Agriculture

In the mid-nineteenth century, agriculture was not simply an occupation for one segment of the community; it was a way of life shared in various measure by nearly the entire population. Judges, businessmen, fishermen, yeomen, widows, and laborers of all types had vegetable gardens, chickens, and the occasional cow, pig, and sheep. Meanwhile, the individuals identified by the census as farmers undertook many additional tasks—constructing buildings, clearing land, harvesting timber, and hauling produce. They often sought supplementary income from tanning and cranberry operations, fishing, carpentry, and other trades.

Farming touched most of society. Where it was concentrated in particular families, it comprised an expansive and diverse enterprise involving many types of often far-flung landholdings. Agriculture was a multifaceted operation of grazing and tending animals; raising diverse vegetables, grains, potatoes, and fruits; contracting and coordinating the exchange of labor; marketing and moving animals and other products to market; and the production of additional commodities for sale, barter, and home consumption. Growing food ran through most lives and the countryside.

Oxen patiently await the pulling contest at the Agricultural Society Fair, West Tisbury.

The pattern of land ownership and farming activities had grown out of early colonial days in Edgartown and West Tisbury when the proprietors (land-owning settlers) were granted blocks of wooded uplands along with salt- and freshwater meadows. As forests were cleared, a distributed geography of land ownership and use developed in which uplands supported pastures, woodlots, orchards, fields of so-called English or upland hay, and tilled cropland; wet meadows yielded salt or fresh hay and seasonal pasture; other freshwater wetlands were developed for cranberries; swamps served as woodlots or were cleared and grazed or dammed; and the coast yielded seaweed for manure and shells for lime. Many families held some of these kinds of land in addition to a fenced yard that might support an animal or two along with fruit trees and bushes, and a robust garden. But true farmers generally assembled a dispersed complex of landholdings containing most of these types of lands from which they would reap the diverse resources needed to sustain a family and yield products to sell to the larger community. It is clear in the allocation of lands to proprietors in February 9, 1681, that the initial divisions of common land were made to balance the needs of all men accorded the privilege of land ownership (which was far from all men or families). "Allen and Mayhew were appointed to lay out swamp or Low land to those men who had none or but little . . . making them Equall as we Could with others."

The Varied Activities of the Rural Life

Journals yield insights into the range of daily activities associated with early farm life. Whether classified as a farmer, tradesperson, yeoman, or other, most nineteenth-century residents raised crops and animals and worried about obtaining adequate wood.

WILLIAM BUTLER, FARMER AND TANNER (1761–1832)

William, like his father, was a tanner by trade as well as a farmer and teacher who lived near Farm Pond.

February 8 . . . a great deal of snow on the ground—employ'd by looking after cattle, sheep &c.
March 14 Employ'd mending fences.
March 16 From Chilmark . . . . brought down from Bassets 36½ lb flax being the sixth part of what came off 2 acres ground.
March 17 Employ'd opening Farm Pond Crick—set it a running
March 24 . . . cutting posts.
March 20 Employ'd getting withs [wood rods] on the Plains
April 4 Employ'd shutting in the meadow.
April 5 This day Peter Norton carried away a half a load Sedge Hay—gave 20 shillings for same.
April 6 Employ'd plowing at home.
April . . . sold 400 wt. hay. Ansall Norton bought 500 English hay
April 21 Up at Squibnocket yarding sheep

GEORGE HASKELL WILLOUGHBY, YEOMAN FARMER (1855–1930)

Born in Lowell, Massachusetts, Willoughby married his first cousin, Maria Ann Downing, and moved to Chappaquiddick in 1879, where they raised vegetables, poultry, and bees, sold eggs, and bartered for other provisions. His entries from 1880–82 highlight the struggle to maintain farm productivity.

January 7 Put board under hen roost to catch droppings. All must be saved; cleaned out poultry house got one bbl [barrel] of nice dressing.
April 28 Helped Isaiah Belain cart one load of Ooze from between the Necks.
May 1 Got liquid manure from G Huxford—said I could have 4 bbl.
May 5 Went to Cape Pogue to work for E Worth—set fence.
May 15 Went down to the house place and spaded over a piece of bushy land.
May 22 Caught 20 Bluefish, took to town and sent off to New Bedford iced.
July 28 Went coaling. Got 138 strap tubs full. Cut hay.
August 13 Went to town to see Geo. Cleaveland about some scrub land.
August 24 Cut part of my meadow.
September 18 Went after pine needles, got 1 bag.
September 24 Ground 2 axes and cut a road through the first swamp.
November 4 Brought 3 loads seaweed—one for pig pen, 2 to bank house.
November 9 Got 6 loads of stone . . . out at Isaiah's farther sheep pasture to build a side wall to the pig pen.
November 17 Went to town to get some corn ground but mill is out of order.
January 1 I have succeeded in living within my income the past year.
January 24 Went after sumac to tan the kidd skin.
April 1 Pulled barberry bushes. Burned whole piece [field].
April 29 Put in potatoes. Put ashes over them.
May 4 Used shark compost in hogshead as dressing [fertilizer].
May 24 Night soil on tomatoes too strong.
April 18 Went to town. . . . When we got home we found Ria [his wife!] in bed with a new born Son. She had been alone since noon.

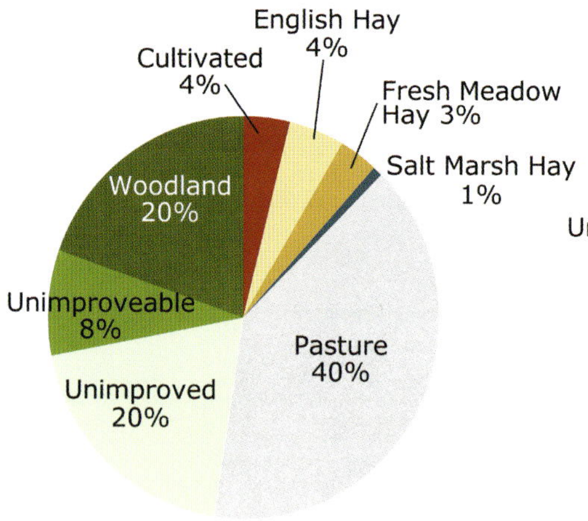

1801

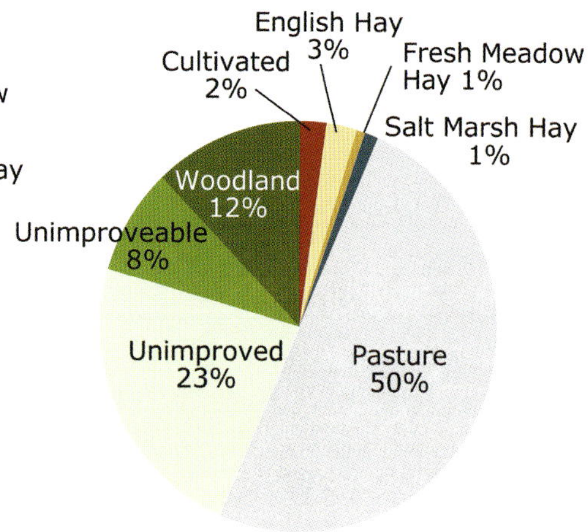

1831

Historical partitioning and use of the land. Census data from the first few decades of the nineteenth century mesh well with descriptions from farmer journals, probate records, and real estate listings through the second half of the century. The dominant land cover was grazing lands (pasture and unimproved/unimprovable) with less woodland. Smaller areas were cultivated for a wide range of vegetables and hay. English hay was largely imported grasses planted on cultivated soils; fresh meadow hay included native grasses, sedges, and herbs on freshwater wetlands; and salt marsh hay consisted of native species.

As noted in the frequent use of agricultural criteria by early explorers in describing the landscape, farming has always been a preoccupation on the Vineyard. Consequently, livestock were among the first imports: cattle arrived by 1651, hogs in 1652, horses and sheep the following year, diverse fowl through the 1650s, and goats by 1660. Over time new animals and breeds were added to maintain genetic diversity and improve the quality of the livestock. The ongoing flow of imports is emblematic of a looming reality: whether at the scale of an individual farm, town, or the entire Island, complete self-sufficiency has always been impossible on Martha's Vineyard. Every family relied on the resources and skills of neighbors and other artisans; different parts of the Island specialized in particular resources and the production of goods lacking elsewhere; and though the Island became an exporter with the first shipment of sassafras back to England in the early 1600s, it has always relied on foods, goods, labor, and fuel from outside.

Though certain farm activities changed over time, the general patterns display great continuity over centuries, even to the present. Indeed, the breadth of land holdings, crops and animals, much of the labor, and quite a few farm-family surnames in the eighteenth- and nineteenth-century landscape would be familiar to people working today on Mermaid, Morning Glory, Norton, Allen, and other farms. This long sweep of activity was punctuated by numerous shocks and declines: the British plunder of livestock in Grey's Raid; droughts, late frosts, and early hurricanes; outbreaks of pests that crippled crops; severe winters that culled herds; and a complex history of tariffs imposed by domestic and foreign governments.

The landscape pattern was shaped by the spatial distribution of different land types and uses and the corresponding arrangement of fencing bounding these activities. Compact gardens lay adjacent most houses and in concentrations near village centers. Larger crop fields sat near or beyond the outskirts of settlements and were cultivated for grains—Indian corn, rye, oats, some barley and flax, and even wheat—as well as lots of potatoes, beans, squashes, and other vegetables. Hard-worked fields lay close by barns and roads on readily tilled lands, generally with the fewest stones and fine-grained soils that retained moisture. Plowed by teams of oxen, cultivated by horses, or hoed by hand, these tilled lands could be any size that neatly fit onto the better soils that a farmer owned. While crops and productive hayfields formed a complex mosaic in any given year, the

Nineteenth-Century Fences

Early on, fences were legislated to control livestock and maintain peace among neighbors. On December 4, 1689, the town of Tisbury voted that "all fences [would be] 4 feet and 3 inches [and] judged sufficient. Any ditches or fences built by casting up panes [spoils]—shall have hedges or rales so equivalent to 4 feet 3 inches." Two men were appointed "judges and savairs of fences," and the position of fence viewer endured for centuries.

In early years the clearing of woods yielded stump fences. More enduring stonewalls were erected in the eighteenth and especially the nineteenth century across Chilmark, Gay Head, and West Tisbury as farms prospered and farmers removed rock from morainal soils. These walls were either built to the required height or capped by wood posts and rails that were eventually replaced by wire. Across the plain where stones are absent, a "ditch and ridge" system was common. Soil excavated from the straight ditch was mounded in a parallel ridge, with each a couple of feet across and a foot or so in relief. Posts strung with rails or wire ran along the ridge in the British tradition of perimeter trenches and banks around in-fields. Ditch and ridge patterns persist today in many woods and open fields from Chappaquiddick to Chilmark along property lines and old field boundaries.

Freeman (1807) provides a contemporary account of this geographic variation in fencing. "Where the land is enclosed, it is in the eastern part [of the Island] entirely fenced with posts and rails, which are chiefly brought from Buzzard's Bay. As many spots [e.g., the Great Plain] however are not worth enclosing, and are destitute of water, they are left in common. In the western part of the island, the land in general is fenced with stonewalls. The stones . . . can be laid in such a manner, as to admit spaces between them; by which labour is saved."

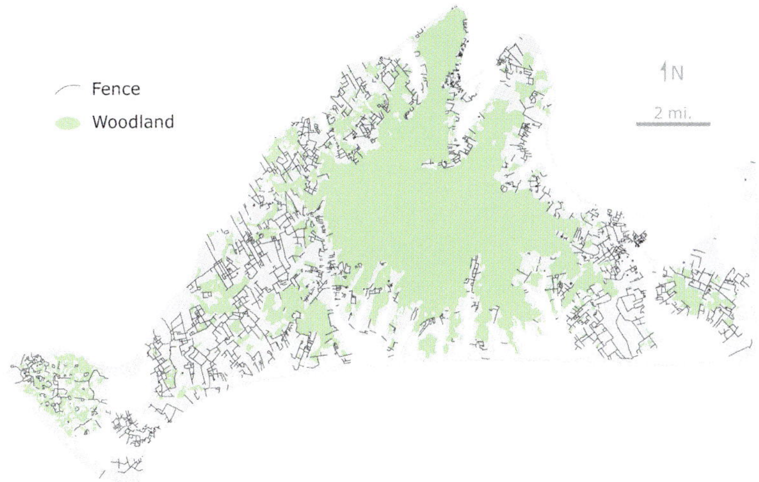

The separation between actively farmed areas and woodlots is clear on Whiting's 1850 map. Areas cleared for agriculture were bounded by fences constructed with stones on the moraine or wood rails set on ditches and ridges on the plain. Continuously wooded areas were seldom fenced.

At an Island scale, the 1850 fence pattern corresponds to areas cleared of forest. Meanwhile, the density and details of the fencelines tell a story of ownership and land use. A tightly reticulated pattern characterized the villages of Holmes Hole, West Tisbury, Middletown, and Edgartown. This pattern opens slightly in adjoining areas of larger gardens, tilled fields, and small pastures. In rural expanses with large pastures the fencing network is coarse, as seen from Menemsha to Makonnikey and across Katama and Chappaquiddick. Some distinctive local patterns appear, including the use of a single fence across many south-shore necks, and the long fence that divided Indian lands on the north of Chappaquiddick from English holdings to the south. A unique pattern appears on Gay Head, where ownership was more communal and land was divided into a few large pastures and many small enclosures protecting crops and gardens from roaming livestock.

The modern affinity for declarative fencing and boundary marking has produced some oddly expensive deviations from ancient styles of stonewalls. These include imported stone, chiseled and tightly fit into ornamental walls; mortared walls; and the construction of massive rock walls on outwash soils that lack large stones and historically supported ditches, ridges, and wooden fences.

specific patterns changed kaleidoscopically over time. Each farmer developed a rotational pattern, switching crops and cover, and inserting animals and their feeding and manuring activity into the mix, in an attempt to maintain productivity and limit damage from insects and pathogens.

Fertility was a preeminent concern as nutrients were removed in harvests and through erosion and leaching into groundwater. Nutrients were returned though the application of wagonloads of organic nutrients—muck from wetlands; seaweed, eelgrass, and shells from beaches and pond shores; potash from ashes; and even leaf litter from forests. Farm waste was composted and spread on fields, with animal (and less frequently human) manure. Bedding of salt hay, pine needles, and leaves soaked up the waste from chickens, cattle, and hogs over winter and was spread the following spring. Animals were placed on land being cleared and on croplands whenever they lay fallow, in order to loosen and mix the soil and convert plant growth into animal products and fertilizer.

The different sources of hay conveniently spaced yields through the summer. In mid- to late July, the uplands were cut of English hay—imported grasses like timothy, red top, clover, fescue, and orchard grass—on fields that had been plowed and harrowed before planting. In August, when water levels were low, freshwater and saltwater marshes yielded native grasses, sedges, and rushes from sites replenished regularly with nutrients delivered by streams, groundwater, and tides. These brackish wetlands reached their greatest extent along Poucha and Sengekontacket Ponds, and scattered great ponds on the south shore. Wet meadows were expanded and maintained by the removal of woody growth, cutting of drainage ditches, mowing, and occasional grazing. Access to coastal marshes was improved by breaching barrier beaches to lower the water levels of the ponds. Conversely, the impoundment of waters like Poucha Pond with dikes to control flow, salinity, and fish populations could flood the marshes and generate conflicts between farmers and groups holding rights to the ponds, fish, and shellfish.

Cattle grazing and transforming the understory of an ancient wood of mixed pine and oak forest along Northern Pines Road, Tisbury.

Freeman provides insights into the varied quality of the different sources of grass that collectively addressed the needs of Island farms.

> Upland English mowing . . . yields about eighteen hundred [pounds] to an acre; the salt marsh, a ton; and the black grass marsh, a ton and a half. . . . Another kind of grass, called creek stuff, grows on the borders of the ponds, and the greatest part of it in the water. It is a coarse sedge, and is worth about one third of English hay. . . . Bordering on the small rivers and brooks, which run into Newtown [Tisbury Great] Pond, there are about seventy or eighty acres of fresh meadow.

Far and away most cleared land lay in pasture, which was concentrated toward the periphery of an active farm and included less workable sites and so-called rough and “unimproved lands.” On the Vineyard, the important grazing lands varied geographically from rolling hills, bouldery slopes, streamsides, and mucky swamps in Chilmark and West Tisbury to the grass-covered sandy necks and pond shores along the south shore, on to the varied terrain of wetlands and uplands on Chappaquiddick and Gay Head. The prominence of pasture shows clearly on historical maps and is well established by tradition. De Crevecoeur (1784) labels Katama as "sheep pasture" and identifies an expansive "Squibnocket Pasture," characterizations that persevere through time. On Gay Head the Wampanoag set off large common pastures—South, Middle, and East pastures—bounded by stonewalls. Annually, tribe members met to decide the allocation of animals to these different areas and the smaller Hog and Fatting pastures. The habit of moving animals from Edgartown to Chappaquiddick across the "Swimming Place" in October and retrieving them in April became a local tradition. Meanwhile, for years the Peaked Hill region provided common summer pasturage for Chilmark and Tisbury.

Across the Island, pasturelands could utilize 90 percent of the open area at any given time, while cropland and upland hay accounted for 5 to 10 percent of the land, and salt marshes added only a few hundred acres. Among towns, Chilmark supported the greatest amount of pasture and hay, while Edgartown led in tillage.

Woodlands

Ask Vineyarders today for a description of the Island during this mid-nineteenth-century height of agriculture, when sheep outnumbered humans five to one, and they're likely to respond that barren sheep pastures extended from Chappaquiddick to Gay Head. While this image suited Nantucket, Block Island, and many of the Elizabeth Islands, the Vineyard presented an entirely different scene. Indeed, the core of the expansive forest covering the Island today lies in a broad woodland that survived through the era of deforestation and farming. That woodland was surprisingly vast—about 24,000 acres in 1850—nearly one-third of the Island. It was, however, as varied as the towns and terrain that adjoined it. And, in its cut-over condition, most of it would be as unrecognizable today as Edgartown would be with its wharves loaded with barrels of whale oil, crowded with ships, and thronged with coarse sailors.

The heart of this enduring woodland stretched from the outskirts of Edgartown across much of the immense Great Plain to the village of West Tisbury and then north beyond Holmes Hole. Though mapped as woodland by Whiting's corps of surveyors, that label would likely raise eyebrows today. The bulk of the wooded plain was a scraggly growth of oak trees and sprouts, scrub oak, and shrubs in an ownership limbo—a quasi-commons of land so unfit for cultivation or habitation that it was described in 1784 by visiting Frenchman Hector St. Jean de Crevecoeur as belonging "to no town" and therefore variously used and abused by people from all sides. Through the years the combination of sandy and nutrient-poor soils, repeated cutting of nearly every stick worth carting away, and occasional runaway fire had reduced this heart of the island to a condition described well by an anonymous mid-century traveler: "Having passed [from Edgartown] . . . into Tisbury, the road lay through what would have been an oak forest, except none of the trees [exceeded] some four feet in height." Though short and scrubby, as if largely composed of scrub oak, this wooded expanse was also filled with tree sprouts, especially white, post, scarlet, and black oak, that could reach forest size when given a reprieve and time.

Toward the Great Plain's periphery—up into the two Chops, scattered in the hills of Tisbury, and in places on the south-shore necks—were true, albeit generally young woodlands, composed of trees that did reach full stature although often sporting multiple stems from repeated cutting. Beech, hickory, and beetlebung, common trees in Gosnold's day, had declined with clearing, cutting, and increased fire, and so oaks predominated. Despite the sandy nature of the soils, pitch pine was patchy, and even surprisingly infrequent. Whiting mapped these pine trees (his was a remarkable map!) in a scattered distribution across the plain. Pine was also described as locally abundant in such locales as the Dark Woods and Pennywise regions west of Edgartown, where it was exploited for its pitch and resin; in what were known as the "Great Woods" stretching across the base of East Chop north of Dr. Fisher's Road; and widely through West Chop.

Nineteenth-century accounts of these woodlands are consistent with regard to their distribution, size, and composition. When it

The Great Plain—a "barren ragged plain of no town"

When Simon Athearn wrote these words in 1698, he articulated a sentiment that would be shared by people over the years. Lacking surface water and dominated by sandy soils, the vast outwash plain forming the Island's center would not support grazing animals or reliable crop production and was unfit for settlement. Though the periphery of the plain, in the villages of West Tisbury and Edgartown and along the coastal ponds, would become dotted with residences, the bulk of this largely featureless landscape remained empty and largely unclaimed until the housing boom after World War II. Until then the plain served to separate communities, provide a backdrop for cross-Island travel, and yield wood, berries, and wildlife to anyone's choosing.

Reconstructing the plain's appearance and vegetation through time is challenging, but one fact is clear—over the past four centuries the entire area remained "wooded" by oaks, scrub oak, and huckleberry. Descriptions, pollen analyses, and Whiting's maps indicate that, despite a speculative statement to the contrary by Nathaniel Shaler that has been quoted over the years, pitch pine was scattered and incidental. The use of the term "ragged" and other pejorative descriptors confirm that tree growth was poor. A long history of drought, timber cutting, fire, and the scavenging of wood by a fuel-starved Island frustrated the full growth of trees and real forests for much of history. Only with the easing of demand for wood in the twentieth century and control of wildfire did the distinctive pattern of forested uplands and scrub oak bottoms become apparent.

Over the years, although the scene likely changed little, the sentiments expressed by residents and visitors varied widely, from incidental and dismissive to delighted and enthralled.

Freeman, 1807: "vast plain of bitter oaks between Edgartown and Tisbury . . . of little or no value."

Daniel Webster, 1849, on driving to Edgartown: "crossed level sandy country covered with shrub oaks."

Anonymous, 1860: "[From West Tisbury] toward the south and east it is a plain, chiefly covered with a growth of stunted shrubbery."

George Willoughby, 1880: "Arose early and went [from Edgartown] to Christian Town. . . . I passed through that wonderfull scrubland. The most part of the undergrowth is about 2 ft high . . . and I could see 2 or 3 miles in all directions and then I couldn't see out of this wilderness of scrub oaks."

Nathaniel Shaler, 1888: "On the sand plain the woods are of stunted oaks and other dwarf varieties of trees. . . . This generally untillable area of Martha's Vineyard has an extent of about thirty-three thousand acres. At present about twenty-five thousand acres . . . is covered by low, scrubby woods. . . . From the saddle . . . the eye ranges above their tops for miles over a billowy sea of deepest green. . . . The extent and unbroken character of the forest is amazing; in one direction we may journey through the woods for ten miles without a trace of habitation or culture. Through it runs a maze of paths."

Hine, 1908: "Another track through this scrub oak wilderness is a half moon known as Doctor Fisher's Road . . . to connect his mill in Tisbury, where water power is to be had, with Edgartown."

Alfred Gross, 1928, on the state heath hen reserve: "an uninteresting waste with a lure and attractiveness that I could not at first appreciate . . . scrub oak for miles, unbroken to the hills of West Tisbury and Chilmark."

Anonymous visitor, 1930: "the Great Plain, thickly grown over with scrub oak that barely reaches the height of an average man. Lonely, wind-swept and haunted by swooping hawks . . . the Great Plain has always been looked upon as a waste place. Scarcely anyone has ever lived there and succeeded in wresting a livelihood from the soil."

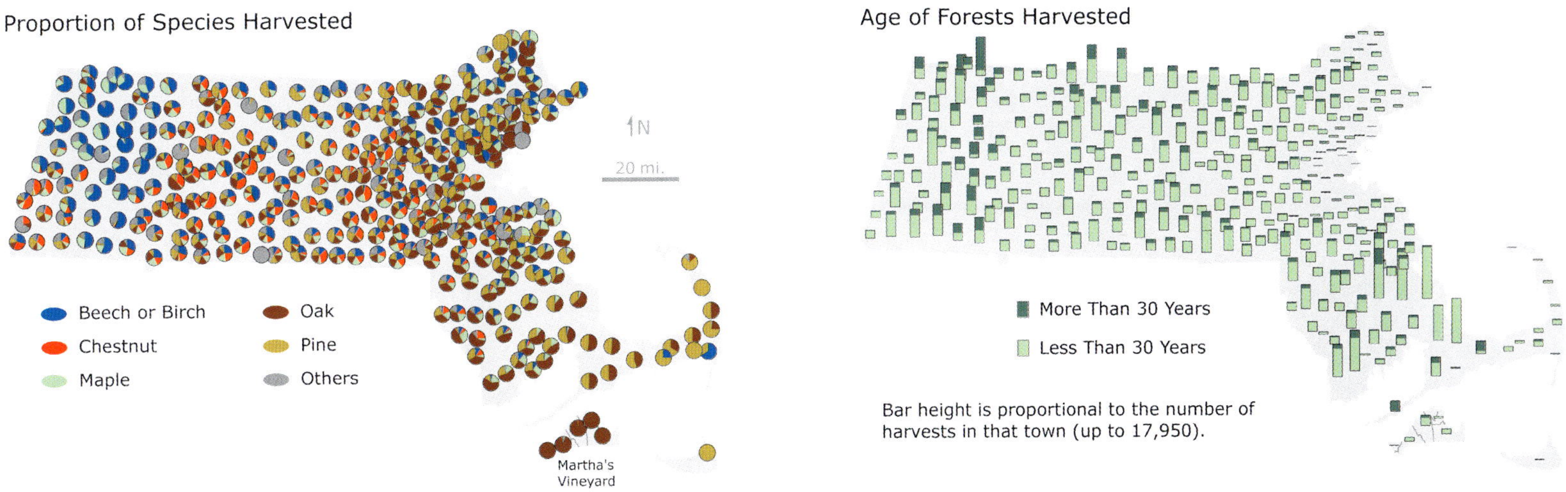

The fortuitous collection of town-based census data on the age, type, and number of forests harvested in 1885 helps interpret forest composition, abundance, and condition during a time of declining agriculture but intensive wood use for heating and industry. *Left:* Species harvested varied from oak and pine in the southeast and east, chestnut in the center of the state, and maple broadly. *Right:* Young forests dominated the coast, the Boston region, and the Connecticut Valley, with older forests in the Berkshires and base of Cape Cod.

On Martha's Vineyard, nearly all forests cut were oak, though the amounts varied. Tisbury outstripped all towns in the amount of forest, based on the number of harvesting activities. Island-wide, most forests were less than thirty years of age.

came to wood resources, Tisbury reigned supreme over Edgartown's expanse of stunted trees, Chilmark's open farmland, Gay Head's thinly wooded swamps, and Chappaquiddick's meager woodlots. Every description also confirms that the demand for wood greatly outstripped the local supply.

> Very little wood land is left in Edgartown and Chilmark: in Tisbury there is more than in both the other townships, about two thirds of the whole island. The trees are principally of white and black oak, and are about thirty feet high: few exceed fifty feet. In Chilmark there is not half fewel enough of wood for the consumption of the inhabitants; and in Edgartown the greatest part of the fire wood which is used is brought from other places, chiefly from Buzzard's Bay, Waquoit, and Coxit: the price of a cord is five or six dollars. (Freeman, 1807)

The need for wood, and especially fuel, pressed relentlessly on the entire Island. In early days the Vineyard exported firewood to neighboring Nantucket. But a large, drafty colonial house easily consumed ten to twenty cords of wood each year in inefficient fireplaces and perhaps half as much once woodstoves became common in the mid-1800s. The progressive clearing of the uplands and conversion of swamps into hay meadows and cranberry bogs yielded a stream of cordwood. But as forest cover declined, farmers husbanded the remaining woodlots with care. Pete Ogden reports that one Vineyard farm family met all of its wood needs for a century from their forty-acre lot. But, for land- or woodlot-poor residents, simply keeping warm every winter was a challenge. George Willoughby's journal from the early 1880s provides some indication of the diversity of fuel sources tapped by families lacking much land or means. George writes that he variously "got one load of swamp wood," "went into my woods for my summer's wood," "got a load of roots for fuel," "went down to Data's swamp and cut wood at halves," "worked [the] day to pay off cord of wood," "went to see folks clearing Mr. Enis wood lot for stumps," and "brought a load of dry cow dung to burn." Despite this effort he concluded one journal entry in April by declaring that "our wood is gone and we are burning whatever we can lay our hands on."

Census data from the last half of the nineteenth century confirms both the characteristics of the forests and their heavy use. During this time, when forest cover was increasing due to the decline in farming, and coal was replacing wood in many homes, trees were still being cut when small, and generally less than thirty years of age. While pitch pine was the predominant species cut on the Cape, nearly all the wood harvested on the Vineyard was oak. In accord with Reverend Freeman's description eighty years earlier, Tisbury accounted for more harvests than all of the other towns combined; it was also the only town with sizable forests exceeding thirty years of age.

On the moraine and especially on Chappaquiddick and Gay Head, where wood was in extremely short supply and cash was scarce to purchase cordwood, some wetlands were harvested for peat. Cut in blocks stacked loosely to allow air circulation while they dried in the sun, the bricks of partially decayed organic matter were eventually stored under cover until needed. Although the practice was not widespread, for at least a few residents across the Island it warranted construction of "peat houses," outbuildings of wood and occasionally brick, which could accommodate the ongoing release of moisture from the wetland fuel.

Among Indian communities a system of common use of wood, peat, pastureland, and other resources in relation to individual and family need persisted into the second half of the nineteenth century. Leavitt Thaxter, in his official role as "Guardian for the Indians and People of colour and overseer of the Poor of said Indians and People of colour," describes the tradition of these "commons" in 1849.

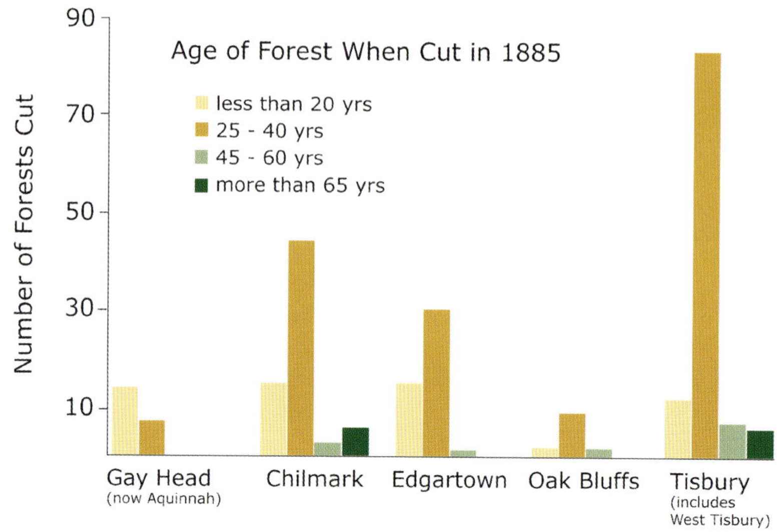

Forest harvesting and variation in the late nineteenth century. Using the 1885 census data, the age of cutting provides a rough indication of the age of the older forests, under the assumption that the older, larger trees would be most desirable. Meanwhile, the number of cutting operations gives a rough gauge of the amount of wood cut and available in each town. Most forests were less than forty years old, and only Chilmark and Tisbury reported forests older than sixty-five years old.

> The Gay Head Indians . . . live on a peninsula, and have little intercourse with the whites. . . . None of the lands are held, as far as we could learn, by any title, depending for its validity upon statute law. . . . The rule has been, that any native could, at any time, appropriate to his own use such portion of the unimproved common land, as he wished, and, as soon as he enclosed it, with a fence, of however frail structure, it belonged to him and his heirs forever.
>
> We have set off in Common (under the following regulations) the Peat Swamps on said Island of Chappaquiddick belonging to said Indians and People of color, viz; Indian Land bounds—581¼ ac. description. Indian Wood Land 111 ac. and 56 rods.

Despite these efforts, shortly after mid-century the Indians on Chappaquiddick had little to no fuel wood.

The sources of the lumber utilized on the Island over time remain conjectural. In early decades, local forests must have yielded boards and timbers for housing. Interestingly, there are no records of water-powered sawmills, indicating that no sustained timber industry ever developed. But references do exist to "saw pits" and "hewing fields" where wood products would have been cut and shaped by hand. The saw pit presented a grueling scene in which one person in a pit below the log and another above used a two-handled saw to cut a series of planks. In the process of hewing, a broadax or adze was employed to square a log into timbers for beams and other uses.

While such labor-intensive activity was surely practiced widely by many farmers and tradesmen, the extensive trade of Islanders and low cost of water transport undoubtedly encouraged importation of the bulk of nineteenth-century building materials. Weekly advertisements for diverse wood products in the *Vineyard Gazette* suggest that the bulk of lumber for construction was imported from Maine. One typical entry from July 11, 1851, highlights the diversity of imported material and helps explain the widespread presence of non-native lumber in many old Vineyard buildings today.

> Just received from Bangor, a large and extensive assortment of Lumber, consisting of [White] Pine, Spruce and Hemlock Boards, of all qualities. From 30 to 40 M. [thousand board feet] of Spruce Timber, of the first quality, free from sap, suitable for frames of largest sized dwelling houses.
> 175 M. of Pine and Cedar, Shaved and Sawed Shingles of all qualities. Clear Clapboards.
> 40 M. of Spruce and Pine Laths.
> 12 M. of Spruce Fencing Rails, 6 inches wide, 1 1-4, and 1 3-8 thick.
> 8 M. Stringers for Fences—2×6, 2×4, 2×3.
> 30 M. seasoned Lumber on hand, of qualities No. 1, 2, 3, 4, from 5-8 to 3 inches in thickness.
> Also, 40 M. Pine Sheathing Boards.
>
> The above will be sold as low as can be purchased in the market.
>
> THOS. M. COFFIN

Other wood-derived resources were also important. Leather, critical for uses from footwear and clothing to harnesses and hinges, required tannins from trees or other plant materials to make hides supple and decay-resistant. The tanning process was arduous, unpleasant, and environmentally damaging as it involved the scraping of hides from butchered livestock and native animals; the collection and grinding of bark, wood, or leaves from oak, sumac, and other species; and soaking of the hides in tannin-rich water for months to more than a year. Across New England nearly every town supported one to three or more tanneries of modest size. On the Island they were distributed widely—near Farm Pond, off Lamberts Cove Road, and at the upper end of Fulling Mill Brook—and always near an abundant source of water. Ads in 1854 for the sale of Charles Cottle's tannery at Lambert's Cove indicate that he processed a thousand hides a year, a typical amount for a small New England operation.

Equally important but even less well documented were naval stores—pitch, resin, and turpentine derived from pitch pine. Details on Island operations are elusive and largely restricted to inference encouraged by place names such as the Tar Kiln Path. It is likely that the bulk of these products were imported from the southeastern United States.

Industry in a Period of Rapid Change

The agrarian era was distinctive but transient. By 1850 New England was rural, broadly settled, and farm-based. Though Boston was an emerging regional hub with 130,000 people, and centers of maritime activity thrived at Salem (20,000), New Bedford (16,500), and Nantucket (8,500), the state's population was distributed remarkably evenly from the coast to the Berkshires in hill towns and valley farm communities of many hundreds to a few thousand residents. On Main Streets across the commonwealth, stately homes and thriving local businesses complemented sprawling rural farmhouses in signaling relative affluence and a secure belief in the enduring qualities of life built on human investment in the land. The Vineyard supported its own version of this successful model with a mixture

The barns, house, and mill at the old paint mill in Chilmark, now the Holly Holm home of Robert and Anne Ganz.

of farming, fishing, and maritime activity supporting 4,540 people, divided among Edgartown (1990), Tisbury (1803), Chilmark (747), and Gay Head.

But larger forces were at work. In the next few decades most rural towns and lands would fall into decline as newly opened Midwestern and western lands became the center of agriculture, and massive brick industrial complexes took shape alongside New England streams and rivers. These changes were facilitated by new infrastructure—canals and especially the railroads—that moved grain, coal, industrial products, timber, livestock, and other resources and by massive demographic changes as easterners relocated and were joined by growing numbers of foreign-born Americans. The ensuing geographical and economic tumult that drove the depopulation of rural villages, invasion of trees and shrubs onto abandoned pastureland, and rapid growth of cities and mill towns across New England brought tensions of cultural change to the Vineyard.

Island agriculture suffered mightily due to competition with farms in the west and the lifting of import tariffs on wool beginning in 1846. The number of sheep continued a long decline, and long-standing traditions fell out of favor and were then lost to memory. The late summer of 1854 marked the final washing of sheep in Shear Pen Pond out on Chappaquiddick. Other economic changes were afoot, including the peak and fall of whaling. While Dr. Fisher's oil and candle factory remained the Vineyard's most lucrative industry into the 1860s, the ships and barrels of oil unloaded on Edgartown docks steadily declined. Nantucket was a larger player in this business, with a less diversified economy, so it experienced greater shocks from these changes. A great fire there on July 13, 1846, destroyed three hundred buildings, left hundreds homeless, and triggered a reduction in whaling and oil production that initiated a long, steady slide in that island's population. From a peak of over 9,000 people in 1840, Nantucket lost a thousand people by 1850 and ultimately bottomed out at 2,797 in 1920. With its diverse towns, population, and commerce, the Vineyard suffered less when a similar fiery tragedy struck Homes Hole in 1883.

Meanwhile, other sources of economic activity did pick up. Most notably, the future of the Island as a great leisure destination began to glimmer with the arrival of such national figures as Daniel Webster (1849) and Nathaniel Hawthorne (1850), both of whom included plover hunting with leading businessman Dr. Daniel Fisher on their itinerary. The construction of small hotels and rooming houses in Holmes Hole and Edgartown, and the growing surge of tent- and then cottage-based camp meetings in Cottage City, extended this new wave of activity. Founded with a small gathering in 1835 and zealously championed by Jeremiah Pease and other leaders, the Vineyard-based meeting grew to a summer gathering of thousands. The climate, scenery, rural splendor, and leisure possibilities that would attract hordes in the future were recognized

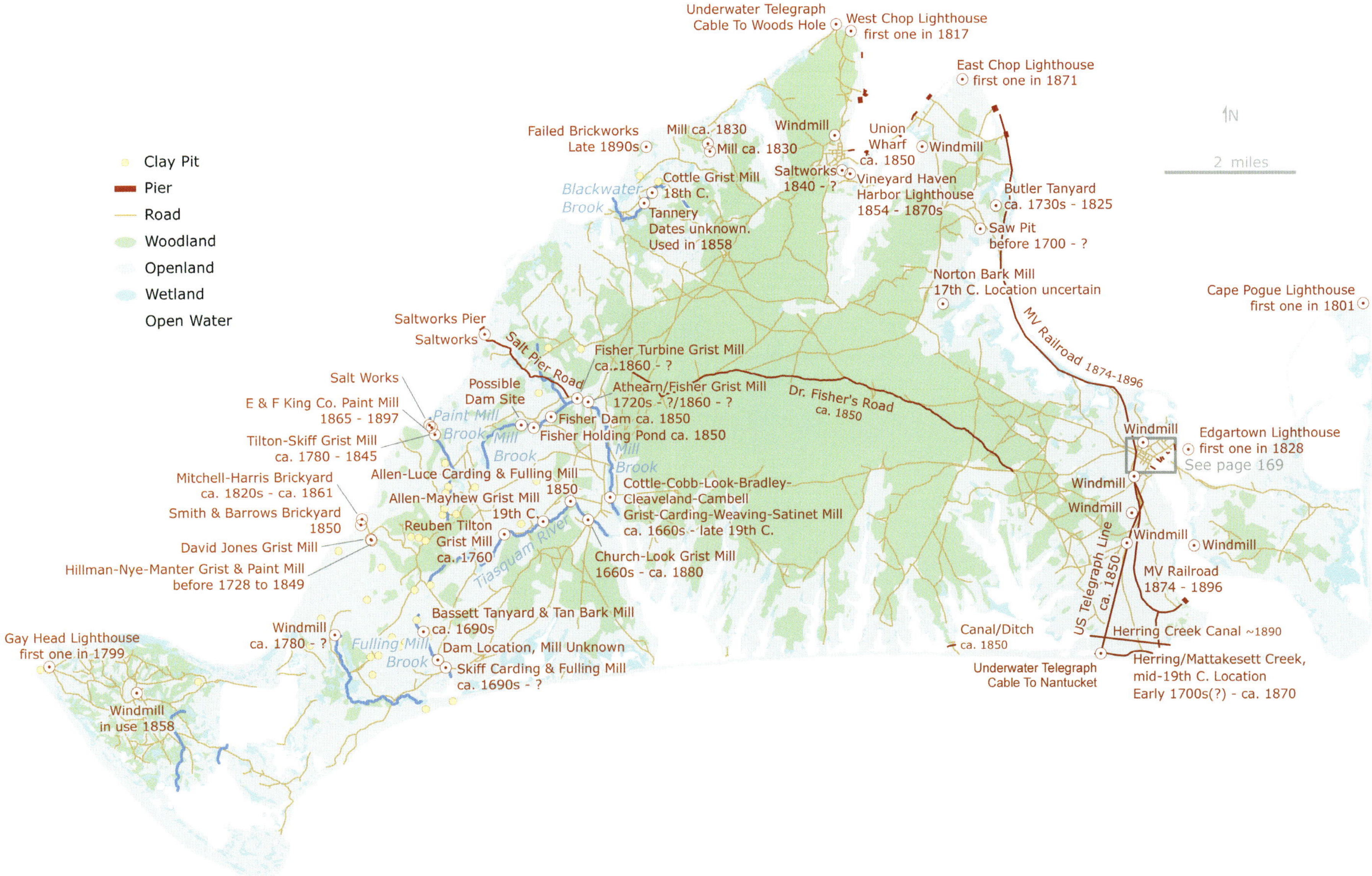

The mid- to late nineteenth century was the peak of industrial activity and associated alteration of steams and waterbodies. Existing gristmills were joined by new mills for carding, fulling, and brick and paint manufacture. Water was the principal energy source, focusing much activity on the up-island region. Windmills also provided power, especially around Edgartown Harbor. Other notable commercial features include saltworks, canals dug for the herring fisheries, the Mattakeset Railroad, and Dr. Fisher's Road. Information from many Island sources including John Flender, historical maps, the Martha's Vineyard Museum, and the *Vineyard Gazette*.

Mill sluiceway and pond that powered the paint mill in Chilmark.

and touted remarkably early in ads and editorial writings in the *Vineyard Gazette,* which started printing in 1846.

Through this period, maritime trade continued, bringing more than twenty thousand ships through Vineyard waters in 1854–55 on their way between New York, Boston, other eastern ports, and Europe, though having less and less direct impact on the island. The keepers at the Island's lighthouses maintained a tally of passing ships, as indicated by a note in the *Gazette* on July 7, 1854: "We learn from Mr. Daniel Smith, keeper of Cape Poge Light, that 6324 vessels were seen to pass the Cape, during the quarter ending June 30."

The mid-nineteenth century also brought modernization in communication, connectivity, and infrastructure that laid the groundwork for patterns familiar today. Successful town petitions led to the construction of Middle Road (1845), North Road (1849), and the Edgartown–Holmes Hole Road (1850). That last effort engaged the surveying skills of the ubiquitous Jeremiah Pease and revealed the entrepreneurial energy and community spirit that characterized that leading Island citizen Dr. Fisher. Having outbid a couple dozen of his fellow citizens on the contract for the first section of the road leaving Edgartown, Dr. Fisher proceeded to subcontract with all of the successful bidders for other sections at their higher bids while charging the town his lower price. As internal transportation was improving, a major step in external connectivity was forged with the laying of the first telegraph cable between East Chop and Woods Hole in 1856.

The northeast's shift from agriculture to industry was experienced on the Island as a modest boom in manufacturing and processing. Though often individually short-lived, each new industrial enterprise was heralded by community leaders and the *Gazette* and left imprints on the landscape along with colorful traditions that often loom much larger in history than the reality of the day. New water-driven mills were erected and old colonial ones were expanded through the hills and small valleys of Chilmark and West Tisbury. Activity peaking toward the middle part of the century included new grist mills at Look's Pond, Fulling Mill Brook (1850,

Dr. Daniel Fisher, 1799–1876

A Google search on his name simply lists his magnificent Edgartown home as a rental space for weddings and cocktail parties. But in the mid-nineteenth century, Dr. Daniel Fisher was a national economic force whose diversified enterprises spanned the Island, where they left many cultural and natural legacies in a landscape he loved.

According to journalist Nathaniel Parker Willis, Fisher was "the tallest, strongest-built, healthiest, and handsomest, as well as wealthiest and most influential inhabitant of Martha's Vineyard." The March 2, 1855, *New Bedford Mercury,* newspaper to the nation's whaling capital, identified the source of that wealth.

> Dr. Daniel Fisher's oil and candles works is the largest in the U.S. and surpasses all in the world other than London. Dr. Fisher is considered the boldest operator in the purchase of oils in the market and buyers are probably as much or more influenced by his opinion in their operations than any other person. The scope of his operation is noted by the fact that he has 28,000 barrels.

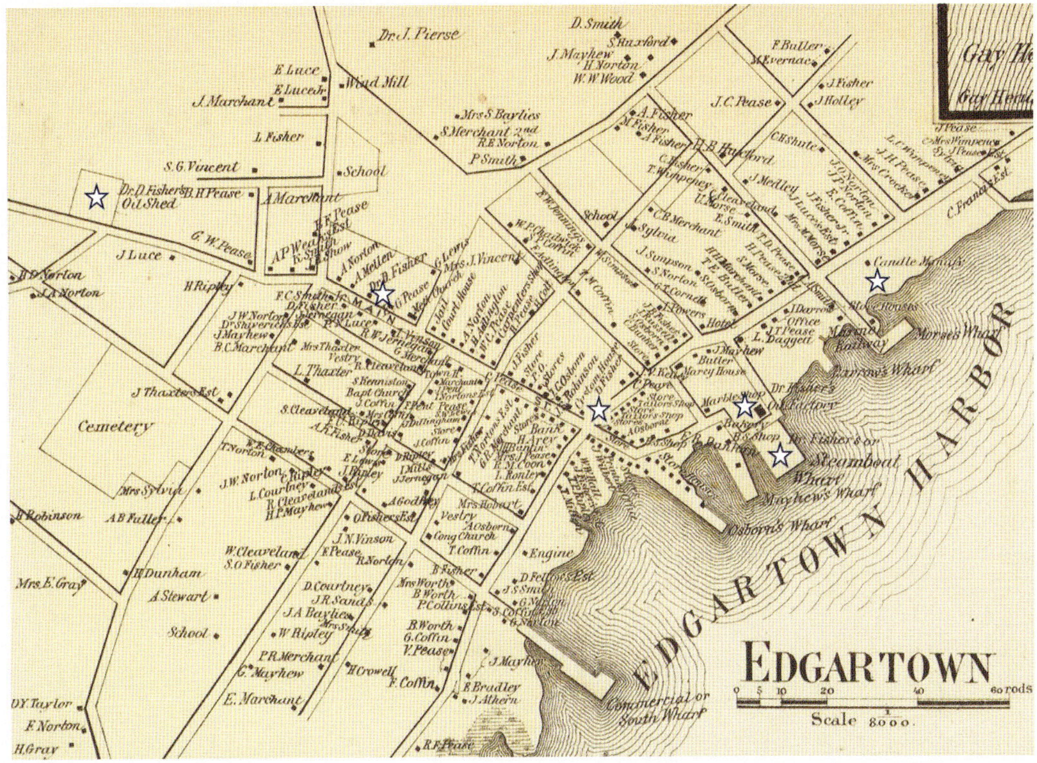

Dr. Daniel Fisher's industrial enterprises. Henry F. Walling's map of Edgartown (1858) shows (*left to right*) the location of Dr. Fisher's oil shed (today the county jail), house, office, oil factory, wharf, and candle manufacturing plant (Gamble, 2009).

Fisher's vast inventory of oil barrels was deployed globally on whaling ships from many ports and stored on the Vineyard behind a sturdy wooden fence in a looming warehouse called the "Fort" on the site of the modern county jail. By 1856, when he founded the Martha's Vineyard National Bank, Fisher's ownerships in whaling ships, Edgartown wharves, and the oil and spermaceti candle factory provided the Island's largest source of employment and annually yielded over 118,000 pounds of candles, 13,200 barrels of refined oil, and $285,000 in income. Many years he won the federal bid as sole producer of oil for the nation's lighthouses, and annually he and Abraham Osborne topped the list of Island taxpayers.

Fisher was as progressive as he was powerful, and he applied his means to support agriculture, industry, and the Vineyard economy while diversifying his own financial enterprise. Beyond his role in the construction and improvement of local roads, he subscribed to efforts to improve regional transportation and communication by bringing the telegraph to the Vineyard, the railroad to the Cape and Woods Hole, and the short-lived Vineyard railroad to aid Edgartown in its competition with Oak Bluffs for summer tourism. In a forward-looking effort toward Island self-sustainability and to provision his ships for their lengthy voyages, he built a hardtack bakery in Edgartown and an up-island flour mill. In developing a substantial milling enterprise, Fisher invested considerable resources and creativity. He constructed or renovated a series of four dams along Mill Brook and outfitted Crocker Pond with "one of the most complete mills in the U.S. for grinding wheat and bolting flour" by importing the best equipment, including a French millstone and expensive English works, and hiring a British miller, Nicholas Martin. He acquired more than six hundred acres of land in West Tisbury to produce wheat raised by local farmers. Though this crop often faltered, with the slack filled by grain from Maryland, the *Gazette* carried ads for Fisher's flour into the 1870s, and the hardtack factory thrived until his death in 1876.

Fisher linked his two commercial centers across a singular ancient way that still bears his name. Dr. Fisher's Road is a delightful but occasionally obscure trail that arcs across the heart of the Great Plain from Edgartown to Middletown. In its day this curious route provided advantages over the Edgartown–West Tisbury Road for four- to six-horse wagon teams. Fisher's track cut north, avoiding the deep hollows that strained animals and providing welcome respite for horses and drivers at Little Pond. The route also avoided the inconvenience of other traffic, pasture gates, and the congested village of West Tisbury. For two decades the thriving enterprise of farms, mills, ponds, path, and factory advanced Dr. Fisher's local and global interests by feeding Vineyarders as they crisscrossed the seas.

Fisher's rise to prominence was facilitated by a creative mind and auspicious marriage. Born in Sharon, Massachusetts, he grew up in Dedham, graduated from Brown and then Harvard Medical School, and completed an internship at Massachusetts General Hospital, before taking a job in 1824 as general physician in Holmes Hole at the advice of an Island classmate. His marriage to Grace Cousens Coffin, daughter of established shipowner Timothy Coffin, introduced him to whaling and set the young family up in the John Coffin house on North Water Street. Intrigued by the challenges of using whale oil in a system of many small lamps to illuminate lighthouses, he invented a single large lamp. This initial success encouraged Fisher to shift his energy to his wife's family business, where he built an enterprise that spanned the globe. With growing wealth, he constructed the "transitional Federal period house with Greek revival portico and intricately carved roof rail" on Main Street where he entertained acquaintances such as John Murray Forbes, Louis Agassiz, Nathaniel Hawthorne, and Daniel Webster. That house, the "track through this scrub oak wilderness," and a series of grand dams, millponds, stoneworks, and buildings along Mill Brook are among Daniel Fisher's enduring legacies.

by Samuel Tilton), Roaring Brook (1849, Francis Nye), and along Mill Brook (Daniel Fisher's dams and mill at Crocker, Fisher, and Athearn Ponds, circa 1860) and construction of the Smith and Barrows brickyard at the outlet of Roaring Brook (1850) and Paint Mill (1847–49) on the brook of that name in Chilmark.

Windmills and sprawling salt-production facilities were distributed along the coast from Gay Head to Holmes Hole and over in Edgartown. At the interface of traditional fishing and industry, a series of fishing companies excavated creeks and constructed dams and buildings to capitalize on the runs of herring and other migratory species along Katama Bay and Chappaquonsett (Tashmoo), Lagoon, Poucha, and Trapp's Ponds.

Legacies for Today

Though more than a century has passed since many pastures were abandoned, rural cartways supported the traffic of wagons and

Dr. Daniel Fisher. From the Collections of the Martha's Vineyard Museum.

livestock, and old grindstones turned harvests into flour and meal, the imprints of those activities remain embedded in the land. The most obvious are the striking physical objects—the stonewalls winding through forests; the deeply eroded ancient ways; the drill holes in and remnants of boulders mined for foundations, dams, and front stoops; and the old cellar holes, clay pits, and hillside gravel pits. Many of our open water bodies—from striking ponds like the chain composed of the Old Mill, Priester's, Crocker, and Fisher Ponds along Mill Brook to countless cow and sheep ponds, marshes, and vernal pools across the moraines—were dammed during that period. From significant waters to minor depressions and wet ditches, these features diversify the land and offer habitats to plants and animals that we cherish.

As pastures and other farmland were worked hard, they were gradually transformed. Though most declined in fertility, some were enriched with nutrients and became more productive. Others were plowed and their upper soils homogenized. Many were stripped of their surface rocks and even gigantic boulders, and most suffered some dimension of erosion as hooves, farm implements, wind, and water took their toll. But as this clear degradation of the originally forested environment proceeded, something marvelous occurred. The landscape was diversified, a range of new habitats was created, and that hard-worked land accumulated a wide assortment of native and imported species of grasses, herbs, and shrubs that expanded the Island's heterogeneity. After decades of heavy grazing, after they were well marked by patchy vegetation, spots of bare soil, rocks, and the eroded trails of hoof prints, rough pastures came to support many plants, insects, birds, and mammals once scarce in the former wooded landscape. Over the centuries a set of distinctive habitats emerged that became familiar to the Island—sandplain grasslands; low maritime heathlands; scrubby shrublands; scrub oak barrens; grassy, sedgy, and ferny fens; and moors. The result was one of the largest concentrations of uncommon habitats and species in the northeast.

Today the Island carries many memories of this past. We can dip into that history by reading the features and characters of the land as effectively as we can delve into the volumes that line our libraries and the papers and maps that fill our archives. Nature itself is a rich archive. If we take the time to read it, we can understand the forces that have shaped it. We can then use those insights from written and landscape history as we decide what we will value in the land today and what we will manage for and how. On adjacent sites across the Island today, we have, on the one hand, old farm woodlots that have never been cleared and that retain the same native species that have grown there for thousands of years and, on the other hand, heathlands and grasslands created by deforestation and centuries of grazing. On Mill Brook we have gorgeous wooded stretches that appear natural and timeless, interrupted by picturesque dams, ponds, and channels constrained by massive block walls. Each of these habitats and scenes supports distinctive species, references to past times, and varied emotional responses. The nineteenth century lends a key to understanding the interplay of cultural and natural forces that give us this rich portfolio of sites.

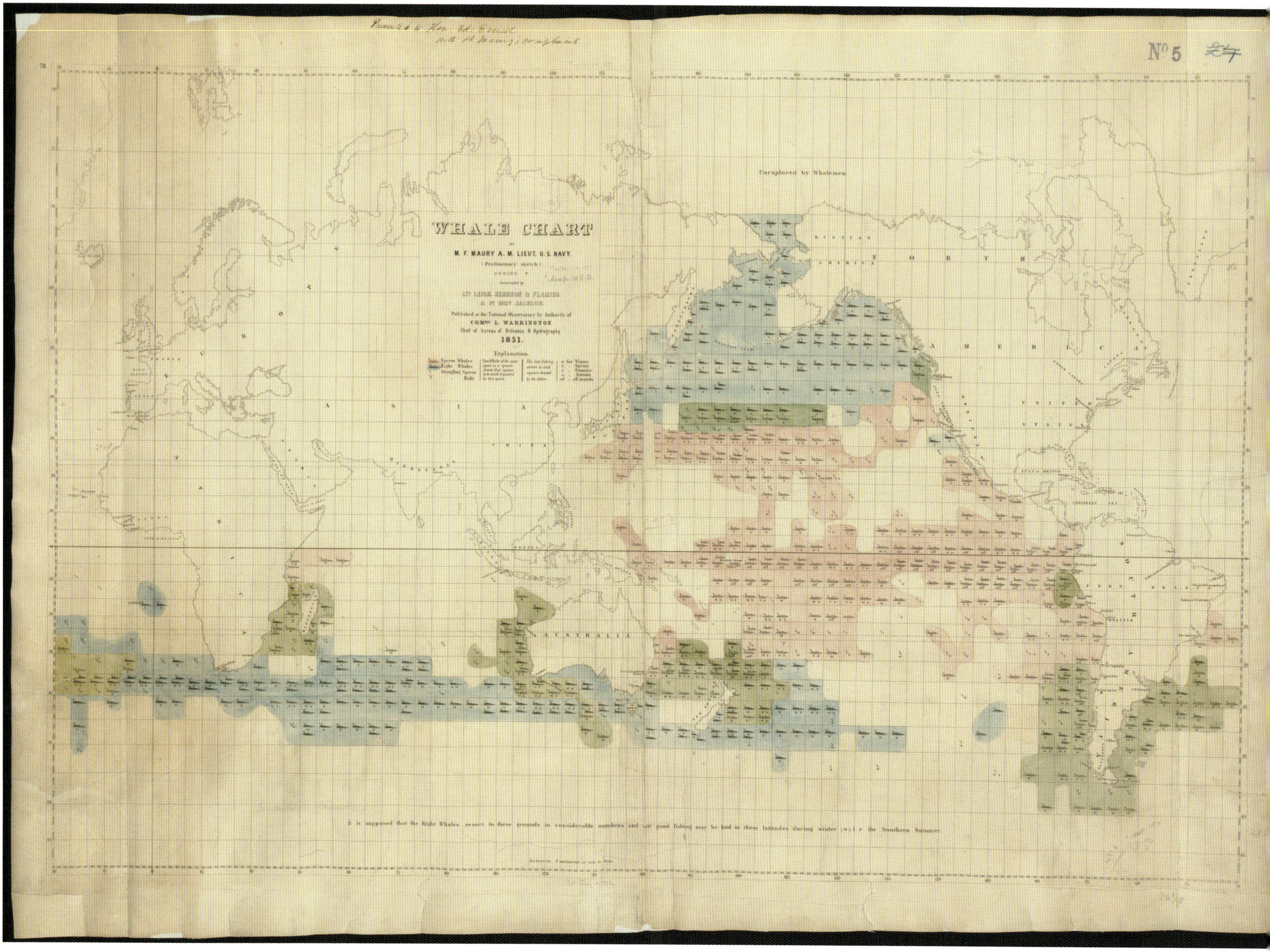

Where did Island whalers go? This extraordinary map by the Bureau of Ordnance and Hydrography of the U.S. National Observatory displays the distribution of sperm and right whale harvest in the mid-nineteenth century. These are locations that men, boys, a few women, and their whaling ships headed for when departing the Vineyard. Map reproduction courtesy of the Norman B. Leventhal Map Center at the Boston Public Library.

Fences. *Upper two rows:* relict stonewalls on the moraines; *third row:* ditch and ridge fencelines on the outwash plain; *fourth row:* aesthetic walls built by modern hands.

Making the Modern Landscape

The abundant and cheap transportation of food-products from these prairies . . . has been so great that, since the middle of the century, it has not only in good part stopped the process of clearing away the Appalachian woods, but in certain districts, especially in New England, extensive areas of land which had been long under tillage have been allowed to return to the forest state. . . . The present writer, in his many journeys through [Massachusetts], has observed many thousand acres of woods with the marks of former tillage—corn-hills, and walls composed of bowlders gathered from the once cleared land.
— Nathaniel Southgate Shaler, 1887

All I see in prospect now is uninhibited development. . . . I can't see any consequence but that the Vineyard will be built up and built up.
— Henry Beetle Hough, 1984

Most persons will heartily agree that it is our bounden duty to transmit the inheritance which we enjoy in the earth unimpaired to the generations yet to be.
— Nathaniel Southgate Shaler, 1887

How do we capture the immensity of the changes that brought the Island from the mid-nineteenth to the twenty-first century and from an agrarian landscape and sea-based economy to a largely forested land fueled by tourism but home to professionals, tradespeople, artisans, farmers, and families that form its backbone? Statistics convey the hard facts. Maps and photographs paint the changing scene. But it is through the words and interpretations of people who lived and cared deeply about these changes that we can capture the essence of the upheavals that altered the pattern of the land and spirit of the place. Two keenly observant but contrasting men provide insightful perspectives. One was a Kentucky gentleman and renowned Harvard professor of science—Nathaniel Southgate Shaler—who founded Seven Gates Farm and in the process became the Vineyard's largest landowner. The other—Henry Beetle Hough—was a newspaperman and writer who spent his early summers at Fishhook, the family place overlooking Cedar Tree Neck and Seven Gates, and devoted his life to the *Vineyard Gazette* and the Island.

The two men never met and lived in different times. Though they differed in profession, background, and means, they shared at least three passions: the Vineyard, exploring the interplay between society and nature, and land conservation. Both wrote prolifically and had the perfect eye to capture the Vineyard era in which they lived and from which it had come. Shaler compiled the definitive description of Island geomorphology in the last half of the nineteenth century, a time when the cutover and deforested condition yielded sweeping views of its physical variation. As he assembled his rambling estate at Seven Gates through the acquisition of declining farms, Shaler witnessed in a personal way the dissolution of the Island's agrarian and whaling life and the obscuring of the land's surface by a thickening film of vegetation. The Vineyard had a unique appeal to this student of the earth who wrote, "No other landscape known to me has so many contracted slopes in an equal amount of profile; . . . nowhere else in this country do I know anything like the variety of scenic effect that is exhibited in this hundred square miles."

From 1896 to 1985 and his domestic vantage points in West

Professor Nathaniel Shaler on Seven Gates Farm. Reprinted with permission of the Harvard Forest Archives, Harvard University, Petersham, Massachusetts.

Tisbury and Edgartown, Hough lived through multiple waves of tourism, exploitation, and conservation. With a historian's eye, he wove the narrative from Shaler's era of early summer folk with their grand hotels, camp meetings, and family places through to his own time of accelerating change. In his popular volumes, articles, and editorials, he depicted and often battled developers. He came to promote a dual path of conservation and growth that balanced the historic and natural with emerging economic forces, all the while extolling the beauty and human spirit of the Island that he seldom left.

Together, the lives of these men spanned many of the major changes in the land and people that helped make the landscape what we have around us today. And, like earlier generations whose relict pasture walls, cellar holes, and crumbling brickworks grace our landscape, both left us marvelous legacies to cherish: Seven Gates Farm, the bulwark of the conserved morainal landscape, and Sheriff's Meadow Foundation, the largest land-owning Island conservation organization.

These men showed that the emergence of the modern landscape is largely the consequence of two opposing forces—the broad-scale conversion of farmland to forests and the punctuated development of fields and woodlands by residential and commercial development. As these contrasting forces clashed, a third influence arose to mediate the interaction—the conservation of woods, farms, shores, and historic structures.

Reforesting the Land

Looking back today, when the sight of stonewalls wending through forests is commonplace, the process of farm abandonment and forest succession is a proud part of our history. Newspapers, books, and documentaries herald the recovery of the New England forest as a grand conservation success. This greening of the east is "the great environmental story of the United States and in some ways of the whole world," according to environmentalist Bill McKibben. Indeed, there is every reason to embrace nature's sweeping recovery. New England has twice as much forest today as it did one hundred and fifty years ago. In most places, like the Vineyard, these forests are growing grander every year. As an ecologist bent on land conservation, I relish this history as a marvelous gift, for it grants New England a second chance to determine the fate of our forests and landscape. It is a chance to conserve the land this time around.

But in the eyes of nineteenth-century New Englanders, the sight of weeds, shrubs, and trees invading rolling pastures was emotionally wrenching, even without the accompanying view of dilapidated farm buildings. The proximate cause—the decline of farming—was a depressing symptom of a failing economy and dissolution of a way of a life that had been forged over generations. The sight of fallen fences, neglected and sagging buildings, and vigorous seedlings becoming forests in old fields made a melancholy scene, not a reason for celebration, for it announced to all who passed that nature was reclaiming the upper hand. The old vitality in the villages and countryside, indeed the character of the residents, appeared weakened if not altogether sapped. Fewer and fewer

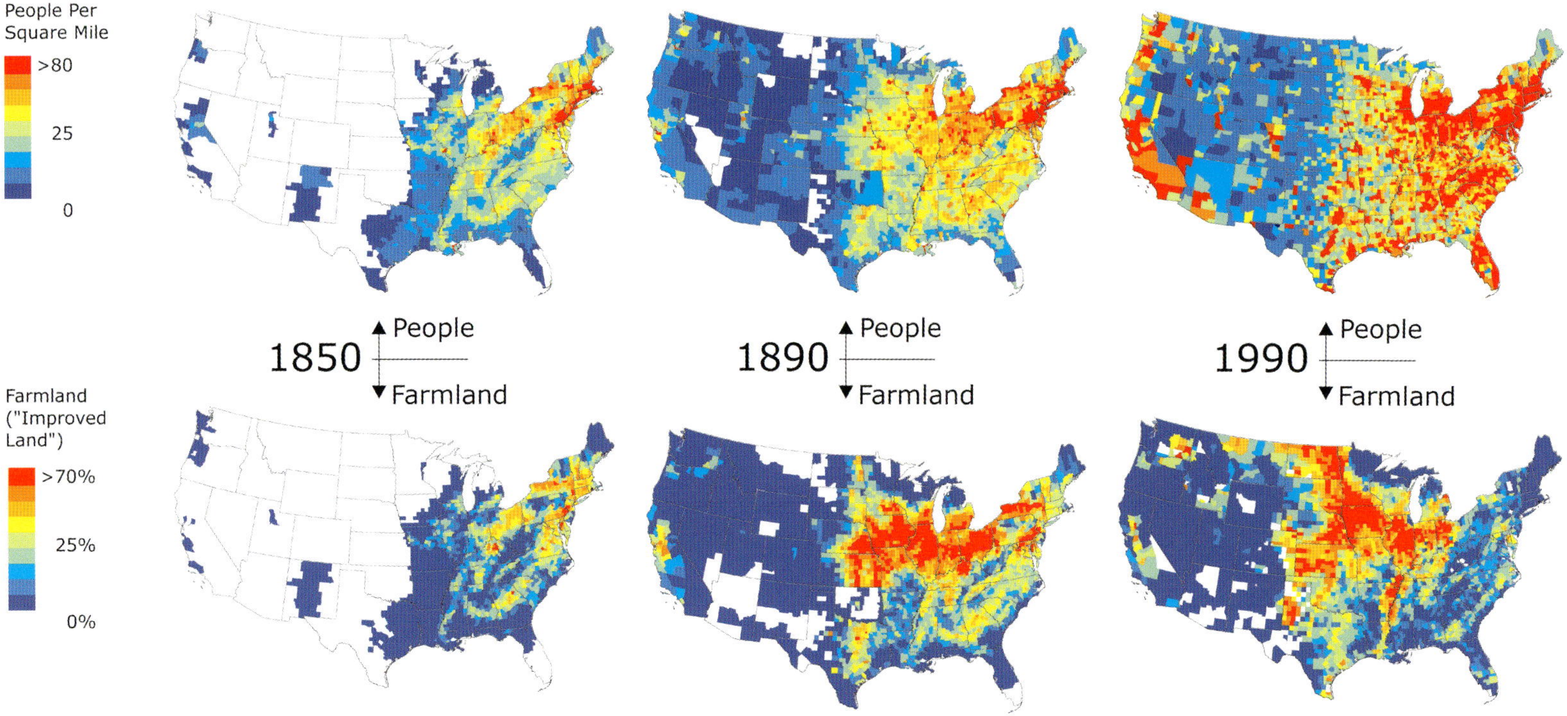

A national shift in people and farming. In the mid-nineteenth century, the United States was an eastern country; the distribution of people and agriculture was closely matched. However, with the opening of fertile lands in the Midwest and west, and growing improvements in transportation by rail, canal, and ocean, the nation's population and food production diverged geographically and eastern farmland reverted back to forest. Data from Pamela Waisanen and Norman Bliss at the U.S. Geological Survey (Waisanen and Bliss, 2002).

people cared passionately for and worked the land. Figuratively, as well as literally, many others were up and leaving it. Something in society seemed to have failed. All the effort expended over previous centuries in mastering the land for productive human uses was coming to naught. The windfall that we have received was their painful loss.

Early on, Daniel Fisher, prosperous owner of the whale oil factory and Edgartown wharf, sensed this change and sought to arrest it or at least alleviate its pain. By modernizing his dams and mills on West Tisbury's Mill Brook, he attempted to bolster both farm and seafaring economies. He employed local farmers to produce grain that was ground in his mill to produce hardtack in his bakery that joined the grueling voyages of outbound whalers that supplied his factories.

Henry Whiting and Leavitt Thaxter, along with other social leaders, also fought the change, by founding the Vineyard Agricultural Society to educate farmers and modernize equipment, livestock, and operations. Gathering in West Tisbury in 1858, they resolved "that the attainment and diffusion of scientific and practical knowledge in the cultivation of the soil, is a subject of such importance as to demand the associated effort of the farmers of Dukes County." A few years later their stated goal was more direct: "to improve the agricultural interests in our county."

Vineyard farmers were late arrivals to a larger movement of state and local agricultural societies dedicated to promoting modern farming practices and fighting the dissolution of a noble profession and established way of life. Aiding this effort were new journals that sought to exchange information, bring science to barn, farmhouse, and fields, and rally spirits throughout the rural countryside. One widely read periodical, *The New England Farmer,* offered articles on equipment, manures, fencing, livestock, domestic practices, and farming methods. It also exhorted its country audiences to change their ways or else lose the rural backbone of the region. In 1856 it zeroed in on the major concern—abandonment of rural life by the region's youth. Two columns captured the emotional range of the times by first acknowledging the stark economic motivation for youthful dissatisfaction with farming, and then exhorting this young generation to reconsider that livelihood.

> That agriculture is unprofitable, compared with other business in Massachusetts, is the practical judgment of farmers generally, deny it as you may, gloss it over as you will; else why do so many of their sons desert the plow, hardly enough remaining at home to take care of the good old fathers and mothers? Have they not seen their fathers and neighbors, hard-working and frugal, farmers till sixty years of age, still relatively poor, while their relatives and equals who have engaged in other pursuits are rich, clad in fine linen, and fare sumptuously every day, with leisure to enjoy life, with means to purchase its pleasures, and comforts, too? Why are farmers willing, even desirous, to have their sons quit the farm, and seek an easier and shorter road to fortune and happiness, than they have trod? The truth must be told, they desire a better life for their children than they have had . . . a man can live in Massachusetts by farming, but only by economy and self-denial, unknown and unpracticed in other pursuits.
>
> YOUNG MEN. Whatever may be your choice of future occupation—whatever calling or profession you may select, there is certainly none more honorable than that of a farmer. The patriarch of the fields, as he sits beside his cottage door when his daily toil is over, feels an inward calm never known in the halls of pride. His labor yields him unpurchasable health and repose. I have observed with more grief and pain than I can express, the visible tokens which appear in all directions of a growing disposition to avoid agricultural pursuits, and to rush into some of the overcrowded professions, because a corrupt and debasing fashion has thrown around them the tinsel of imaginary respectability. Hence the farmer, instead of preparing his child to follow in the path of usefulness himself has trod, educates him for a sloth; labor is considered vulgar, to work is ungenteel. . . . How melancholy is this delusion, which, unless it be checked by a wholesome reform in public opinion, will cover our country with wreck and ruin!

Nathaniel Shaler did his part to maintain the noble rural occupation. As he began purchasing ailing farms in the 1870s, he was not simply capitalizing on an opportunity to develop a personal estate and "a piece of the moraine." He and his successors at Seven Gates, especially his son-in-law Willoughby Webb, who assumed control when Shaler died in 1906, sought to hold a solid chunk of the old farm landscape together in an active and even exemplary working condition. Shaler employed staff and neighbor farmers, grazed the hills with sheep and cattle, and famously strode his pastures, uprooting weedy thistles while he oversaw his land and planting daffodils along the way. He laid the groundwork for the farm experimentation pursued by Webb and extending in modest fashion to the present.

When Shaler turned his most optimistic eye to the Vineyard landscape, as in an 1874 article in the *Atlantic Monthly,* nothing surpassed the up-Island scene of farm and nature blending together.

> Finally we emerged from the enveloping woods into the central valley of the island, where lie the villages of North and West Tisbury. It is a beautiful valley, of broad grass and grain fields with overgrown hedge-rows and fences covered with vines. On the east is the interminable forest [the Great Plain]; on the west and north a range of hills rugged with the vast piles of huge, gray bowlders, and dotted with pastures and forests. . . . To the south the valley sinks in widening fields. . . . The beautiful brook which gathers its pure waters in the hills to the westward, and becomes in a few miles a little river, expands into a great pond with irregular shores and a narrow channel, through which it escapes into the sea. . . .
>
> The people get a comfortable subsistence from the broad pastures, well stocked with cows and sheep, and the rich meadows beside the stream. The land is of an excellent quality, quick to answer any legitimate demands upon it, and not readily worn out. Some of the fields of maize and wheat are as good as one finds in the Connecticut Valley. I have never seen better ground for the gardener. Strawberries grow as in southern France; roses have a glory unattainable anywhere else in New England.

But such glowing reflections were rare for this keen observer of nature and man. Increasingly, and with a strong tinge of remorse, Shaler viewed the evidence of neglect in most directions.

> All along the road from the town of West Tisbury to the Menemsha Bight stand empty farm houses in the midst of overgrown fields. Here a little cottage clings to the precipitous hillside, big houses stand in valleys, all with paneless windows and shattered chimneys. Wild stuff is creeping over once cultivated fields. . . . The call of the city has depopulated once prosperous farming communities. . . .
>
> Agriculture here, as everywhere else in Massachusetts, is in decay. One never sees a field newly won from the forests, while on every side are signs of the gain of the woods on the fields. There are many deserted houses and every little while there is a little pile of crumbling brick, or an old well, to mark where once stood some house which has been pulled down for other uses.

As he explored the Island, Shaler noted evidence of decline extending from the coast to mill valleys.

> Along the south shore there is a number of great ponds, with vast reaches of upland plain by their side; these were once cleared and cultivated, but now the fences are falling away, and a few sheep that browse on them are all there is to mark the presence of man. . . . [In the village of West Tisbury] there are a few stores looking only half alive. . . . A little mill sits astride the dancing brook with a business-like air, but the grass and bushes of many years growth gathered around its doors suggest anything but work.

As he sought explanations for this downward trend, Shaler declined to pin it on the residents or quality of land. Like the editor of *The New England Farmer,* he recognized that the root cause was a cultural and economic shift sweeping the eastern United States. Many forces—jobs and opportunity on the one hand, and disillusionment with the staid farm and Island life on the other—were drawing people away. Most dispiriting was the loss of the Island's young. More and more children were departing for mainland cities, western lands, and other promises.

Shaler also noted the other major challenge facing the Island. As agriculture weakened, the whaling and maritime industries that had lent vibrancy to the ports of Holmes Hole and Edgartown began suffering a series of contractions. The Civil War, losses of ships to the rebels, a calamitous winter in the Arctic ice, and a shift from biologically to geologically based fuel reduced the whaling fleet to a handful by the 1870s.

> It is worth another ride over the eighteen miles of road along the south shore to see the pretty village of Edgartown, at the easternmost end of the island. In a commercial sense it is a place far advanced in decay: of all its whale-ships, which got from the sea the hard-earned fortunes of its people, there is but one left.

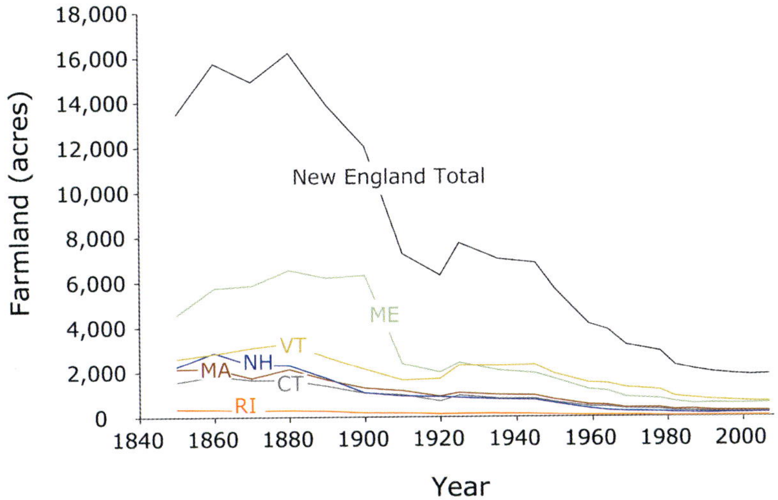

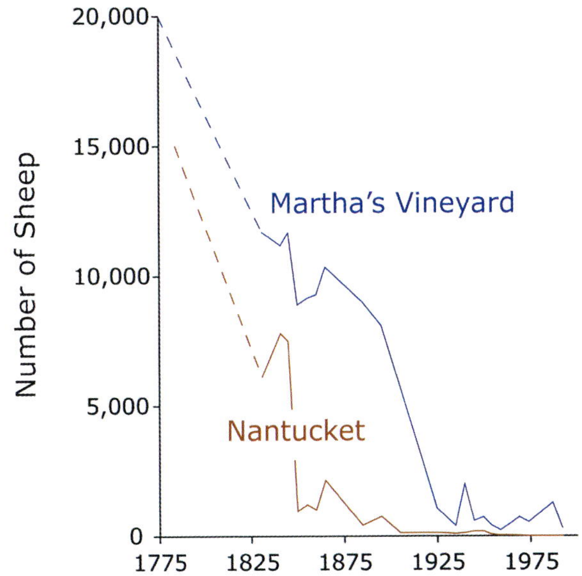

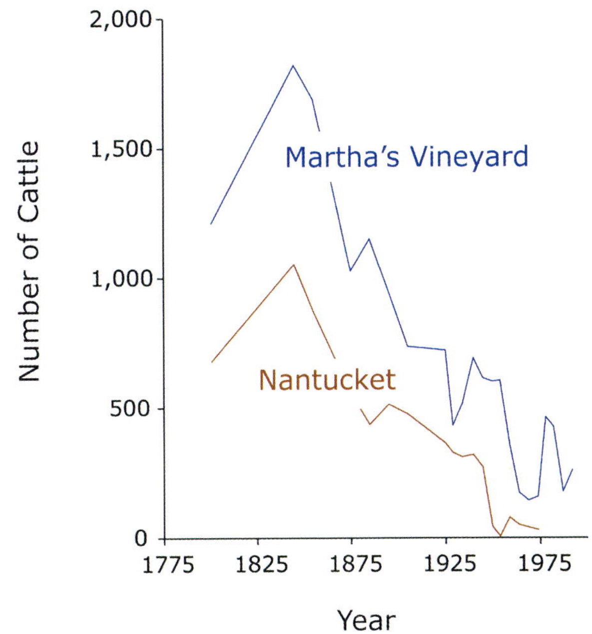

A decline in New England farms and farm animals. Farmland data from Sarah Sutton and Brian Donahue at Brandeis University (unpublished). *Left:* Beginning in the mid-nineteenth century the amount of land in agriculture declined. *Right:* On the Vineyard and Nantucket, the farm decline is noted by the sharp drop in sheep and cattle.

This lies upon the ways, stripped of its rigging, looking like a mere effigy of a living craft.

While farm outbuildings were collapsing, the bustle on wharves in Edgartown and Holmes Hole was subsiding, and the handsome "big square, roomy houses, which shall always be ship-shaped and well painted" grew dingy, legacies of former wealth, "comfortable homes [that] . . . mark a period of prosperity which has passed never to return."

Rural decline invited nature's return through a two-step recovery process: the spread of woody plants across abandoned pastures followed by the growth and succession of the various species. As farmers retired their scythes and reduced their herds, the weeds and trees moved in. In less than a century after the peak of agriculture, this process was largely complete. From Gay Head to Chappaquiddick, a rank growth of shrubs, scrub, and

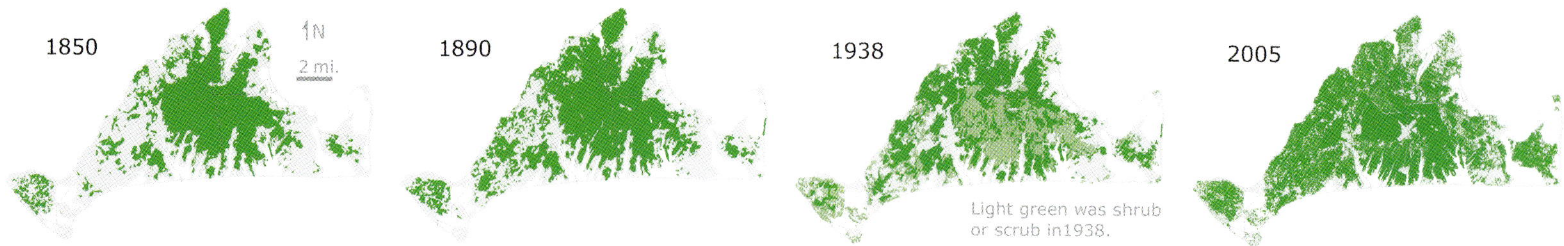

Making of the modern forest cover. From the peak of farming around 1850, farm abandonment and woodland expansion proceeded piecemeal. Through 1890 changes were concentrated across the moraine in Chilmark and northern West Tisbury. Woodlands expanded from Tashmoo to Cedar Tree Neck, around Roaring Brook and down to Waskosim's Rock, on to Cape Higgon, and then south across Kings Highway and the Tiasquam River in disconnected fashion to Quansoo. Large farmlands persisted through the early 1900s along the south shore and on Chappaquiddick. By the late 1930s many details had changed through infilling and expansions of woodlands. The latest maps highlight a Vineyard that is solidly forested but perforated by development and isolated farms.

sprout woodlands dominated the landscape. But the expansion of nature did not yield a magnificent growth of trees overnight. As noted by ecologist Pete Ogden, the scrubby thicket covering most of the Island as he was finishing his thesis in 1958 was far from majestic. The tidy agrarian landscape disappeared rapidly, but true forests emerged through an agonizingly slow process that varied considerably across the Island.

The moderate coastal climate can support prolific tree growth. But once the forest is stripped away, the wide-open environment is hostile. Winds carrying salt and sand sweep the land, and icy winter blasts tear at the emerging growth, damaging shoots and thwarting upward extension. Prostrate, contorted, and multi-stemmed individuals emerge as the first cohort of what can barely be called trees. Adding to the challenge, the old woodlots and newly formed woods continued to be cut heavily into the early twentieth century. On the Great Plain, great wildfires knocked back the growth on this perennially neglected terrain. And, on the Island's extremities, where trees had been eliminated for centuries—the necks, Squibnocket, Cape Poge, much of Aquinnah, and Chappaquiddick—it was challenging for heavy tree seeds to spread and for saplings to push through the tangle of grass, low shrubs, and vine.

Succession was so slow it appeared suspended. Many of the scrubby pastures and thickets changed so gradually that they seemed permanent. People who had never read Gosnold's tales of "tall-timbered" forests interpreted twentieth-century grasslands, heathlands, and bizarrely stunted forests as the natural and enduring condition. It was deceptively easy for people to doubt Ogden's assertion that forests had been and would become the natural cover.

But trees did emerge from the scrub, and this first windswept generation provided shelter for a second wave of taller growth. The next generation escaped the environmental extremes witnessed by their predecessors and grew straight and taller. With time, trees also spread along the necks and Moshup Trail, across Wasque, and even on Cape Poge. Across the Island, at a variable pace controlled by exposure and the impacts of storms, fire, and the ax, woody growth advanced. By the 1980s, Pete Ogden's prediction was vindicated. True forests were widespread, and the measured growth of individual trees was astonishingly good. The Vineyard was found to be a great place for trees and forests. As scenic views became obscured, heathlands declined to a rarity, and fields shrank alongside towering oaks and pines, the reality became clear. Forests were taking over the Island. Today there is little doubt that the Vineyard once was, and will again be, a largely forested land. That is, if we allow it to. For nearly two centuries, other forces besides farming have been working against it.

Victorian cottages in the Martha's Vineyard Camp Meeting Association, Oak Bluffs.

The First Boom, 1835–1900

As Shaler watched the abandonment of farms and trees spreading across the land, he became aware of another growing cultural phenomenon of which he also played a part—the arrival of summer people. While the countryside was losing old farmhouses, and the stately homes in Edgartown were graying, a new type of house was cropping up on formerly unsettled sites. The first were established by the Camp Meeting Association amid fields and oaks north of Farm Pond. Laid out in 1835 by Jeremiah Pease, the seasonal settlement of Wesleyan Grove began with a cluster of tents. Over time it grew in size, permanence, and most surprisingly, economic importance. By Shaler's earliest visit as a Harvard undergraduate in 1860, there was a defined settlement of tiny Victorian cottages tucked among the trees. With little but cart paths linking the grove to the Island's two established harbors, visitors accessed their emerging village via a wharf at Eastville.

As this seasonal Methodist settlement grew, it provoked a series of developments variously fueled by secular and religious motivations. Certainly the most successful was the village center just to the east, conceived in the late 1860s by a group of on- and off-Island schemers. The mastermind was local businessman Erastes Carpenter, but the plan was drawn up by landscape architect Robert Morris Copeland, lifelong friend of Henry Wadsworth Longfellow and designer of Sleepy Hollow Cemetery in Concord, Massachusetts, and estates for Samuel Colt in Hartford and Frederick Billings in Woodstock, Vermont. One of the first planned residential communities in the country, the plan embraced nearly one thousand small lots, curving streets, and what became ten parks, all situated between the campground, the Nantucket Sound waterfront, and Farm Pond. Initiating a custom of coining folksy names to aid the marketing of new developments, Copeland's ambitious village was christened Oak Bluffs. The parent investment company was the Oak Bluffs Land and Water Company.

Growing recognition of the economic benefits of tourism encouraged a series of competing and complementary development proposals associated with the coastal villages of Cottage City (Oak Bluffs), Holmes Hole (Vineyard Haven), and Edgartown. These plans varied in their planning, financial backing, and ultimate success. Through the last three decades of the nineteenth century, a dizzying array of formal and conceptual plans emerged for settlements and grand hotels across the down-Island region, starting at East Chop, extending along the east side of the Lagoon, and across to the Tisbury side, the harbor at Holmes Hole, and on to West Chop. Meanwhile, Edgartown business interests and town fathers felt their loss of prominence to the upstart village to the north and devised their own schemes. The most ambitious eyed the waterfront along Mattakeeset and Katama bays and sought to link the towns by rail. By 1873 more than two thousand acres of subdivisions were planned Island-wide, accompanied by a litany of delightful if redundant names: Vineyard Highlands; Bellevue, Sunset, Prospect, Ocean, and Lagoon Heights; Forest Hills; Sea View, Bay View, and Ocean View; Oak Point and West Point Grove; Grovedale; Lookout Mountain; Central Place; Cedar Neck; Engelside; and Oklahoma.

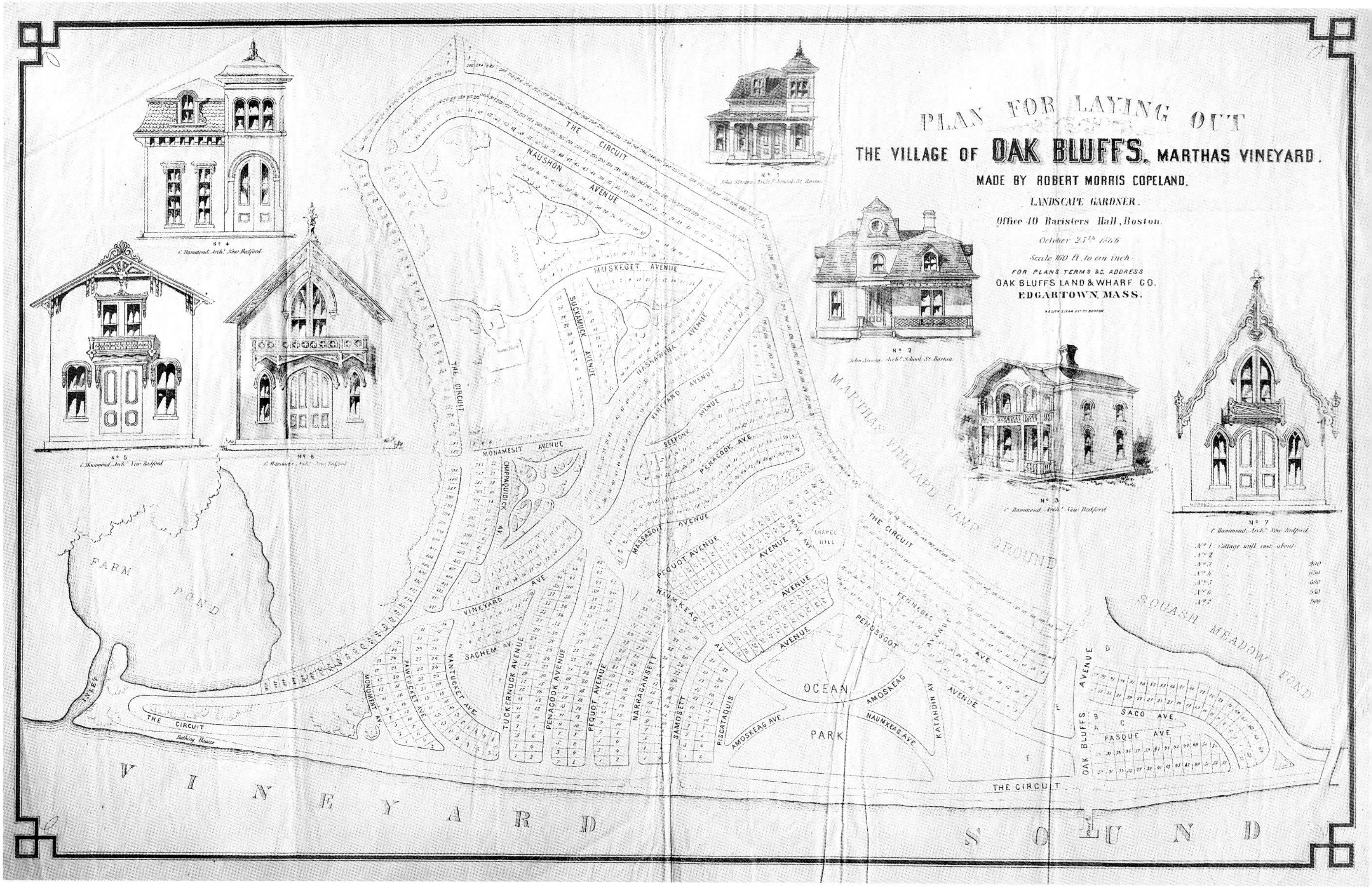

Robert Morris Copeland's design for one of the first intensively planned communities in the country, in Oak Bluffs. Map reproduced courtesy of the Norman B. Leventhal Map Center at the Boston Public Library.

Stereoscopic images of Mattakeset Lodge at Katama Bay, ca. 1890.
From the Collections of the Martha's Vineyard Museum.

At least eighteen plans centered around Cottage City, which by the late nineteenth century had been transformed from shaded cow pasture and a seasonal collection of tents to the Island's most prosperous village and innovative economy. With sound concept, financial backing, and a prominent location, Copeland and Carpenter's design sprang fairly rapidly from plan to execution; others, like Oklahoma and Lagoon Heights, blazed briefly as concepts, faltered, and eventually sputtered to life in altered form years or decades later. A few, such as the railroad from Cottage City to Edgartown and associated Mattakeeset Lodge, along with the Sea View Hotel, Highland House, and Innisfail House, burst into action with great fanfare only to collapse years later due to finances, fire, or both. And quite a few never progressed beyond the original idea or sketch.

The railroad was a grand enterprise that, like larger transportation projects to Hyannis and Woods Hole, attracted Island investors with its promise of connecting to mainland populations, revitalizing the local economy, and modernizing the Vineyard. Opened to great fanfare in 1874, it peaked in 1882 with 21,142 passengers, and shut down in 1896 following a run through precarious financial times and a barrier beach subjected to overwash and erosion. Like many subsequent paper developments proposed elsewhere, the five hundred houses and grand hotel were never constructed. The hotel guardhouse was completed and served as a perfectly elegant lodge for years before being carved up and carted into town.

A stunning coincidence of fires marked the demise of many grand hotels, including the Seaview (1892), Highland House (1893), Prospect House (1898), Innisfail House (1906), and the Wesley House (1894). Though the fire at Wesley House was extinguished, the owner later confessed ignominiously to arson and went to jail.

The explosion of activity brewed major cultural tensions within and among towns. Most community leaders were solidly supportive, the *Vineyard Gazette* spewed boosterism, and newcomers delighted in the growing outlets for entertainment; but many of the old guard and rural citizens were wary or disgusted. The patrician Shaler had disparaging remarks for these cultural developments, aiming his harshest comments toward the growing throng of seasonal visitors to Cottage City, who brought little appreciation for Island history and character. In Shaler's caustic view, the new structures and settlements were as incongruous in style and location as their owners' appreciation for the land was thin. He conveyed sentiments that future generations would share toward subsequent waves of newcomers.

> So far the new-comers have displayed the admirable lack of discrimination so characteristic of those who haunt the shore in summer. . . . Oak Bluffs, where oaks and bluffs are both on the average less than ten feet high, has grown to be a pasteboard summer town capable of giving bad food and uneasy rest to . . . thousand[s of] people. We want the good reader to have the best opinion of Martha's Vineyard, so we will turn ourselves away from the huddled roofs of the new-made town which looks out of the bushes, the aforesaid "oaks" of the name, and journey towards the central part of the island. . . . This ten miles of growing forest

Plan for Lagoon Heights, ca. 1873, with a panorama of the Lagoon, Cedar Neck, Holmes Hole, and sprawling Cottage City. Details show the bath and boathouses, Saratoga and Forest Parks, the Wing Cottage, Prospect House, and 949 lots. From the Collections of the Martha's Vineyard Museum.

is, for us fortunately, a waste in the eyes of the good cits. . . . It has no stage effects to tickle their dulled perceptions; besides, they get lost in it.

Like others associated with the Island's colonial roots, Shaler worried about the shifting of power from old villages to new. "Old Town [Edgartown] was fading and a dense tinsel town city of shacks [in Cottage City and elsewhere] was taking its place."

The motivations of the new class of developer were also suspect. Shaler complained, "The speculator or investor is almost as constant in the landscape as the tree." While more than a few grand schemers were local or from familiar Boston circles, interest began arriving from afar. One of these was O. G. Stanley from Kansas City, who formed the Tashmoo Springs Water Company with an eye toward providing water and safety to the village of Holmes Hole after its devastating fire in 1883. Purchasing a square mile encompassing the springs and shoreline, Stanley was simultaneously land speculator and engineer for the massive water works with its capricious twin steam engines. Though his development dreams for lower West Chop never materialized, the bold vision Stanley espoused in 1887 has haunted Islanders ever since: "One vast resort from Edgartown to Gay Head!" That year more than fifteen thousand summer residents arrived in Cottage City.

The expanding population and prosperity on the north end of the Island, with its ready access to steamboats from New Bedford and the rail terminal at Woods Hole, cultivated growing divides within the vast townships of Tisbury and Edgartown. Each covered a good third of the Island and, more important, extended from the protected waters of Vineyard Sound to the open ocean on the south shore. In 1872, the rail line was completed to a new wharf at Woods Hole with support from Erastus Carpenter, the Oak Bluffs Land and Wharf Company, and investors including Daniel Fisher. Long tied to New Bedford and reliant on ships from across the eastern seaboard, Island eyes increasingly turned to "the Hole" for ferry service, telegraphic communications, and a steady flow of goods and visitors.

Tourists walking from Pilot's Landing at Gay Head, ca. 1890. From the Collections of the Martha's Vineyard Museum.

The interests of the growing population in Oak Bluffs tipped strongly toward Holmes Hole and Woods Hole. This bond strengthened in 1871 with completion of the beach road and bridge across the top of Lagoon Pond. This long-discussed project shortened the arduous drive between villages by three-quarters and increased Oak Bluff's independence. Fearing further decline and the growing financial strength of its tourist-based village, Edgartown built its own beach road to Oak Bluffs the following year. Yet, the fissures widened. The inevitable split occurred in 1880 with the incorporation of the new town of Cottage City.

As two towns emerged on the Island's east side, tensions grew between the rural families in West Tisbury and the mercantile population in Holmes Hole. Financially, politically, and culturally, the northern and southern parts of Tisbury were distinct. The physical distance and span of unoccupied plain separating them was emblematic of the divide and reinforced it. This case was clearly stated in the petition for division presented by West Tisbury to the state legislature: "[This is an] appeal on the part of a parent

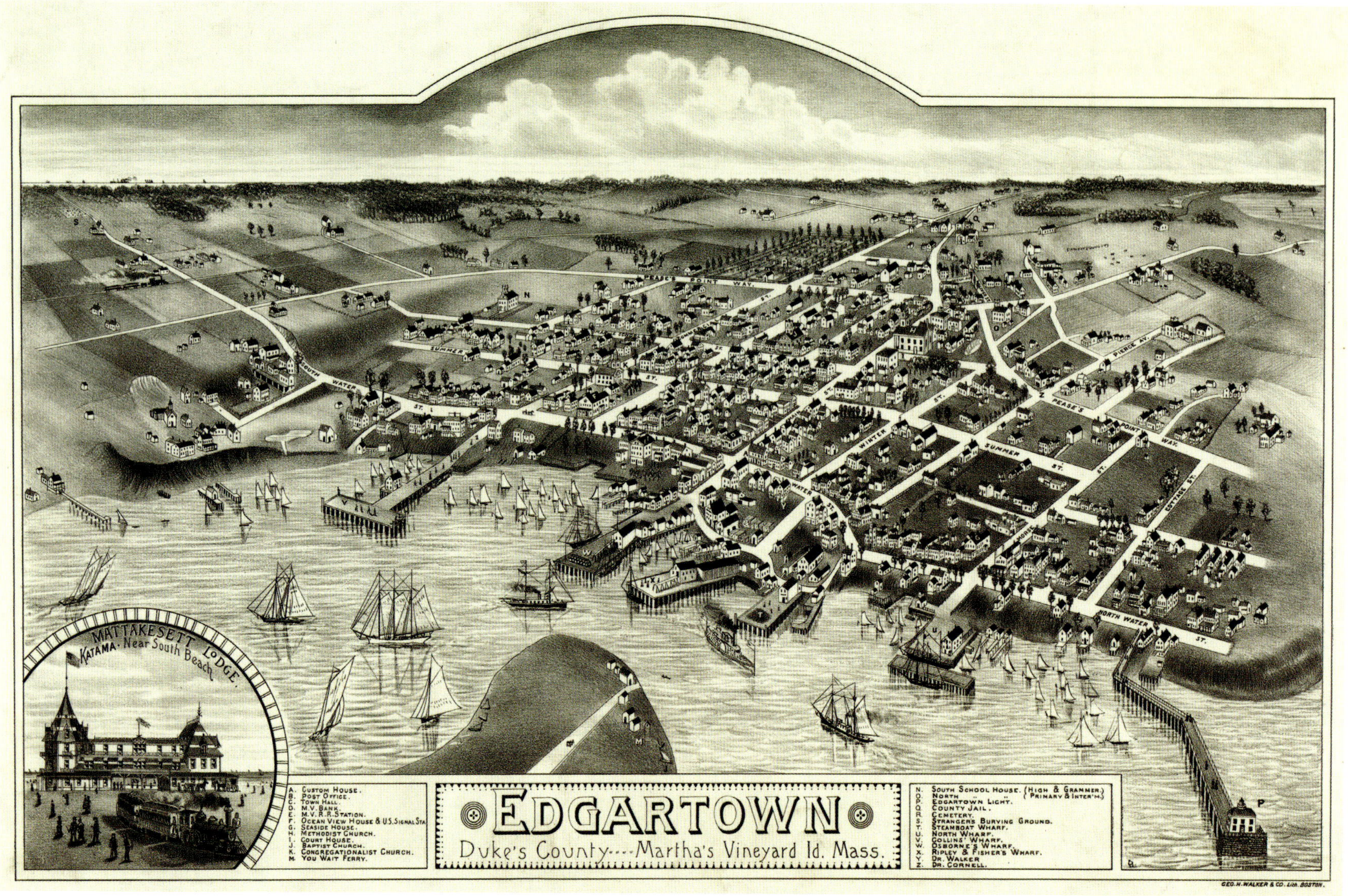

Bird's-eye view of Edgartown, 1886. The harbor, wharfs, and town center are bordered by an agrarian landscape of fields and woodlots. Note the railroad to Mattakesett, the wharfs organized more for steamship, ferry, and schooner traffic than whaling, the lighthouse on the end of the pier, a detail of the Mattakeset Lodge, the Whaling Church identified as the Methodist Church, and the "You Wait" ferry on Chappaquiddick. The Dr. Fisher warehouse is now the county jail. Map reproduction courtesy of the Norman B. Leventhal Map Center at the Boston Public Library.

for protection . . . [from an] offspring [Holmes Hole] . . . on behalf of a farm people to be set off from a people whose tendencies are 'citified' . . . and to be relieved from . . . the taxes which were perfectly proper for a sea port and summer resort but were heavy burden on farmers and fishers."

The original settlement of West Tisbury remained a rural New England village with modest colonial houses, community buildings and church, shops, and cemetery embedded in an agrarian setting. Local leaders were committed to maintaining the lifestyle rooted on farms, Mill Brook, and Tisbury Great Pond. In sharp contrast, Holmes Hole lacked farms and was a tight collection of houses, shops, small industry, and community buildings focused on and hugging the harbor. In the 1870 bird's-eye lithograph of the village, the street just above and parallel to Main Street is intriguingly and accurately named Forest Avenue. Despite two centuries of village expansion, the road cuts through a forested expanse all the way to West Chop.

Eight years after the first petition for separation, Tisbury was formally divided in 1892. In supporting the political division, Shaler acknowledged the real and cultural distances between the town's two worlds. "The division is one by nature. . . . The people at the port are extremely prosperous . . . and can afford to pay large taxes. After you leave town . . . you pass through waste and desolation until you get to . . . North Tisbury. . . . You cannot expect to reconcile a people divided by a diversity of interests and by nature and distance. On the one hand are a people who make money easily and spend generously, and on the other a people who can only live by the exercise of the greatest thrift. . . . A separation is essential."

Shaler himself was part of a small but important wave of new Island landowners. This group presaged a growing role for outside wealth in acquiring property for personal enjoyment, development, and conservation. Like his friend Henry Whiting, Shaler loved the Island for the combined pursuit of pleasure, land improvement, and science. Shaler's initial purchase came in 1887, when he acquired the first of "24 or 26 abandoned farms with houses that were unoccupied and dilapidated." His real estate explorations were aided by Everett Allen Davis, who joined the Whiting family through marriage and assembled his own collection of up-Island properties, often discovered through his position as judge of probate and insolvency. To Shaler, Seven Gates Farm offered "a second childhood" that has remained an exclusive retreat for its owners through ensuing centuries.

Two years after Shaler's initial acquisition, Mrs. Lucy Stanford purchased 270 acres to begin an estate that came to be called Windy Gates on Chilmark's south shore at Wequobsque. In 1873 William Hart, inspirational president of Stanley Works, acquired five lots from the Oak Bluffs Land and Wharf Company to initiate a family compound in Cottage City that grew to eight residences. By the early 1900s, Hart shifted his focus south of Farm Pond, where he bought parcels of "worthless marshland." In 1913 these holdings stretched continuously from the end of the Oak Bluffs seawall to the second (south) inlet of Sengekontacket Pond. Forming Hart Realty, the family developed an enclave of houses and harbor known as Harthaven. Alongside these and other notable individuals acquiring large tracts, there came many others who picked up an old farmhouse, built a small summer cottage, or settled permanently. The transition to a new economy and landscape continued.

Summer Resorts, 1900–1950

The early twentieth century brought a flourishing era of land purchases and vacationing for families from Boston, the east coast, and beyond. While some of these eventually became the nuclei for conservation reserves, others morphed through the years into high-end developments. Notable in both categories were the hunting clubs that acquired sweeping landscapes on the great ponds with access to marshes, pastures, woodlands, and the waterfowl and upland birds these attracted. The first of these—Watcha Club—was established in 1903, extending from Scrubby Neck to Oyster Pond. This area has come down today as Scrubby Neck Farm and the twenty or so immense homes off the peculiarly S-shaped Oyster-

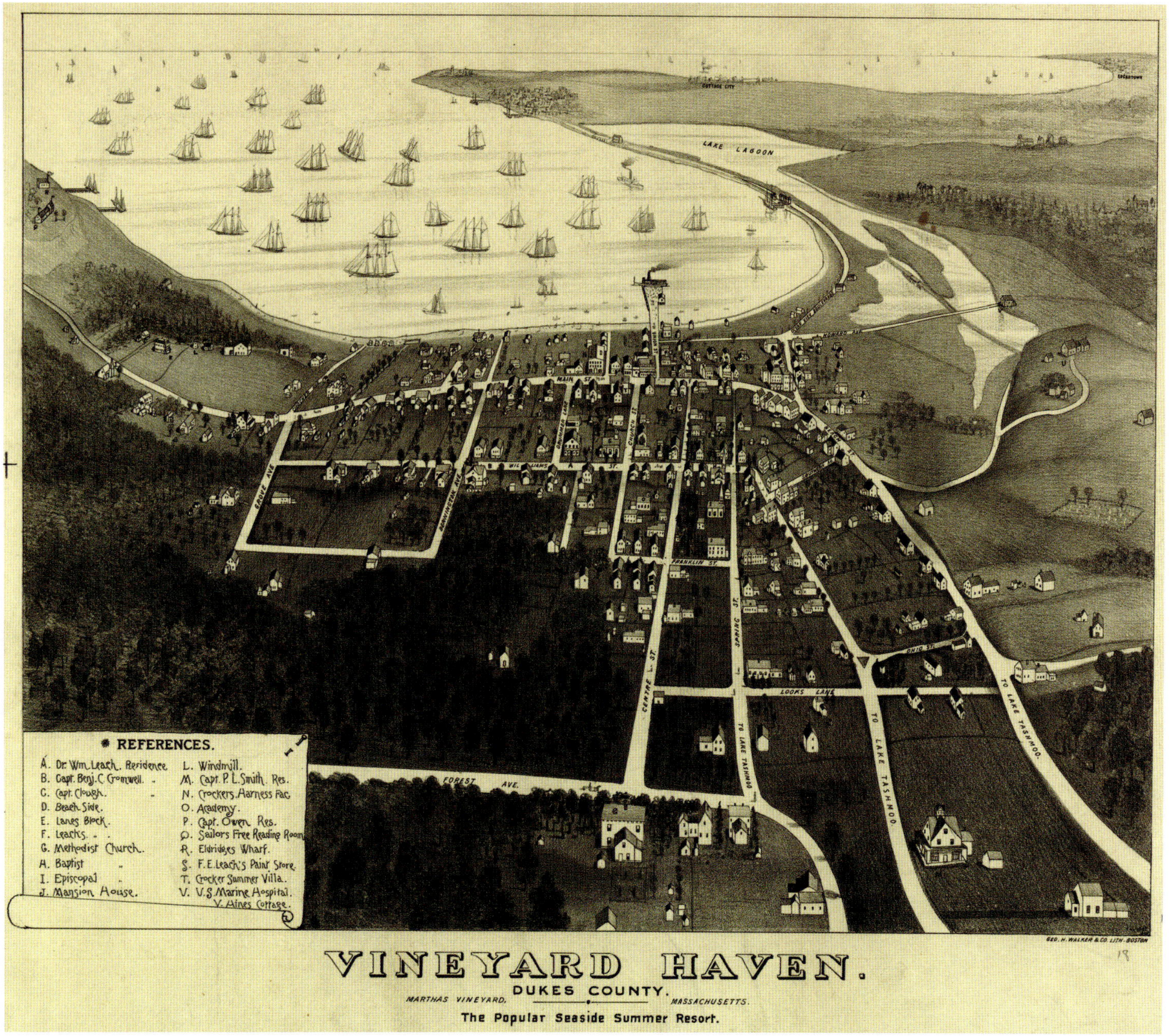

Vineyard Haven (1887) with tightly clustered houses, a harbor full of ships, gardens, a windmill, steamboat at the wharf, "Lake Lagoon," the completed beach road and bridge to Eastville, a small highland area, and Cottage City. The expansive woods surrounding the village and extending across West Chop is indicated by the name Forest Avenue for modern-day Franklin Street. Map reproduction courtesy of the Norman B. Leventhal Map Center at the Boston Public Library.

Beef cattle graze a pasture at Seven Gates Farm.

Watcha Road. In 1912, partners in the Tisbury Pond Club purchased 470 acres reaching across Long Point and erected a clubhouse near Middle Cove Pond. Upon the club's termination in the late 1960s, and with backing by former partners, the Long Point Reservation was permanently protected by The Trustees of Reservations.

Just to the east, George Daniel Flynn arrived from Fall River in 1903 and began to acquire the old Samuel Smith farm and adjoining properties in the Pohogonot region between Oyster Pond and Job's Neck Cove. The family holdings would peak at over 1,500 acres and ultimately yield such diverse offshoots as the Boldwater and Swan Neck's "limited developments" and the 800-acre Pohogonot tract of the State Forest. Over on Chappaquiddick, Bob Welch from Marblehead visited with duck and goose hunting in mind but ultimately purchased a couple hundred acres along Poucha Pond that encompassed a stunning array of farm fields, small woodlots, and some of the Island's greatest salt marshes, punctuated by a series of oak-covered islands. Out on the last island he established a duck blind and camp, while back near the barn and house, he fashioned a skeet field on what he called Pimpneymouse Farm. All across the reforesting Island, bird hunting flourished seasonally on both water and land. By the late 1930s, nearly twenty areas were annually stocked with quail, while more than a dozen areas released pheasants.

While men dominated the fall hunting scene, many families came for the entire summer, often moving wholesale with kids, belongings, and even a cow to set up house. As recounted so delightfully in the books *Last Farm on Chappaquiddick* and *Chappaquiddick, That Sometimes Separated but Never Equaled Island,* by Bob Welch's daughter and granddaughter, Edo and Hatsy Potter, the families immersed themselves fully in Island life from farm chores, fishing, and shellfishing to sailing and horseback riding.

The first half of the twentieth century also brought the first golf courses and yacht clubs, often accompanied by distinctly different attitudes toward the land. Oak Bluffs hosted the first golf course in 1910, followed by Edgartown in 1926, and Mink Meadows on West Chop in 1928. The Edgartown Yacht Club emerged from the Home Club in 1927, and a year later the Vineyard Haven Yacht Club was formed. Through this period more schemes for large developments continued to arise, though thankfully few came to pass. There was an effort to establish a 775-lot community—Chappaquiddick by the Sea—on Wasque (1913), a proposal to carve sixty house lots out of the woods alongside the Mink Meadows Golf Course (1928), renewed construction on the neighborhoods of Oklahoma and Lagoon Heights, and a scheme in 1937 for 240 lots spread across more than 400 acres on the headland above Lambert's Cove called Makonikey Heights.

This lengthy post-agrarian period highlighted the profoundly changing relationship between people and the Island. Rather than living directly off the land and drawing resources and their paychecks from the bounty of the sea or fields, more and more people arrived to simply enjoy themselves. Yet, even as fewer

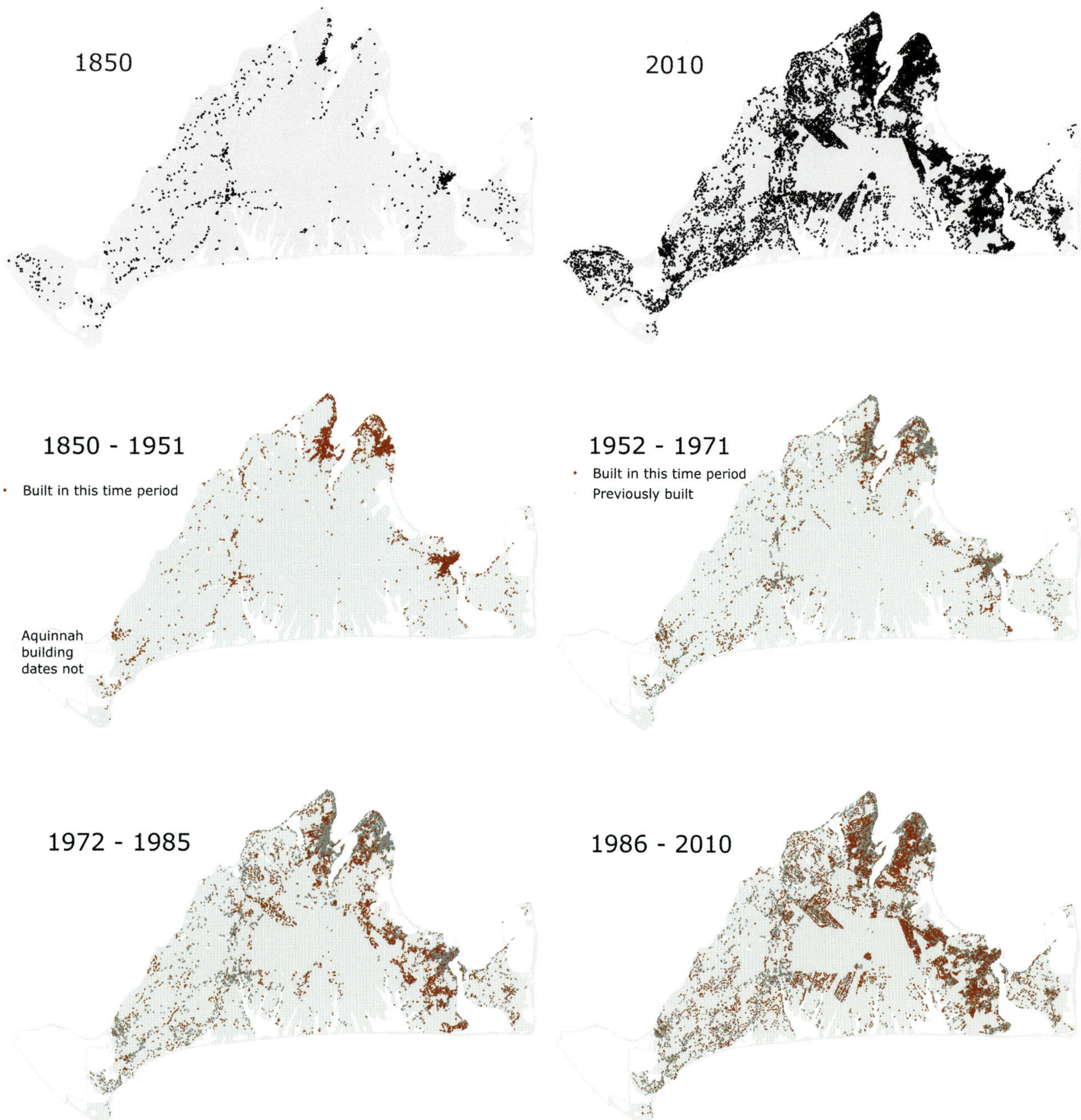

Making the modern building pattern. Maps depicting buildings in the nineteenth and twentieth centuries and the changing building patterns during key periods. Note: this map excludes all buildings less than three hundred square feet for simplicity and those in Aquinnah due to a lack of data.

Whiting's 1850 map captures a widely dispersed population with centers at Edgartown and Holmes Hole and smaller settlements at Eastville, Middletown, West Tisbury, and Gay Head. The Central Plain and Katama were largely empty, while scattered buildings occupied the south shore. Today, dense settlement extends from West and East Chop on to Katama and surrounding the State Forest. The density of buildings outlines many conservation areas, large bodies of water, and golf courses.

Although the population grew substantially through 1950, new construction reinforced the established pattern with exceptions: Menemsha, where dredging allowed the development of the small harbor in the 1920s; around Cottage City, down to Harthaven and along Vineyard Haven to West Chop and toward Lagoon Pond in the 1930s; and at Ocean Heights along Sengekontacket Pond and along stretches of Lagoon Pond in the 1940s. After 1951 a different pattern emerged, including, by 1971, the first development on the Great Plain, across Katama Plain, at Abel's Hill, and on Chappaquiddick. The boom years of the 1970s, 1980s, and 1990s brought concentrated developments near the plain, on Katama, south of the State Forest, off Lambert's Cove Road, and across West Chop, while dispersed development continued through Chilmark and West Tisbury on large building lots mandated by new zoning laws. Building dates provided by the Martha's Vineyard Commission.

and fewer full-time or seasonal residents pursued the old ways, and more turned to servicing the summer folk, a strong, though minority, tradition of working and understanding the land persisted. For most newcomers and seasonal folk, the relationships to historical practices and the land were different. The fields, woods, and water provided backdrop for leisure. Activities that sustained the original residents and a handful of their successors—fishing, hunting, tree-felling, oystering, sailing, and farming—morphed into engaging pastimes for the visitors rather than necessities or a way of life. Henry Beetle Hough captured the sentiment of many who had fished the sea and farmed the land for a living when he quoted the aging Captain Nathaniel Jernegan: "Our summer population is a community of idlers, and they grasp at everything that offers which is calculated to vary the monotony of existence."

Looking back at the time when he arrived with wife Betty to take ownership of the *Vineyard Gazette* in 1919, Hough wryly recognized that "there is no stranger social organism than a summer resort."

Hunters with a day's worth of ducks along a south-shore hunting club at Homer's Pond. From the Collections of the Martha's Vineyard Museum.

Third hole on the Farm Neck Golf Course, overlooking Sengekontacket Pond and Beach Road in the distance.

The large land purchases, house construction, golf courses, and proposals for new developments triggered the first well-articulated and widely distributed call for conservation. Specifically, Henry Hough began to speak forcefully on the subject from his prominent Island podium, the editorial page of the *Vineyard Gazette.* In a prescient May 1930 editorial, Hough foresaw that increased building and growth could reach "a point . . . beyond which the Island will begin to lose instead of gaining." He concluded his argument by issuing an early call for balance: "We must treat conservation as of equal importance with development." This appeal, and even a desire to tip the balance the other way, would become the mantra of the newly emerging environmental and conservation communities.

Peter Rosbeck's Deep Bottom Pond development, with green and eutrophic artificial pond and horse pastures, was built in the early 1990s, along the West Tisbury–Edgartown Road.

Hough treaded an intriguing path as editor, Island advocate, and author. While he emerged as a leading voice for conservation, his prolific nonfiction books and articles frequently extolled the Island's cultural and environmental qualities, enhancing its reputation as a seasonal destination and fueling the very impacts that he opposed. In works like *Martha's Vineyard Summer Resort, Soundings at Sea Level,* and *Country Editor* (which earned him bestseller status and an enduring reputation in the nation's journalism schools) and pieces such as "Escape in August to Martha's Vineyard" in *Town and Country* magazine, Henry Hough became a prominent promoter of the Island that he sought to conserve. Historian Arthur Railton captured Hough's dual role and inner tension: "There was probably nobody who did more to make the Vineyard a nationally known summer resort than Henry Beetle Hough. And he came to regret it."

But it was not simply the elite and newly arriving voices that lamented the impact of new developments on the Island. Many natives raised early concerns over the loss of nature and beauty. In the late 1920s, trees fell in West Chop as the vast woodland that had covered the broad peninsula for thousands of years suffered its first clearing. The project launched by New York banker Robert

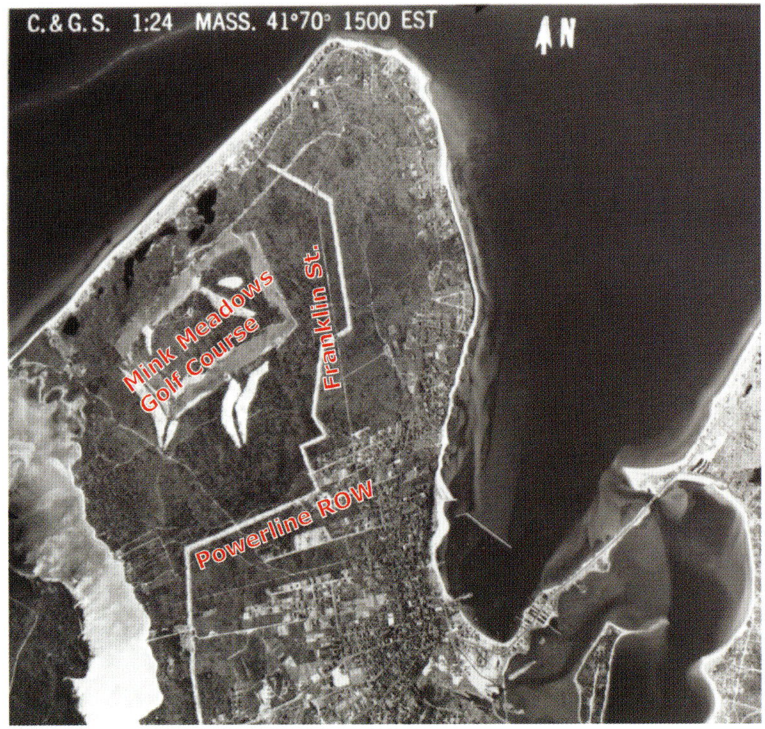

Despite lying adjacent to a bustling port, West Chop remained heavily forested until the early 1900s. The photograph (May 2, 1948) shows the first substantial deforestation by the peculiarly rectangular fairways of the Mink Meadows Golf Course, newer fairways (subsequently abandoned), the electrical right-of-way from Woods Hole (wide light-colored line), and buildings extending south from the West Chop Club along Vineyard Sound and the harbor. The sole road is Franklin Street, although Old Lighthouse Road and other ancient ways are visible. From the Harvard Forest Archives, Harvard University, Petersham, Massachusetts.

Bigelow to develop Mink Meadows Golf Course employed young Edwin Athearn on a crew clearing the woods. As Athearn ran a small Creatrac tractor to haul logs and brush, this Island-born jack-of-many-trades cringed as "the best forest land anywhere in the island" fell to axes and saws. As he recalls poignantly in his delightful *Chronicles of an Islander,* the scene stuck with him, and his complicity in their destruction left a sense of lingering guilt. "Four lanes will be cleared to form a rectangle. . . . [Bigelow] had hired a crew of woodchoppers armed with axes and two-man saws to destroy hundreds of big, beautiful trees."

This was the beginning of the end for one of the most majestic ancient woodlands remaining on the Island. Over ensuing decades, houses would puncture the surrounding woods, and a series of developments would fill the peninsula from town center to the outlet of Lake Tashmoo and north to the West Chop Club and lighthouse on the northern tip. In the end, only a small but majestic trapezoid of grand pines remained. West Chop Woods, owned by Sheriff's Meadow Foundation, is encircled by houses and slashed by a corridor carrying high-power wires and electricity to the Island. There, scattered in view of Mink Meadow's fairways, and throughout many West Chop neighborhoods, grow some of the largest and oldest pines on the Island.

In 1939 the Island's permanent population stood at 5,700, growing to 40,000 through the summer. In this latter number, some saw success and promise for a grander future. That year, an Island booster wrote, "The resort industry of the Vineyard appears to have a firm foothold!! . . . The writer knows that the following invitation, if answered, will assure the Vineyard of a means of Livelihood. 'By land or sea or air . . . we hope you will make the Vineyard your summer homeport. We wish you a good voyage. We offer you the old whaling captains greeting. . . . "Come aboard."'"

Resort and Boom Times, 1950s to Present

As Henry Hough later opined, the real shift to the modern era of summer resort, with its crush of development and stream of tourists,

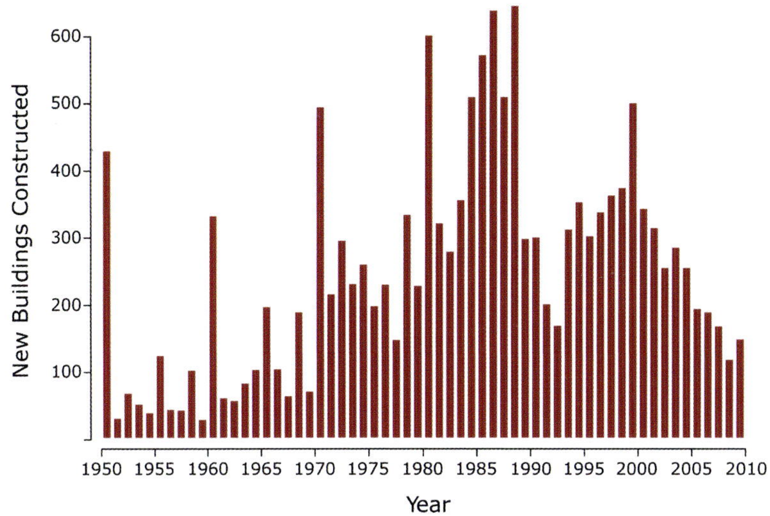

Fairly consistent growth in new construction from 1950 led to a peak in the late 1980s, a second peak in the late 1990s, and decline with the recession until recently. Building dates provided by the Martha's Vineyard Commission.

began around 1950. Housing data confirm this, but show that like other geometric processes, the explosion of development that came to shock the Vineyard emerged from modest beginnings: 700 new houses in the 1950s, 800 in the 1960s, more than 1,600 in the 1970s, 2,700 in the 1980s, and a peak in the 1990s, when nearly two thousand houses were built in just the first five years. Numbers have dropped recently, but at least 200 houses were built each year in the 2000s.

As historian Railton put it, after World War II "being a summer resort became a full time job." But the nature of that job, the pace and seasonal duration of the activity, and the associated changes to the Island shifted markedly over time. What began as a comfortable summer migration of families to old farmhouses and unheated cottages crescendoed in a few decades into a flux of day visitors and workers, weekly turnover of vacationers, mushrooming of construction from condos to castles, and steady growth of year-round housing. The housing boom had the greatest impact on the

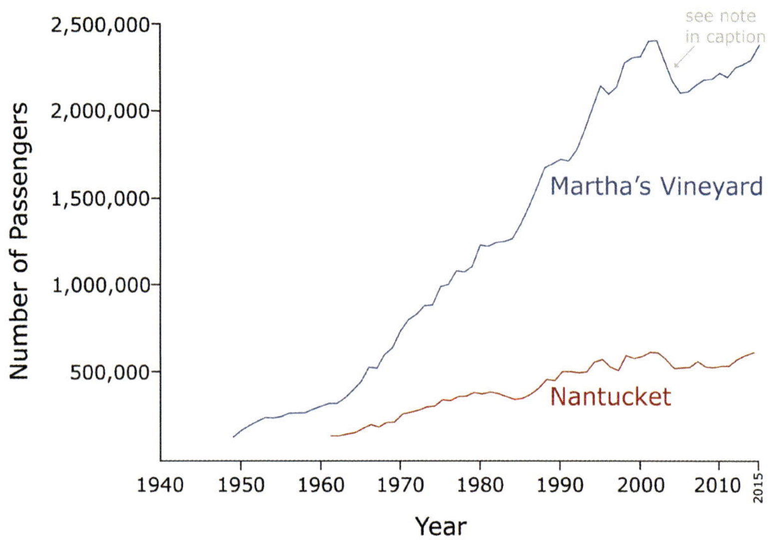

Ridership on Steamship Authority ferries has increased steadily until recently. The drop after 2000 marks a shift in record-keeping as other ferry lines initiated separate records. Data from Steamship Authority annual reports.

land, but it was paralleled by a frenzy of fishing derbies, JawsFests, concerts, carnival-embellished agricultural fairs, and fund-raising galas that altered the feel of the place. Especially noticeable was the traffic, noise, and congestion. And all of this was accompanied by a gradual extension of the busy tourist season beyond that defined by school vacation and Labor Day. Gradually the season extended to Columbus Day and Thanksgiving. Then, Christmas and even New Year's became busy. Now, the only truly quiet period, when more lights are out and few tire tracks wind down the long dirt drives, is the dead of winter, from January through early March. Nonetheless, the quietest month, February, still witnesses a stream of eighteen-wheeled deliveries of lumber, sheet rock, windows, and furnishings as contractors in a fleet of pickups and white vans tackle repairs, upgrades, and new construction, all in preparation for the Island's next tourist season.

The Vineyard's international reputation has grown with news coverage of regular vacations by Presidents Clinton and Obama. *Above:* Police and crowds congregate near Farm Neck Golf Club while the president was on the course, August 2009. *Left:* A First Family sculpture was trailed around the Island to presidential venues by artist Jay Lagerman.

Conserving the Vineyard

All the houses are within a mile or two of the sea coast: the internal parts of the island will probably always remain without inhabitants.
— Reverend James Freeman, 1807

Development and conservation lie in opposition in more than one way. While development is a declarative and transforming land use, conservation is understated. Both begin with plans, deeds, and surveys, and often involve staking out colorful flagging in the woods. Development then proceeds with chainsaws, excavators, and cement trucks followed by construction crews and septic systems. Natural processes are disrupted and the land is altered in ways that are, in general, irreversible. In stark contrast, land protection is often a monumental non-event. No matter how bold the headlines or how great the sums of money exchanged, the grandest finale leaves the land, the shore, and the views just about the way they always were.

As a consequence, during a rural landscape's early days it is development that captures the attention. Lay train tracks across a sandy plain to a grand Victorian hotel, or construct an obscene house and monument to one's ego on an ocean bluff, and the world can't help but notice. But purchase a thousand-acre chunk of that same sweeping plain or take that bluff and declare them "conserved in perpetuity," and there is little to mark the change save peace of mind and (hopefully) the public's ability to explore a place that they likely trespassed on beforehand. The plants, the animals, the flows of water, nutrients, and air, all remain as they have been.

It is only as human demands on the land multiply that conserved lands really begin to stand out vividly and be recognized for the values they represent. At some critical turning point in regional development, conservation emerges visually and culturally as a locally dominant and controlling use of the land. At that juncture, conservation joins fully with development in determining how nature and people flow in the land. It is also a juncture where new threats to conservation emerge—when the inviting open space "just sitting there doing nothing" becomes the target for the siting of "indispensable" public benefits like schools, sewer treatment plants, and solar arrays or the relocation of houses and public infrastructure retreating in the face of sea level rise. View a satellite image of the northeastern United States at night and the immense successes of

Hayfield, pasture, pond, and open-grown oaks on the Keith Farm, Chilmark.

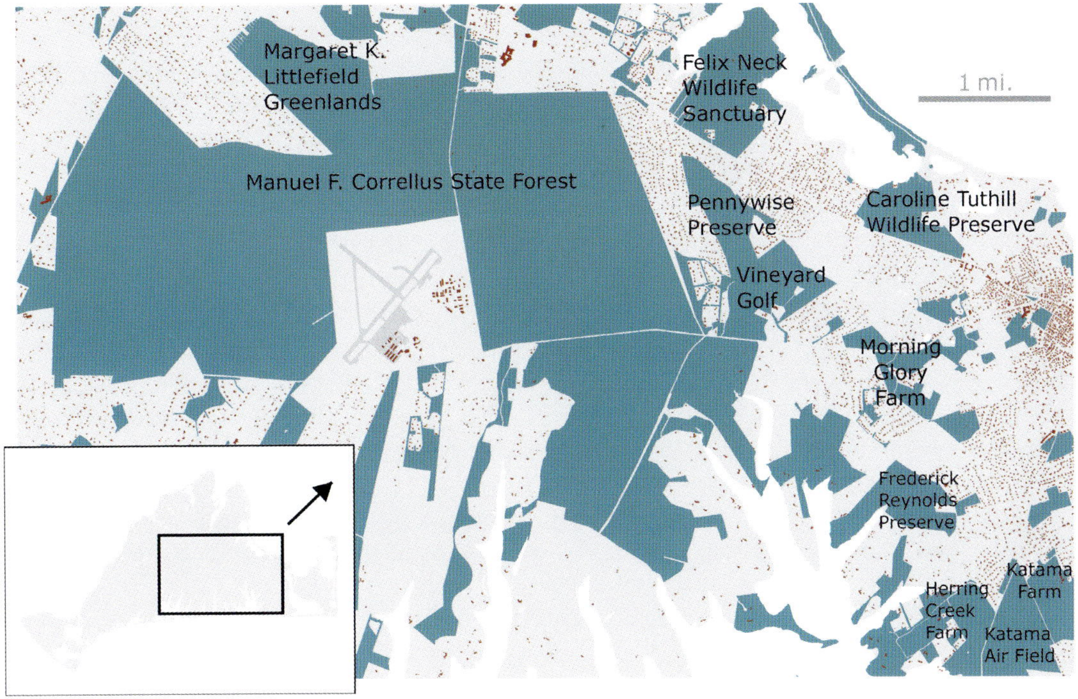

Development defines conservation. Map of building rooftops outlining the major conservation lands in the eastern part of the Island.

the region's conservation history jump out—the Adirondacks and Catskills in New York, and in Massachusetts the protected watershed of the Quabbin Reservoir and its regional conservation landscape, the Holyoke Range north of Springfield, or the Cape Cod National Seashore. Map the houses on the Vineyard today and conservation jumps out structuring the land. Many notable conservation lands are framed in development—Katama Farm, the State Forest, Southern Woodland Reservation, West Chop Woods, Pennywise Area, and others. It is when the pressures on the land become overwhelmingly intense that the decades of prescient effort and the leading role in conservation of the Vineyard, Massachusetts, and the northeastern United States are fully appreciated. At that point it is too late to begin to make such transformative differences from scratch; rather, it is time to dig in and complete the job that others with great insight thankfully started decades ago.

It is the stealth quality of conservation—the fact that it only really stands out when much of the die is cast—that makes us beholden to so many people from the past. To understand and appreciate the land that we enjoy today, we need to recognize this legacy of commitment by many groups and special people. The story is lengthy and complicated, as Massachusetts and the Vineyard have been world leaders in conservation and support an unequaled number and diversity of conservation organizations. The world's oldest of these groups (The Trustees of Reservations) and the largest (The Nature Conservancy) are active on the Island. The Vineyard's unique qualities and challenges have also spawned distinctive Vineyard-grown groups and approaches to conservation. Theirs is an inspirational history, which is fortunate, because it is our propensity to become inspired and committed to carry that effort forward that will determine how the story plays out.

Like most conserved New England landscapes, the Vineyard supports a heterogeneous collection of lands and waters assembled by public entities, private organizations, individuals, and, distinctively for the Island, a Native American tribe. Motivations for conservation among these groups vary—recreation, biodiversity and habitat protection, aesthetics, human health, historical and spiritual values, and essential resources such as drinking water, shellfish, and agriculture. The types and level of protection also vary, although most conservationists strive for "in perpetuity" as their gold standard. The seemingly haphazard nature of the resulting geographical patterns of conserved lands arises from these diverse motivations, from the vagaries of opportunity, and from an early lack of coordination among groups with different constituencies engaged in quiet competition for supporters, funds, and occasionally, projects. Since 1997, the Conservation Partnership of Martha's Vineyard (Vineyard Conservation Society, The Nature Conservancy, The Trustees of Reservations, Sheriff's Meadow Foundation, and Martha's Vineyard Land Bank) and the emerging central role of the public Land Bank have advanced greater coordination and steady effectiveness in land protection.

The seeds of Vineyard conservation were sown in the late nineteenth century when owners like Nathaniel Shaler purchased large tracts for privacy and for their natural and cultural values. However, despite often-admirable intentions on the part of their early owners, the land of Shaler, the hunting clubs, and other large holdings of Island farmers and off-Islanders did not become permanently conserved until the past few decades. In the intervening period, changes in family circumstances or ownership have led to the development of many tracts purchased years ago by conservation-minded owners. It is this potential for unanticipated change, even with a history of good intentions, that drives the focused effort to conserve all of the remaining sizable parcels today.

Formal land protection began in 1908 with the Commonwealth's acquisition of 612 acres of the Great Plain to create the Heath Hen Reservation. Like much conservation, this one was triggered by an emergency—the looming extinction of a game bird whose entire range had shrunk to the Vineyard. This purchase was also typical of many early public lands in that it involved the acquisition of relatively "worthless" unoccupied tracts, which in this case was not necessarily the prime habitat supporting this species. Therefore, it was a relatively simple transaction that lacked the complications of modern conservation deals that may tap multiple sources of funding, involve a combination of land purchases and easements, require extensive efforts to clear titles and engage multiple landowners, and include existing residences or the potential for limited development. The Heath Hen Reservation was also emblematic of much land protection: the tract yielded future values that were unappreciated at the outset.

Kiteboarder under the watchful eye of a Trustees of Reservations ranger at the eastern tip of Norton Point, which at this time extended in front of the southwest corner of Wasque, Chappaquiddick, December 2013.

Sadly, the reservation failed in its original mission, as the heath hen went extinct two decades later. But over the years, additional lands were added to this public holding, ownership was transferred to the state agency of forests and parks, and the tract gradually

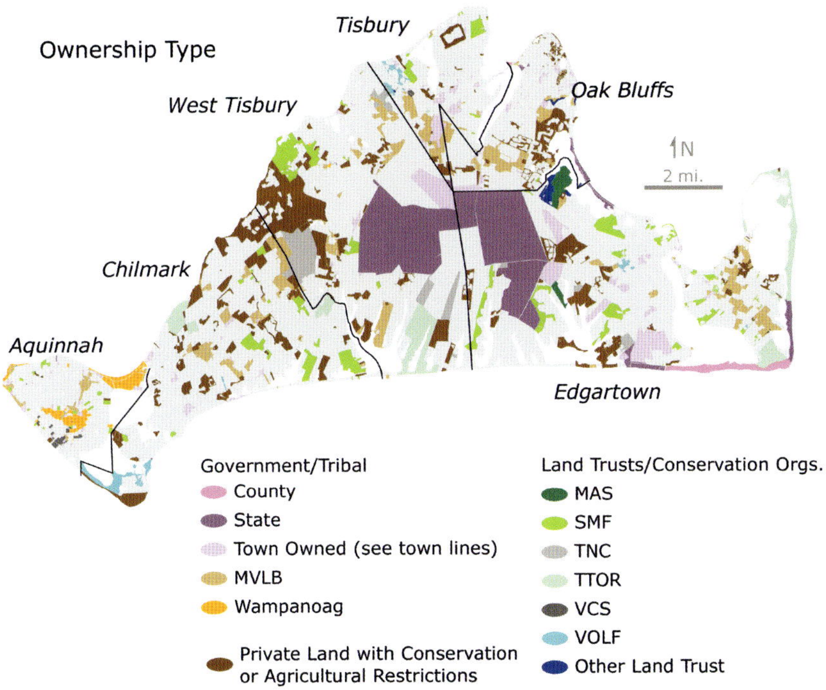

assumed its position as the literal core of Island conservation. At 5,350 acres, the Manuel F. Correllus State Forest is both the largest conservation tract and a globally significant and surprisingly intact sandplain landscape. Due to its large size and historically low intensity of land use, the area remains an amazingly wild and distinctive landscape despite the rectilinear pattern of fire roads that partition it.

Emergency and seized opportunity characterized one of the next significant activities and heralded the emergence of Island-grown conservation. In 1958 Sheriff's Meadow, a small pond, marsh, and field close to Edgartown's center, was threatened by development. Neighbors Henry and Betty Hough used the sale of the magazine rights from Henry's book *Once More the Thunderer* to purchase the land. Unable to locate an organization willing to partner on such a small, isolated tract, the couple established the Island's first land-holding conservation organization—Sheriff's Meadow Foundation (SMF).

There were other catalyzing moments when conservation acted to define parts of the Island. Many were triggered by the threat of development, others were delivered through the simple generosity of thoughtful landowners, and a few arose through shrewd conservation planning. As development pressures built in the late 1950s and 1960s, momentous land conservation deals brought The Trustees of Reservations (TTOR) and Massachusetts Audubon Society (MAS) to the Island. Tracts acquired include iconic landscapes: Cape Poge Wildlife Refuge (TTOR, 1959), Cedar Tree Neck Sanctuary (SMF and MAS, 1960s), Felix Neck Sanctuary

Top: Lands permanently protected from development vary in distribution and ownership among conservation organizations, public entities, the Wampanoag tribe, and privately owned lands with conservation or agricultural easements. *Bottom:* Full public access lands are generally open with limited restrictions, such as time of day. Limited public access includes recreation lands restricted to town residents. Areas with no public access include a portion of the privately held lands with conservation or agricultural easements. Conservation land data provided by the Martha's Vineyard Commission and the Harvard Forest.

East beach, salt marshes, and oak islands in Poucha Pond comprise a complex of conservation lands maintained by The Trustees of Reservations, Martha's Vineyard Land Bank, and Pimpneymouse Farm.

(MAS, 1963), and the Menemsha Hills and Wasque reservations (TTOR, 1966). The presence of two of the nation's oldest and most well respected conservation organizations added expertise in land protection, education, management, and research. Importantly, it also expanded the base of conservation support.

The mid-1960s attempt in Gay Head to pave a state road, build houses, and erect a wharf on the west side of Menemsha Harbor triggered one landmark battle. This episode might be forgotten were it not so beautifully chronicled in Anne Simon's *No Island Is an Island.* That successful effort by a handful of passionate pioneer activists led to the permanent protection of a distinctive peninsula of coastal heath. It also helped to codify a formal though ancient role for the Wampanoags as stewards of important conservation land, and launched the formation of the Vineyard Conservation Society as a leader on environmental policy and action.

In the 1970s Ted Kennedy advanced the Island Trust Bill in the U.S. Senate, seeking to conserve the Vineyard and Nantucket comprehensively, through a heretofore unsought partnership of federal, county, and town governments. The effort was greeted by many as "one last chance for the [Island's] rescue" and by pro-development opponents as "a major blow to the Island economy." Though the bill was ultimately defeated, the contentious debate it ignited was fought over kitchen tables, through news coverage, on Beacon Hill, and in Washington, D.C. Ultimately, Kennedy's provocative move had positive repercussions. By igniting Island-wide consciousness, the fight and ensuing state legislation stimulated the formation of a regional planning agency—the Martha's Vineyard Commission (MVC)—and eventually the Martha's Vineyard Land Bank. Ultimately, it generated ripples of thought and action that led to such far-reaching efforts as the MVC's Island Plan.

Along the way there were notable victories over development—the rejection of a McDonald's restaurant (1977) with a strong message to national chains; defeat of Ed Redstone's proposal for the Nobnocket bank and supermarket complex in Vineyard Haven (1985); rejection of commercial expansion at the airport, and development of a golf course along Edgartown Great Pond; and repeated rebuff of aggressive housing and golfing proposals by developer Cory Kupersmith for the Southern Woodlands section of Oak Bluffs. This last region provides the Island's best example of magnificent pinewoods after West Chop Woods; its partial conservation by the Land Bank provides clear evidence of progress in land protection.

There were many other milestones in the tense interplay between development and conservation. In 1962 the Island was stunned by the designation of the Edgartown–Vineyard Haven Road as a federal secondary highway, which required "the cutting of great swaths of trees and bulldozing to meet federal standards of 66 feet wide, ten-

A forest perforated by houses, Oak Bluffs.

foot shoulders, and a minimum distance of twelve feet from a tree." Over the previous century, this road laid out in 1851 had become pleasantly overhung by trees. In Henry Hough's words, the road upgrade was one first step in a direction Islanders have resisted thereafter—a narrowing of the distinction between the Island and the mainland: "Now cars whiz along the short improbable highway from one sister town to the other . . . and people begin to forget what a quiet island road looks like."

In 1971 a comprehensive study commissioned by the newly formed Dukes County Planning and Economic Development Commission (DCPEDC) shocked the Island with its prognosis for environmental disaster. Though the report's most strident section was initially suppressed, the symptoms identified were widely recognized. A local newspaper—*The Grapevine*—spread the news and *New York Times* editor John Oakes wrote a front-page article in the *Vineyard Gazette* calling attention to the report's diagnosis. Both argued for increased land protection.

As development accelerated, concern broadened. The 1984 Boldwater subdivision plan shocked Edgartown with 32 lots on 444 acres of land belonging to the Flynn family. The grandeur and layout were unprecedented, as houses were positioned strategically for isolation and water views, and consequently inflicted maximum perforation on the landscape adjacent Edgartown Great Pond. Four years later, the *Vineyard Gazette* headline "Island in Troubled Water" accompanied the release of a survey overseen by Louis Harris, the nationally renowned pollster and a seasonal resident. Though highlighting dire issues, the poll had one silver lining: it confirmed a surprising unanimity among residents and summer people against "unbridled building."

During this period of overwhelming growth, Henry Hough's successor at the *Gazette,* Richard Reston, maintained the conservation drumbeat. "The island is for sale and nothing is safe or sacred. . . . It is time for the Vineyard to wake up and act. There is a war going on here. The barricades are up and they stand at South Beach." Adopting Winston Churchill's resolute posture and words, Reston declared the barrier beach the Vineyard's Dunkirk. "We shall defend our island, whatever the cost may be. We shall fight on the beaches . . . we shall fight in the hills: we shall never surrender."

And yet, the pressures continued and the relentless streams of day visitors and seasonal residents were spurred by celebrity appearances and scandals, along with promotions such as this one from the 2013 Dukes County website: "Martha's Vineyard. Miles of free public beaches and other recreational opportunities, coupled with exciting nightlife and exquisite natural beauty, give the Island its world-class reputation. The many celebrity homeowners and visitors are evidence that the 'rich and famous' consider Martha's Vineyard a premier playground."

Over the years, the pressures and concerns triggered approaches coupling development and conservation. The Vineyard Open

A winter walk of the Vineyard Conservation Society heads across the Allen Farm to Chilmark Pond.

Land Foundation (VOLF) promoted an early model of limited development that retained the countryside's visual quality, if not its intact nature, by keeping road frontage unfragmented. The result often protected moderate-sized areas with conservation easements or deed restrictions while enabling housing across the backland at a density that afforded privacy to individual residences but consumed broad swaths of land. VOLF figured variously in many notable housing projects—Meetinghouse Road in Chilmark, Nat's Farm and Cranberry Acres in West Tisbury, Sweetened Water Farm and Katama Farm in Edgartown, and Pilot Hill Farm in Tisbury. The group also assisted in land division plans that maintained spectacular views across the Keith Farm at the Chilmark end of Middle Road and along the Howland curve on State Road at the Aquinnah line.

A second innovation targeted the affordable housing crisis that emerged as prices skyrocketed. By the 1980s the Vineyard had (and continues to have) among the state's highest housing costs and lowest average family income, making the residents who kept the Island economy and seasonal homes intact increasingly unable to afford to live there. Among the projects advanced by organizations promoting permanently affordable housing (Island Housing Trust, Dukes County Regional Housing Authority, Habitat for Humanity, Community Preservation Committees, town boards) are Island Cohousing, Eliakim's Way, and Sepiessa Point projects in West Tisbury; Fairwinds, the Pennywise Path/Morgan Woods, and Watcha Path developments in Edgartown; and the Middle Line neighborhood in Chilmark.

As monstrous houses came to disrupt neighborhoods, scenic vistas, and historic landscapes, the Vineyard Conservation Society provided leadership in seeking to regulate their location, construction, and impacts. In 2013, when the MVC declined to intervene, Chilmark approved the first bylaw regulating house size. Meanwhile, an unanticipated threshold in the real estate market was realized. On spectacular large rural parcels with breathtaking views,

Estate construction consuming ancient woods on Watcha Pond along the West Tisbury–Edgartown line, October 2011.

Access

As I went walking I saw a sign there
And on the sign it said "No Trespassing."
But on the other side it didn't say nothing,
That side was made for you and me.
— Woody Guthrie, "This Land Is Your Land," 1940 variant

The grandmother explains that ordinarily they would not go on someone else's property but the "No Trespassing" sign, being bad manners, practically invited them.
— Henry Hough, *Far Out the Coils*, 1985

Thus, the following things are by natural law common to all—the air, running water, the sea, and consequently the sea-shore. No one therefore is forbidden access to the sea-shore . . . for these are not, like the sea itself, subject to the law of nations. . . . [These] cannot be said to belong to any one as private property.
— Justinian (482–565 A.D.), first articulation of the Public Trust Doctrine

Over time and across the Vineyard, access to the landscape has been an important issue that private landowners have dealt with in diverse and occasionally contentious ways.

One of the most direct ways that conservation makes a huge difference on the Vineyard is through access. On an island in which land prices are astronomical, beaches *are* privately owned, and some landowners defend their property with glaring eyes, vigilant caretakers, and motion-triggered cameras, the simple conveyance of an easement or title to the Land Bank or a conservation organization can open remarkable resources to everyone. Thus, we revel in spectacular landscapes such as Wasque and the Cross-Chappy Trail, beaches at Long Point, grasslands at Quansoo Farm, solemn and ancient forest at West Chop Woods, beech and beetlebung groves at Cedar Tree Neck, and the bike paths and trails traversing the State Forest. And yet, of the approximately thirty-five thousand acres of the Island's conservation land, more than 40 percent is not widely accessible. There is interesting history and some considerable debate behind that fact and the larger issue.

Access remains one of the most contentious and critical of Vineyard issues. The Land Bank expends great resources acquiring trail easements, beaches, and connector parcels to open the Island's natural treasures to all; association members pay tens to many hundreds of thousands of dollars for exclusive beaches; no end of imaginative and occasionally hostile communication is employed by landowners to assert their right to exclude others; letters to the editor debate whether landowners should receive tax deductions or reduced taxes on conserved lands not open to the public; and towns tightly regulate passes to beaches, creating a pecking order in many minds regarding the location of one's residence or rental property (Chilmark tops most lists). Anyone who enjoys exploring the Island landscape has been astonished by the contrasting welcome received when crossing (inadvertently or not) a landowner's space—everything from a warm handshake and open sharing of knowledge to expletives and threat of constabulary action. There was an old saying on the Vineyard that with the departure of the summer folk, the Island returns to the natives. With a longer "summer" season and changes in attitudes among residents, this is less true. Nonetheless, there is no question that winter is the time for boundless exploration.

The tension around access is grounded in a few basic facts. Massachusetts law allows private ownership of beaches down to the low tide line, an artifact that traces to a 1640s ordinance of the Massachusetts Bay Colony that sought "to encourage private wharf construction and maritime commerce." As a consequence, shoreline access can be purchased and the public barred from private holdings. Having invested dearly in their property, most landowners guard it jealously. So, signs abound, caretakers are vigilant, altercations arise, and debates rage concerning the issues.

- Should the Land Bank purchase property through a third party to disguise the intent to thereby open a previously exclusive pond to public access?
- Are public values adequately served by conservation easements on private lands that exclude the public?
- Should the legislature modify the seventeenth-century statute and assert the Public Trust Doctrine, affirming that unique natural resources are held in trust for the general public?
- What happens to the ownership of exclusive beaches when they migrate inland over publicly owned coastal ponds?

Many questions have simmered beneath the Vineyard surface for decades, with little open debate beyond the courts. Fortunately, the Land Bank has provided a flow of new opportunities to all in recent years. Simply along the coast and great ponds, new access includes: Tisbury Great Pond and Beach, Chilmark Pond and Beach, Moshup Beach, Great Rock Bight and Wilfrid's Beach on Vineyard Sound, Edgartown Pond and beach, Duarte and Blackwater Ponds in West Tisbury, Little Duarte's and Doug's Cove in Tisbury, and Quammox and Hickory Cove on Chappaquiddick.

And more will come. While the debate simmers, many private landowners, towns, and conservation groups work steadily to make more of the Vineyard open to one and all.

the appraisal standard of "highest and best value" shifted from subdivisions with many houses at extraordinary prices to limited development of a few houses on large tracts at astronomical values. In 2013 two lots of forty and fifty-four acres restricted to three houses each were listed on the Kennedy family's Aquinnah estate for $20 million and $25 million. In West Tisbury 314 acres on Homer's Pond with five house lots were advertised for $118 million.

These extreme real estate listings represent the latest in a growing trend. In 2001 the Flynn family (Pohogonot Trust) sold 81 acres in four parcels on the east side of Oyster Pond for $15.5 million, at that time the Island's highest single residential sale. The largest of these tracts (34 acres restricted to two houses and one guesthouse) sold for $13.4 million. That same year the 215-acre Herring Creek Farm on Edgartown Great Pond was purchased for $64 million

Cedar Tree Neck Sanctuary with Daggett Pond, West Tisbury.

in a complex deal brokered by The Nature Conservancy (TNC) that allowed limited development on more than half of the area with the rest conserved (62 acres to TNC and 40 acres to the Farm Institute). According to realtor Tom Wallace, "The market has come to the point where larger-sized properties with reduced density are of higher value." Thus, elements of the historic nineteenth- and twentieth-century era of large ownerships have been retained. Like the early period of grand estates, most large tracts held by individuals or families today preclude public access even when conserved through public support, tax-reducing strategies, and partnership with nonprofit organizations. At the same time, efforts by public and private conservation entities have resulted in more public open space, including beaches, than at any time in Vineyard history.

Over more than six decades, the relentless push for development and the inescapably finite capacity of the Island often produced an air of resignation that the prediction by O. G. Stanley for one resort from Edgartown to Gay Head might come true. In the 1970s Richard Reston editorialized, "We are going to have more people and more business regardless of what anybody says. No editorials in the Gazette, no matter how eloquent are going to stop that trend." In 1995 James Lengyel, director of the Land Bank, made a similarly pessimistic prediction paired with a gloomy assessment of his own agency's role, one that fortunately he and we can reject out of hand today. "The die has been cast. The Vineyard in 20 years is going to feel like any other suburb . . . the Land Bank's impact on the Island is 'trivial.'"

Testimony in late 2012 by the director of the Vineyard Conservation Society to the MVC lent concerned support for Lengyel's prediction. In a plea to private landowners to grant more conservation restrictions on their land, Brendan O'Neill's assessment of the struggle to control development was somber. "We know that the conservation groups working to set aside more open

The southeast corner of Wasque and Chappaquiddick at sunset near the breach on Norton Point beach, November 2014.

space won't have a significant impact in reducing that future density [of housing]. . . . The rate at which land is being developed is far outpacing the rate at which land is being conserved through gift and purchase. No foreseeable combination of public and private dollars will stop that."

Indeed, in the discussions of conservation there has been a strong sense of loss and shifting baselines. Humans have a phenomenal capacity for accommodating change. Each person tends to evaluate current conditions and rates of change through the narrow historical window of his or her own experience. As older generations pass and new people arrive, the baseline of acceptable conditions and practices that fit with the notion of Island culture and aesthetics shifts. Values change and what was once inconceivable becomes tolerable and then the norm. This point was underlined by Martha's Vineyard Commission member James Athearn, whose Island roots extend to the seventeenth century, during a 2005 decision regarding affordable housing on Middle Line Road in Chilmark.

> The selectmen have argued that this proposal [for three duplex apartment buildings and six one-acre home sites on twenty-one forested acres] is protecting a rural standard. But I submit that it's actually a suburban standard that's been generated in the last 30 years . . . it is a housing style that people have grown comfortable with. But I don't think it looks back far enough in history or forward enough into the future to protect what we think of as the rural character of Martha's Vineyard.

Not surprising, given his long-term perspective and role in leading the Commission and its Island Plan, Athearn was seeking Island-wide perspective. He and many others sought to employ intelligent planning and the permanent protection of farms, forests, and shorelines as an antidote to development.

The ratcheting increase in second-home development across the Cape and Islands argued for a coordinated regional conservation solution. Though one response, Senator Ted Kennedy's Island Trust Bill, was ultimately rejected, the associated analysis urgently

Fishing the rip on the eroding end of Norton Point, Chappaquiddick, September 2012.

underscored the need for accelerated land protection. In its regional scale and zoning, this legislative effort and related studies laid the groundwork for Island-wide conservation planning. Through the turmoil of the 1970s Sheriff's Meadow Foundation grew into an Island-wide force in conservation by establishing the Caroline Tuthill Preserve through gifts that were among the largest in the organization's history. The Trustees of Reservations also increased its south-shore holdings and public beach access through acquisition of old hunting club lands at Long Point.

One defining moment in Island history transpired in the mid-1970s with modest fanfare. Motivated by a legacy of conservation and desire to resolve mounting estate, governance, and financial concerns, the successors of Nathaniel Shaler worked quietly to conserve his old holdings permanently. On Shaler's death, the land and residences had been placed in trust "for the purpose of providing summer homes in the manner of [Shaler's] own

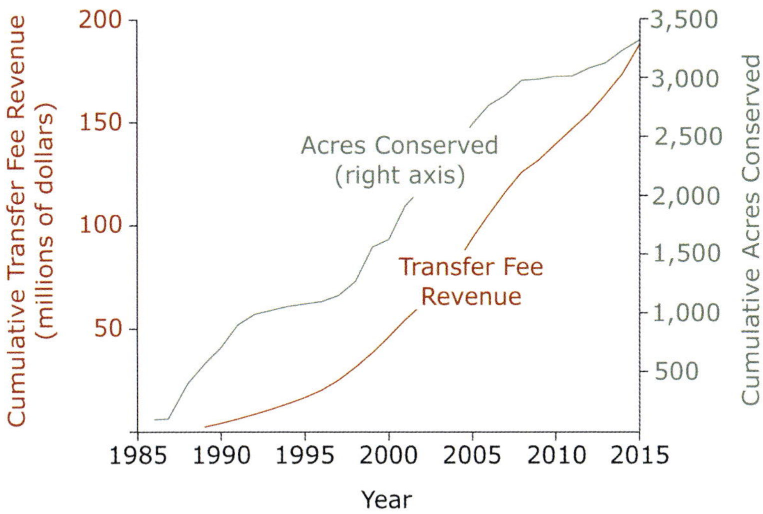

Role of the Martha's Vineyard Land Bank. The Land Bank steadily advanced its land protection activity through 2008, when the rate began to slow, whereas revenue streams remain steady. Land Bank funding comes through a 2 percent fee on most residential estate transactions, with totals varying among towns due to real estate values and activity. Data compiled from Martha's Vineyard Land Bank annual reports and the *Vineyard Gazette* for 2015.

conception." By 1921 Shaler's son-in-law Willoughby Webb had persuaded other families to join in "an experiment in communal living—sort of an idealistic utopian idea." The corporation established thirty-six private building sites of five acres or more. Then, after forty years with no permanent protection of the vast surrounding holdings, the families constituting the Seven Gates Farm Corporation worked with The Trustees of Reservations to conserve more than one thousand acres running from the Mill Brook to Vineyard Sound. The conservation restriction prohibited development but allowed forest management and agriculture and made Seven Gates Farm the Island's second largest tract of conserved land. But, in striking contrast to the larger, publicly held State Forest, the lands amassed by Shaler remain restricted to the exclusive use of members and guests.

Skyrocketing building of the 1980s brought fear, frustration, and then action. Pursuing an Island-wide mechanism to advance conservation that respected town-level priorities, Island leaders worked with the state legislature to establish the Martha's Vineyard Land Bank in 1986. A public institution administered by a commission of seven people (one representative of each town and one designee of the Massachusetts Secretary of Environmental Affairs) and coordinating with advisory boards in each town, the Land Bank acquires, holds, and manages land and interests in lands that may include water resources, habitat, scenic views, agriculture, and passive recreation. Funded by a 2 percent fee on the purchase price of most property, the Land Bank rapidly emerged as the dominant land protection force on the Island. Its first purchase was of historic and strategic significance—8.6 acres of woods and sacred ground in Christiantown surrounding the tiny replication of the first chapel erected there by the Christianized Indian settlement. By 2015, the Land Bank had expended more than $180 million to protect in excess of 3,100 acres, or 5 percent of the Island, through purchase or conservation restrictions.

Other important, lower-key activity also transpired during this period. Through support from Mary Wakeman, who led the protection of Wasque, the major conservation organizations—VCS, SMF, TTOR, VOLF, and the Garden Club—developed the Wakeman Conservation Center Trust (1984). This modest complex off Lamberts Cove Road houses organizational offices and is close to the Island headquarters of The Nature Conservancy and The Trustees of Reservations. About the same time, descendants of Henry Whiting boldly donated a conservation restriction over the 511-acre Frances Newhall Woods Nature Preserve. Tying together Seven Gates Farm, Polly Hill Arboretum, and Waskosim's Rock Reservation, the Woods property was the largest individual conservation restriction in state history. When 485 acres of the land were ultimately donated to The Nature Conservancy in 2013, it constituted one of the most valuable gifts in the organization's history. Over time other entities entered the scene as meaningful players in holding and managing conservation land, including town conservation commissions

Edgartown Great Pond and the south-shore barrier beach, October 2011.

Beech woods and bridge on Three Ponds Reservation and the Cross-Chappy Trail.

(Sailing Camp Park, Katama Farm, Katama Airfield, Margaret Littlefield Greenways, and property on Moshup Trail) and the Island Grown Initiative, which acquired Thimble Farm in Oak Bluffs and the Dunkl property in Chilmark.

The Vineyard Conservation Society has been notably instrumental in warding off incursions on existing conservation lands, such as attempts to build a waste treatment facility on West Chop Woods (1986), utilize the State Forest for state police facilities and a county jail, or advance developments on environmentally sensitive lands. The latter include three major efforts to establish golf courses: the Down Island course, which eventually became the Southern Woodlands Preserve; the Meetinghouse course along a half-mile of shoreline on Edgartown Great Pond, which was protected through conservation restrictions and limited development; and the Vineyard Golf Club, in which a 148-house development morphed into an environmentally managed golf course that accommodates sandplain habitat through partnership with the Sheriff's Meadow Foundation.

The Modern Conservation Landscape

Today, somewhat more than twenty thousand acres of land, close to 40 percent of the Island, is conserved in a mosaic that is diverse ecologically as well as in ownership, use, and public access. After seventy-five years, the Commonwealth of Massachusetts remains the single largest owner of conservation land, with about a quarter of the total. Though the bulk of this lies in the State Forest, other heavily used state lands include South Beach, Joseph Silvia Beach, and a portion of Leland Beach out on Chappaquiddick.

Whereas all state conservation land provides public access, one of the next largest categories of conservation holdings—private lands owned by individuals, families, and trusts—stands in contrast. These more than four thousand acres (20 percent) of land occupy dozens of scattered parcels. Since the land remains private, public access is typically not granted. While some tracts are notably large, such as Seven Gates Farm, the Woods Preserve, portions of the Ganz and

Open shrublands along Moshup Trail looking southeast toward Squibnocket.

Kloss properties on Vineyard Sound, and Squibnocket Ridge and Beach, many tracts are less than twenty acres. The bulk of this land (93 percent) is protected from development through conservation restrictions overseen by private conservation organizations, the state, or towns. The argument that allows public funds and tax benefits to support these lands is that it poses little burden on the local tax base (such as schools and police) and brings significant benefits to Island residents and society through the protection of water, biodiversity, farmland, and scenic vistas.

Collectively, the Island's major nonprofit conservation organizations are the second largest landowner, with more than 4,500 acres (22 percent). These groups' activities and operation differ but are complementary. The leading landowner, Sheriff's Meadow Foundation (1,941 acres), owns properties across the Island, is committed to public access and active use for farming, education, and forestry, and holds more than 600 acres of conservation restrictions. The next largest, The Trustees of Reservations, owns four large properties (1,340 acres) that are heavily used by the public (Cape Poge, Wasque, Long Point, and Menemsha Hills) and holds conservation restrictions on 1,800 acres (Seven Gates, Ganz, Kloss, and Squibnocket Associates properties). The state's largest owner of agricultural land, TTOR also owns 99 acres that it leases to the Grey Barn Farm, and has recently merged with The Farm Institute on the town of Edgartown's Katama Farm. The majority of The Nature Conservancy's 950 acres have restricted access due to habitat or gift restrictions. TNC also holds conservation restrictions on more than 600 acres of land.

Steady acquisitions by the Martha's Vineyard Land Bank make it the third largest conservation owner. True to its governing policies, Land Bank holdings are spread across the Island and towns, with many parcels of ten to more than one hundred acres. Examination

Barrier beach and dunes along Squibnocket Pond.

of a map highlights the intentionally strategic role that these lands play in connecting and expanding the effectiveness of other conservation holdings, forging trails and linkages, and providing public access to key natural, scenic, and recreational resources. The Land Bank is also the largest holder of Agricultural Preservation Restrictions that support local farms (e.g., the Allen, Athearn, Sterling, Thimble, Morning Glory, and Flat Point Farms). One vision underlying these efforts is an Island-wide network of hiking trails. While this is achieved in a "virtual" fashion every spring through a cross-Island walk hosted by the Land Bank and has been recently accomplished for Chappaquiddick in collaboration with the Chappaquiddick Open Space Fund, it remains a goal articulated by Land Bank director James Lengyel. "It would be a north-south-east-west, coast-to-coast pedestrian hiking trail . . . an archipelago where along that trail are beautiful, sizeable pieces of conservation land."

Although town-owned lands have multiple uses, over 1,900 acres are devoted to conservation. Edgartown owns more than double (863 versus 419 acres) the landholdings of the next closest town (West Tisbury), whereas Chilmark holds more than 300 acres in conservation restrictions. In Aquinnah, the hundred acres of town-owned land is bolstered by more than 450 acres owned by the Wampanoags (or held in trust by the Bureau of Indian Affairs) that provide conservation, cultural, and other values at West Basin, at Gay Head, and along South Road.

Flying the Vineyard these days, or employing Google Earth to scan the coast, moraines, and plain, the distinctive episodes and processes advanced by many people meld together in a landscape mosaic of natural and cultural features. This land is far from static, as major processes—forest growth and ecological dynamics, development of housing and human infrastructure, and the conservation of remaining open space—continue apace. We shall come back to assess the future of these processes, but first we need to delve into the land to see what all of this history has wrought.

IV Shaping the Present and Future

Exploring the Natural and Cultural Landscape

No one has yet described for me the difference between the wild forest which once occupied our oldest townships, and the tame one which I find there today. It is a difference which would be well worth attending to. The civilized man not only clears the land . . . and cultivates open fields, but he tames and cultivates to a certain extent the forest itself. . . . The sun, air, and perhaps fire, have been introduced, and grain raised where it stands. . . . The most primitive places left with us are the swamps.
— Henry Thoreau, *The Maine Woods*

Given the human impacts the Vineyard has experienced in the past four hundred years, Thoreau's question helps focus our exploration of the modern landscape. The Island may well be one hundred and fifty years into a process of reforestation and, surprisingly enough, rewilding, but to what extent are the new groves like those that greeted Gosnold in 1602? And, how do those that have developed secondarily on abandoned farmland differ from ancient woods that came through history without being cleared, albeit used as woodlots for centuries? These queries can inform a walk in every part of the Vineyard, and their answers yield a framework for understanding the variation in vegetation across the Island.

European settlement brought the most sweeping transformations to the land since the glaciers. Following their initial declarations of awe at nature's resources, the colonists set about taming the land with a focus on agricultural productivity. Forests were felled and cleared, wetlands were drained for pasture and crops, and thousands of animals were set out to graze the land. While the ax, fire, and plow exerted terrific impacts, nothing transforms a landscape like the incessant work of the domestic animals that follow these—cattle, sheep, horses, hogs, goats, and fowl. Plants are survivors, equipped with multiple mechanisms to endure disturbance and stress. But animals are relentless. Their repeated browsing eliminates seeds, seedlings, and sprouts and gradually converts forest to scrub, then heath, and finally grass. Across uplands, wetlands, streambanks, and pond shores, humans and their animals exerted enduring consequences on the land.

This agrarian rampage appears starkly in pond records. Charcoal levels rose sharply as fires followed the ax in clearing forests. Tree pollen crashed with an accompanying explosion of grass and the weeds of civilization. First came a peak in sorrel, that innocuous red-green herb of rough grazed land and poorly managed pastures. With farms and their crops expanding there came an eruption of ragweed, an aggressive annual that thrives in the disturbed soils of fields, barnyards, and roadsides. The biggest winners were the grasses that came to dominate as the hills and plains became pasture and fields grew thick with hay, as well as oats, rye, barley, and other grains. The relative abundance of the pollen of grass, herbs, and weeds versus trees provides an index of deforestation and land-use intensity. Extremely low in prehistory, this index peaked in the mid- to late nineteenth century and has declined ever since.

The crescendo in human activity and wide-open landscape is paralleled by a dramatic shift in one of the simplest, most informative, and sobering analyses that we conduct on lake sediments. "Loss-on-ignition," or LOI, is the value calculated when small samples of sediment are burned in a laboratory oven to determine the relative proportion of combustible organic material versus mineral material that remains as residue. As the amount of sand and silt increases, the percentage of sediment lost through burning declines. This technique helps to pinpoint erosion around the lake or in the watershed. The sharp decline in LOI and increase of mineral material coincident with the crash in tree pollen signals that colonists did more than remove trees; they fundamentally altered major ecosystem processes on the land and in every stream and pond. The mid-nineteenth-century nadir of LOI links this evidence for ecological degradation with the peak in grazing. With much forest removed, soils regularly disturbed, wetlands trampled and ditched, and animals mucking shorelines and streambeds, the sands, silts, and clays that were liberated were carried downslope and downstream to collect in valleys and the bottoms of Island ponds. In few sites has the value for LOI returned to that of the undisturbed pre-European condition.

Nearly as abruptly as it flourished, agriculture peaked and fell. Following the mid-nineteenth century, nearly every measure of farm activity declined: number of animals grazed; acres mowed, plowed, and tended; farm buildings occupied; and, of course, families farming. Firsthand descriptions paint a more colorful dynamic picture: grasslands became coarse and then woody; clones of huckleberry and blueberry expanded across pastures; trees crowded the margins of fields and then forayed across the open hills; and wet meadows, stream banks, and pond shores healed and began to sport delightful wildflowers and overhanging shrubs and trees. Gradually, in Pete Ogden's words, "bedraggled looking and depauperate woodlands spread across the land."

Zonation of vegetation from salt marsh to shrub wetland and oak and pine forest adjacent a small pond and inlet on Katama Bay, Chappaquiddick.

The consequences of this history can be read at two levels. At the broad scale, the recovery of natural vegetation is remarkable. The forests that regrew broadly include the same native species that reigned for millennia, and the persistent expanses of scrub oak, heathland, and grassland have a wild appearance despite their origins in forest clearance and grazing. But in detail, the landscape was transformed and the vegetation exhibits many subtle legacies of its agrarian history. Overall, the recovering forests are notably simpler than their predecessors. There has been an increase in those species that sprout, tolerate dry conditions, resist fire, spread readily across open sites, and are able to persist in the face of the ax, hoof, flames, scythes, and teeth that prevailed for three centuries. The fine pattern of these effects varies across the landscape with the details

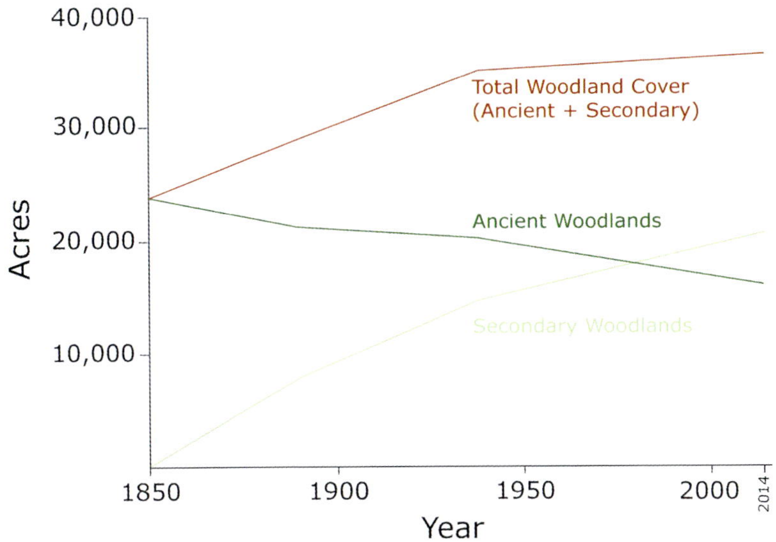

Patterns of changing forest cover. In 1850 nearly twenty-five thousand acres remained continuously wooded—"ancient woodlands" that were never cleared. Since the mid-nineteenth century, abandoned fields have reforested to form "secondary woodlands." Local deforestation accompanying development since 1850 has reduced the total amount of ancient woodlands and dampened the increase of secondary woods.

Hudsonia ericoides, or false heather, on a bluff in the Menemsha Hills Reservation, Chilmark. Restricted to open and chronically disturbed sites when the Island was heavily forested before European settlement, *Hudsonia* spread widely with deforestation, farming, and other forms of human disturbance.

of land-use history. Across the moraine, there has been a substantial decline in beech, beetlebung, and hickory, and an increase in oak and (to a much lesser extent) pine and red maple. The abandonment of open fields across the moraine allowed shrubs and vines to engulf large areas. Catbrier, grapes, bittersweet, Virginia creeper, poison ivy, viburnum, and blueberry grow thickly on reforesting old pastureland, where they compete strongly with the trees. On the great plain and eastern moraine, the dominance of oak has been strengthened in the old woodlots, whereas pine has increased on abandoned fields and where it was planted. Across the Island and throughout the coastal region, the reputation of these landscapes as dry, scrubby, fire prone, and dominated by oaks and pitch pine has emerged and been reinforced by the history of intensive land use.

As we interpret this modern scene and seek to understand the patterns of vegetation and individual plants and wildlife in detail, we need to consider the three processes that link the landscapes of the past, present, and future. First is the *sieving action* of human activity. The tree felling, land clearing, grazing, burning, hunting, and other activities since the days of the Mayhews strongly shifted the balance among native plants and animals. Weedy species thrived; resilient species persisted and increased; and those weak in the face of disturbance, repeated stress, or persecution declined. Some even disappeared. Early on, the balance swung toward opportunistic native species. Over the years a growing trove of exotic species—transported inadvertently or specifically introduced by man—became increasingly important.

By the second half of the nineteenth century, the flora, fauna, and relative abundance of species were starkly different than in Gosnold's day. In the woods, resilient oaks thrived due to their ability to sprout and out-compete most species following repeated cutting. Scrub oak exploded with tenacious growth wherever the tree canopy was removed. The big losers were beech, hickory, and beetlebung, whose lower tolerance of stress and preference for moister morainal soils laid them vulnerable to cutting, fire, and land clearance. Across continually grazed, mowed, and planted lands, openland species of grasses, herbs, birds, and butterflies thrived. Meanwhile, at the margins of woods, along roadsides, fencelines, and other neglected areas, weedy native species (red cedar, pitch pine, black cherry, catbrier) accumulated with a growing number of imports (black locust, bittersweet, honeysuckle).

The sieve of history edited more than species. It also eliminated important structures and substituted others. Gone were the immense trees and the distinct habitats they proffer. Towering canopies create great vertical structure and deep shade, qualities missing from the thirty- to fifty-foot-tall forests that dominate many sites today. Massive trees display deep cavities and sharply furrowed bark that provide refuge and footholds for insects, mammals, birds, and other plants. In death, their standing or fallen carcasses add structure to woods, to streams where they dam the flow to form deep pools and eddies, and to ponds where they break up the shoreline and form diverse habitats beneath the water surface. When large trees fall, their roots rip up mounds of earth and leave wide, deep pits. These and many features that characterize old-growth forests were largely stripped from the Vineyard and are just beginning to reappear as our woods age.

As large trees were edited from the scene, many *legacies* were added. Most notable are the unusual tree forms. Harvesting of ancient woodlands yielded forests of multi-stemmed trees. Most sprouts have two or three similar-sized stems extending upward at slight angles. But frequent coppicing by axe or fire induced oaks to form distinct rings of four to eight stems that gradually expand outward with successive cutting. Equally distinctive are the broad spreading limbs of trees that once grew in open pastures or abandoned fields. The branches spread widely to capture light, while pruning by wind and winter ice kept the growth low. On sites like Quansoo and Menemsha Hills, true forests of straight, tall trees are today engulfing these spreading monarchs. Over time, these relicts of the past century will succumb to age and shade.

Other biological legacies abound. Most notable are the plants introduced through human activity. Plantations of spruce and pine extend deeply into the State Forest and are spawning descendants that line the Edgartown–West Tisbury Road. Elsewhere many pitch pine forests are derived from nineteenth-century planting by landowners seeking to stabilize soils, provide windbreaks, or convert "worthless" land into a future crop. With modern landscaping, exotic species have proliferated, selected for character, beauty, or privacy. Other plants persist as legacies of past eras. Lilacs, honeysuckle, Japanese barberry, myrtle, and lily of the valley expand from old home sites and cemeteries now in the woods, whereas garlic mustard, mugwort, and bishop's goutweed spread from gardens or soils moved through careless landscaping.

Much less obvious is the hidden legacy of seeds from prior vegetation lying dormant in our soils. Extended dormancy is a notable trait that allows many weeds to lie buried for decades awaiting sunlight, warmth, or disturbance to trigger their germination. The ancient pasture history of West Tisbury and Chilmark is occasionally revealed by the startling appearance of field weeds in intact forests. Across the Woods Preserve and Seven Gates Farm, where the oaks were widely killed by moth defoliation after 2008, many sites took on the reddish-green hue of sorrel. The death of the eighty-plus-year-old trees triggered the reappearance of this widespread plant from the intensive sheep-pasturing days of Nathaniel Shaler. Many species of grasses, shrubs (for example, raspberry), and the various species of cherry have long-lived seeds that enable them to reappear in our woods whenever conditions are right and long after any memory of their past presence.

Given the centuries of farming and industry, many physical legacies persist across the Island on sites that otherwise appear quite natural today. These include ditches cutting through vast salt marshes and towering swamps of red maple and beetlebung; crumbling dams along cranberry bogs, small mill streams, and

Leaves of post oak, a regionally uncommon species that is common across the Great Plain.

herring runs; cellar holes and foundations from long-abandoned barns and houses; and ancient fencelines marked by stonewalls on the moraine and ditches and ridges running through the plain.

Inertia refers to those ecological processes that play out in enduring fashion. Many involved in the recovery of our lands and woods from past forest clearance and farming will unfold for centuries. As farming subsided and wood chopping declined, many plants scrambled and competed for new footholds. In the woodlots, white and black oaks, scrub oaks, huckleberry, and blueberry tenaciously held their place. Open fields were invaded by a collection of native and exotic species whose seeds thrive in full sunlight and are readily dispersed by wind and birds. They joined existing pasture plants and were gradually supplemented by woodland species that disperse more slowly, along with pitch pine planted by enterprising landowners. Across the abandoned pastures, hayfields, meadows, and cranberry bogs, plants battled through the processes of dispersal, establishment, growth, and competition. While native species that had occupied these sites in prehistory were often successful, their proportions differed in the new forests. Aggressive shrubs and vines flourished. Catbrier, wild grape, poison ivy, and the introduced bittersweet, Japanese barberry, tree-of-heaven, and honeysuckle came to join oak, sassafras, beetlebung, black cherry, oak, and beech in many secondary forests.

While trees now cover much of the Island, the process of forest growth and recovery remains far from finished. On the Island's extremities and across its most challenging upland environments—Moshup Trail, Squibnocket, parts of Chappaquiddick, and the scrub oak bottoms of Quansoo, Pohogonot, and the State Forest—seedlings and saplings of trees continue to struggle to gain a foothold. The future of forests there and on remaining open sites relies on human indifference, for only active effort will prevent the eventual success of trees.

Where trees and woods have taken hold, they are thriving and getting older, larger, and more majestic. As the canopy expands, shrubs thin and the structure and composition of the woods

Ancient woodlands loss through time. Although 40 percent of the original forest landscape of Martha's Vineyard survived intact (but was harvested for wood) through the peak of clearance in 1850, these ancient woodlands have since declined through development.

continue to evolve. New tree stems are single and straight. Early successional trees—red cedar, gray birch, black locust, black cherry—gradually decline. Oaks solidify their hold and then begin to lose it as more shade-tolerant species gradually expand. These include beech, the native species that declined most during the colonial period, and many non-natives—white pine, white and Norway spruce, and Norway maple. The Island will see much more of all of these species in the future.

The full maturation of the Island's forest will take centuries, if nothing else changes. But natural and cultural forces are far from quiescent on the Vineyard. Storms will uproot swaths of forest and initiate new bouts of recovery; insects will episodically weaken and kill many trees; warmer temperatures and longer growing seasons will alter the competitive balance among species; and landowners will incessantly tweak their woods in minor and more abusive ways. But processes initiated since 1850 when Henry Whiting completed his defining map will continue to reverberate through the land.

The Moshup Trail landscape of shrublands and woods on sandy uplands and intervening wetlands. The dominant plants on the sandy soils are little bluestem grass (*Schizachyrium scoparium*) and false heather (*Hudsonia ericoides*).

The Modern Woods

Two dominant factors control our modern woods—the local environment and the history of land use. Soils in the morainal hills from West Tisbury to Aquinnah in general have greater variation

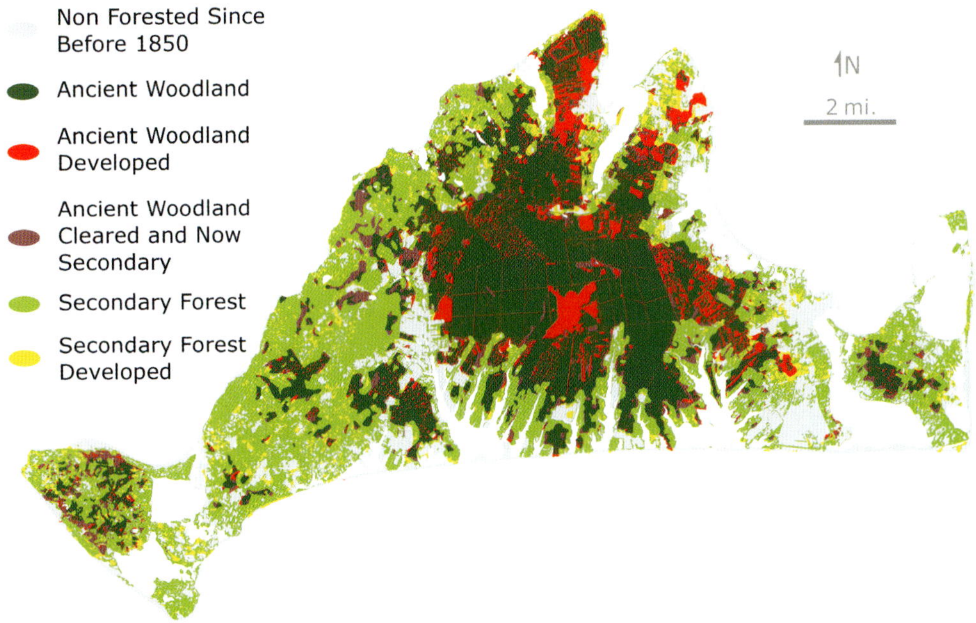

Woodland history in the modern landscape. Since 1850 the net result of forest regrowth on abandoned fields and the destruction of forests by development yields the modern pattern of ancient woods, secondary forests, and developed areas of each. Ancient woodlands have been extensively developed at the airport, the eastern half of Edgartown, much of Oak Bluffs and Tisbury, and around the State Forest in West Tisbury. Secondary woodlands have been perforated Island-wide. Development has also occurred in many agricultural fields.

than those on the flat and sandy Great Plain. As a result, they support a greater diversity of plants and vegetation. But those relatively productive morainal soils were good for agriculture and so were broadly cleared. When these areas were abandoned, they were gradually occupied by what we call *secondary forests,* which develop on sites that have been cleared of their native vegetation. These forests support a great mixture of weedy and non-native species and bear less resemblance to the Island's original forests than do the old woodlots.

In contrast, the historical woodlots on the moraine and much of the Great Plain are *ancient woodlands* (sometimes called "primary forests") that were never cleared and have therefore remained relatively intact. Soils in the ancient woodlands escaped the plow, hoof (in any great intensity), and fertilizer; much of the woodland's original flora remains; and the continuity of vegetation has proven surprisingly resistant to the invasion of weedy and exotic species. Though these woodlots were cut repeatedly in ways that altered their structure, these ancient woods retain continuity with the forests of prehistory. Given the strong connection between history and ecology, we can go a long way in our interpretation of modern forest conditions by "simply" mapping the distribution of ancient versus secondary woods. When we also take into consideration soil moisture, we can account for much of the forest variation across the Island.

The modern distribution of ancient woodlands is the product of processes that interrupted the continuity of forests elsewhere, especially agricultural clearing into the mid-nineteenth century and development in the past century and a half. By 1850, forests had been reduced to 40 percent of the Island. The heart of the Great Plain, from the heads of most ponds up through West Chop and much of East Chop remained as woodland, albeit often a low growth of scrub oak with trees. Substantial woodlands also remained

Distinctive horizontal branching and short shoots on a grove of beetlebung in the uplands of the Menemsha Hills Reservation.

on Chappaquiddick south and west of Sampson's Hill, on South Ridge along Kings Highway, and in the swampy headwaters of the Tiasquam River. Across the moraine, remaining woodlots were small and scattered.

From 1850 forward, and even as forest cover has expanded overall, ancient woodlands have continued to be destroyed. Through the nineteenth century, the process was a nibbling around the edges as small areas were cleared near houses or fields. By the 1950s, larger chunks were eliminated by land clearance—around Oak Bluffs and Vineyard Haven, near the head of the Lagoon and Duarte Pond, and at Mink Meadow. Since then commercial and residential activity has perforated broad areas of ancient woodlands. The two Chops lost most of their ancient woods, save for West Chop Woods and Southern Woodlands. The airport and commercial areas cut a gaping hole in the heart of the Great Plain's vast ancient woodlands, and major developments encroached into it from every direction. Since 1850, the ancient woodlands have declined by nearly eight thousand acres, and most intact areas have been substantially compromised.

Tree Ages

In 2005 we set out to examine how the history of the ancient and secondary woodlands had shaped their characteristics across the Vineyard, from their composition and appearance to tree ages. As we aged trees by examining cut stumps and extracting thin cores of wood from living trees, we were particularly interested in finding the oldest trees, clarifying the range of forest ages, and examining whether the ancient woodlands and secondary forests differed in age. In this endeavor we built on previous studies, and consultation with many naturalists, conservationists, and landowners. Coring more than 1,200 trees revealed many surprising insights concerning the Vineyard's ecological history.

As expected, many trees are old, but despite their gnarly shape, few exceed one hundred and fifty years in age. Many have survived well more than a century, but the average age of the larger trees in most woods is between seventy-five and one hundred years. Intriguingly, there is scant age difference between the ancient woodlands and secondary forests. Apparently most old woodlots continued to be cut heavily into the late nineteenth century, overlapping the period when farms and fields were abandoned and secondary forests became established. Wood use, which had been critical for home heating and industries like the brickyard, decreased

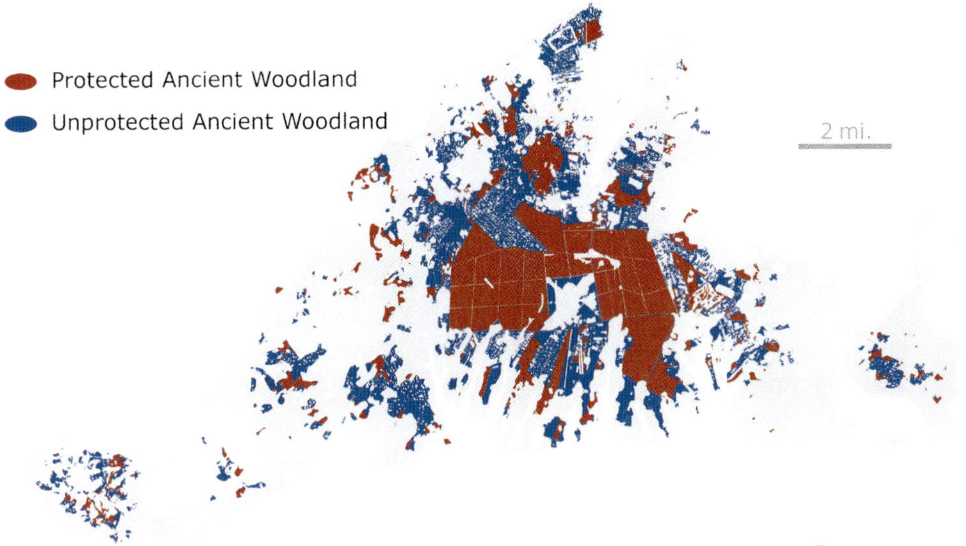

Land protection status of the Island's ancient woodlands. The largest continuous area of conserved ancient woodlands lies in the State Forest and adjoining Littlefield Greenways. Substantial unprotected blocks also persist north of the Southern Woodlands, adjoining Wapatequa Woods, northwest of the State Forest, and toward the south shore.

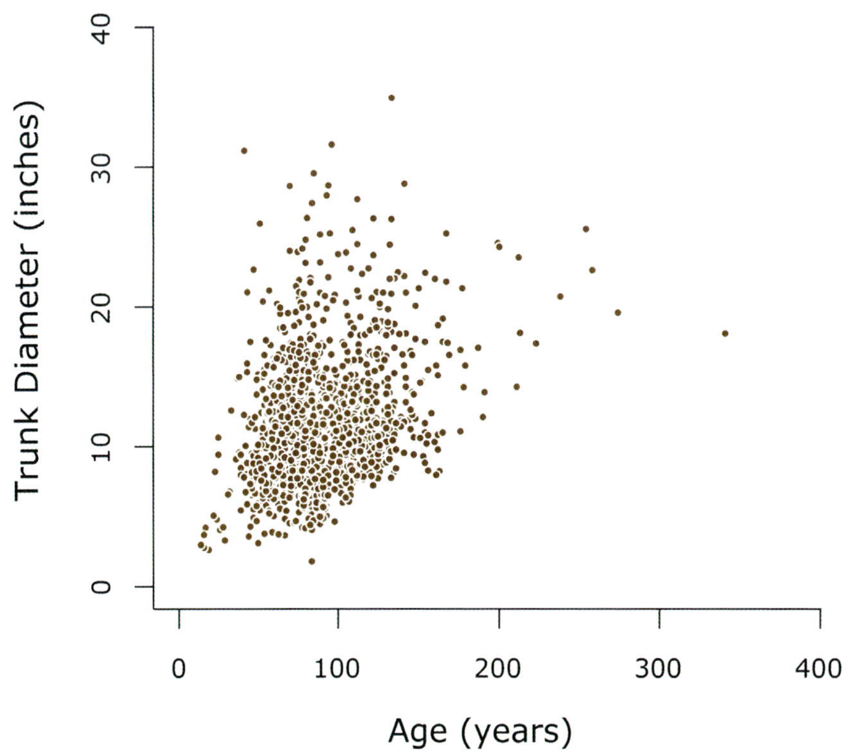

Relationship between tree age and size. The annual growth rings of trees from across the Island display little relationship between tree trunk diameter and age. Most trees are 50 to 150 years and vary from five inches to more than two feet across. While the oldest trees tend to be larger than average, they generally are not the very largest. Trees more than 150 years old range from eight inches to two feet in diameter. The largest tree sampled (a three-foot-diameter white oak) grew slightly more than one-quarter of an inch a year over 134 years. The oldest tree (a 343-year-old white oak under a foot and a half in diameter) averaged just 1/20th of an inch a year.

Brian Hall using an increment borer to core a white oak on Brookside Ridge, Chilmark. By extracting a thin dowel of wood, it is possible to count the annual growth rings and determine the tree's age and history of growth.

in the late 1800s as coal, oil, and electricity became increasingly available. Since then, less forest harvesting has been spread across an increasingly larger area, allowing many trees to grow older.

The size and age relationship among trees is notable. While broad generalizations hold—the smallest trees are often young and the oldest trees are fairly large—there is no clear relationship between tree age and size. Trees that are one hundred and fifty years old can range from less than eight inches to nearly forty inches in diameter, and trees twenty inches in diameter may be three hundred years old or barely over fifty. As a consequence, it is simply impossible to estimate the age of most trees by their size. Indeed, the largest trees are often not terribly old; they simply grew fast by sitting on good soils with lots of light and protection from winds. Examples include the large wide-spreading trees that grow in fields such as the Keith Farm, Quansoo, and Peaked Hill. Benefiting from full sunlight, these trees grew fast and came to support an enormous girth and a surprising spread of branches. However, despite their impressive qualities, most are about average age—a century old.

When we examine individual tree species, some distinct relationships emerge. Red maple, beetlebung, sassafras, and beech follow a general trend of increasing size with age. On the other hand, white oaks and pitch pine exhibit almost no pattern. One partial explanation for this difference is soils and moisture. The first group tends to occur on moist sites where growth is fairly regular. In contrast, oaks and pine occupy a broad range of sites, from moist to extremely dry and forest settings from scattered open pastures to dense woods. Under such variable conditions, these species display a wide range of growth patterns.

All truly old trees occur in old woodlots. The oldest was a 343-year-old white oak near Witch Brook on Seven Gates Farm. This age compares favorably with the oldest trees on the coast, including some 350-year-old white oaks we cored on Naushon Island. But we also located a 275-year-old beetlebung at Waskosim's Rock, a 259-year-old beetlebung at Cedar Tree Neck, and a 239-year-old white oak on the John Presbury Norton preserve. The location of some old trees was also surprising. West Chop Woods, just north of Vineyard Haven, has among the oldest pitch pines, including many about 180 years old. The Caroline Tuthill Preserve on the outskirts of Edgartown has a 212-year-old beetlebung.

Having remained permanently wooded, the ancient woodlands have retained more wild character. The Island's oldest trees occur here, and human artifacts are scarce. No stonewalls, ridge and mound fencelines, or old cellar holes appear, though ancient ways are common and some woods bear evidence of small quarries for rock, gravel, sand, and clay. Many trees are multi-stemmed. Native species dominate and early successional trees are scarce.

Walking through some of these forests, it's not difficult to project a bit of additional size and age to the trees and envision a Vineyard landscape four hundred years ago. Across stretches of West and East Chop and much of the Great Plain and Chappaquiddick, white oaks dominate over a continuous cover of knee-high huckleberry. In scattered locations, towering and platy pitch pines join the oaks, especially toward the north of the Island. Hickories and beech are

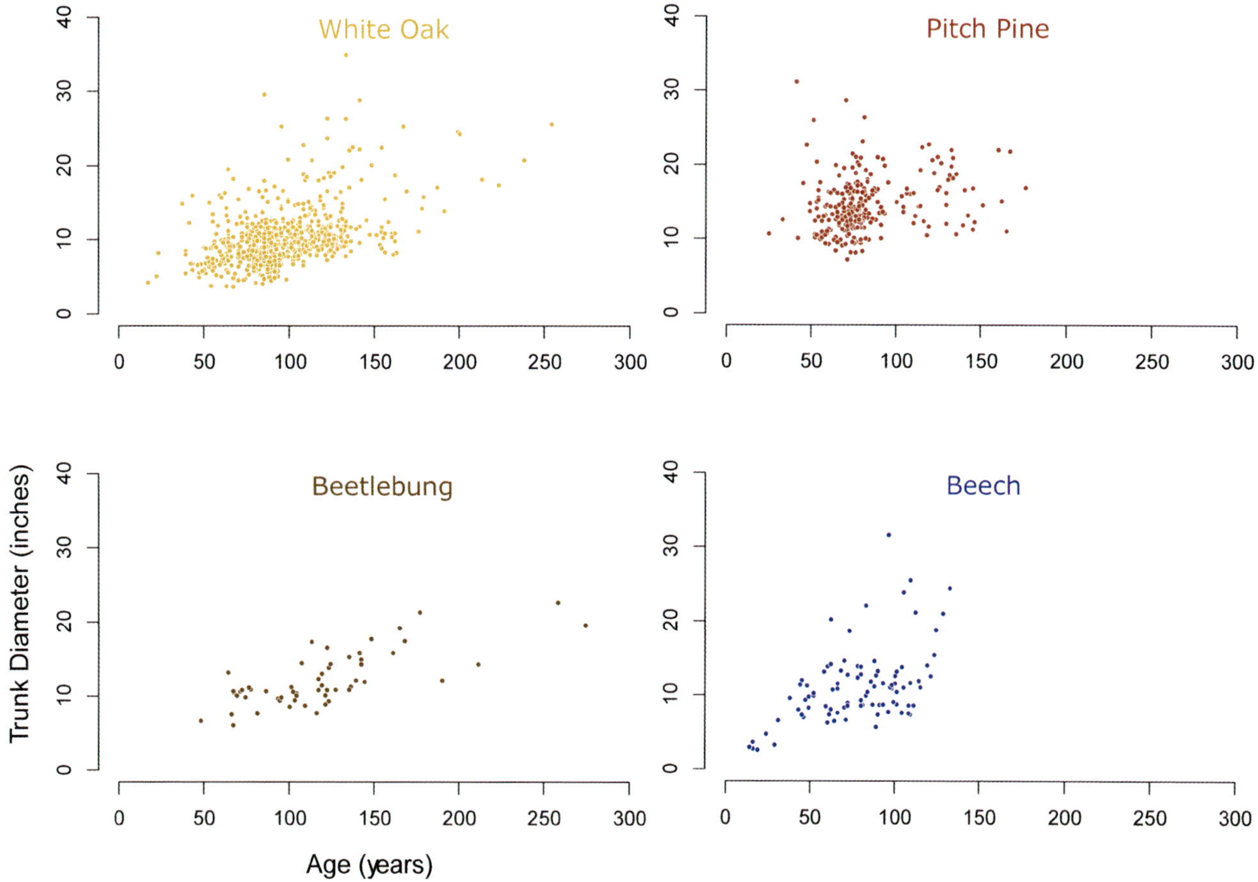

The major tree species exhibit considerable variation in growth rates, size, and age. Pitch pine shows no relationship between size and age. The relationship is weak in white oak, slightly stronger in beech, and strongest in beetlebung, which exhibits a consistently slow rate of growth.

more prevalent on rolling and dissected sites, and scrub oak is widely spread through understories on the Great Plain and toward the south shore. On the moraine, richness increases with more hickory, hop hornbeam, and especially beech, and with pockets of large beetlebung and red maple. On rich, moist sites the understory is full, tall, and more varied than in the oak woods.

The Vineyard's forests can be delineated into several broad forest types with fairly distinctive distributions.

Oak Forests

Oak forests dominate across more than twenty thousand acres but are far from uniform. Clear differences in the shrubs and understory relate to geography and land-use history. Toward the Island's center, on the Great Plain and extending to the south side of Chappaquiddick, the landscape supports ancient oak woodlands with a scrub oak or huckleberry understory. Scrub oak, actually two shrubby species, is a good indicator of forest continuity, and nearly 80 percent of forests and scrublands containing it are ancient woodlands. These are concentrated on the State Forest and south

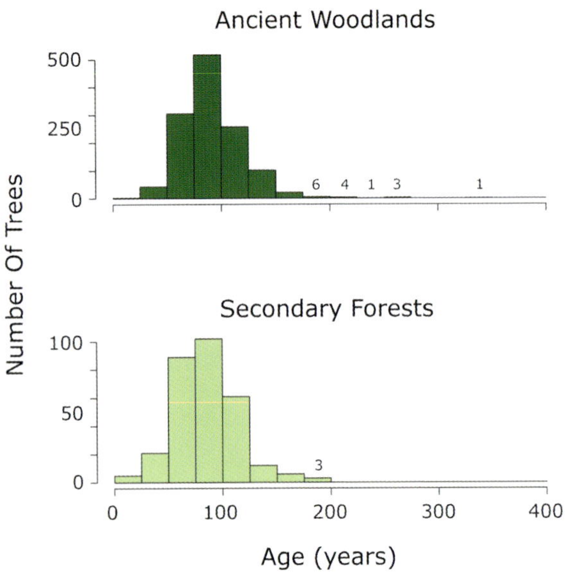

Ancient woodlands and secondary forests exhibit little difference in tree ages and a broad and rather symmetrical distribution centering on seventy-five to one hundred years old. All trees exceeding two hundred years in age occurred in ancient woodlands.

between Edgartown and Tisbury Great Ponds, where it occurs in bottoms, impenetrable expanses, or as an understory element below white oak. Extending out from this concentration of scrub oak in the Island's center, huckleberry and blueberry dominate the shrub layer in nearly ten thousand acres of forest comprising white, black, scarlet, and post oak. These woodlands are often open and easily traversed.

Oak Heath Forests

Forests of oak, huckleberry, and blueberry are the most widespread type of ancient woodlands. In areas of the Southern Woodlands and Ripley's Field Preserve, they seamlessly add pitch pine, whereas across the State Forest, Margaret Littlefield Greenlands, Pohogonot, and the Watcha Ribbon Lots, they grade into scrub oak types. Given its extensive distribution, good views of this open forest of modest-sized white and black oak over continuous low huckleberry are common along many of the Island's well-traveled roads. These include State Road from the Tashmoo Overlook to North Tisbury;

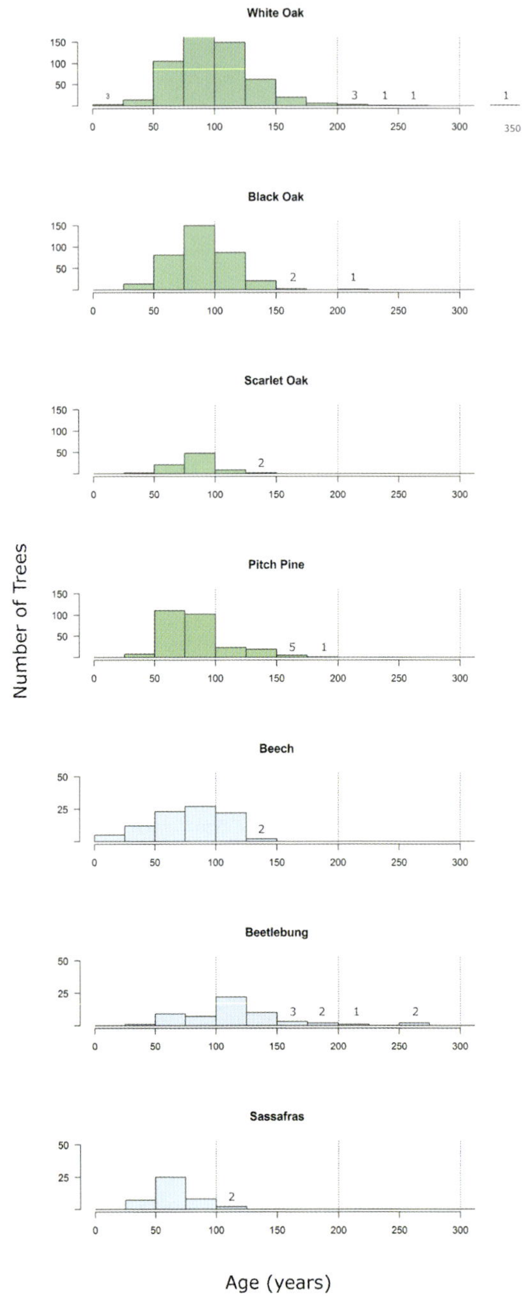

Within the general trend of tree ages there are some important differences among the species. White oaks tend to be older than other oaks, conforming to the interpretation that it has been the most abundant oak species historically. While a few pitch pines in ancient woodlands are very old, relatively few exceed one hundred years in age, and the majority are fifty to one hundred years old, suggesting that they established following field abandonment, disturbance (harvesting or fire), or planting. Beech also tends to be young, as it is actively reproducing and spreading into many forests to become a larger part of the Island landscape.

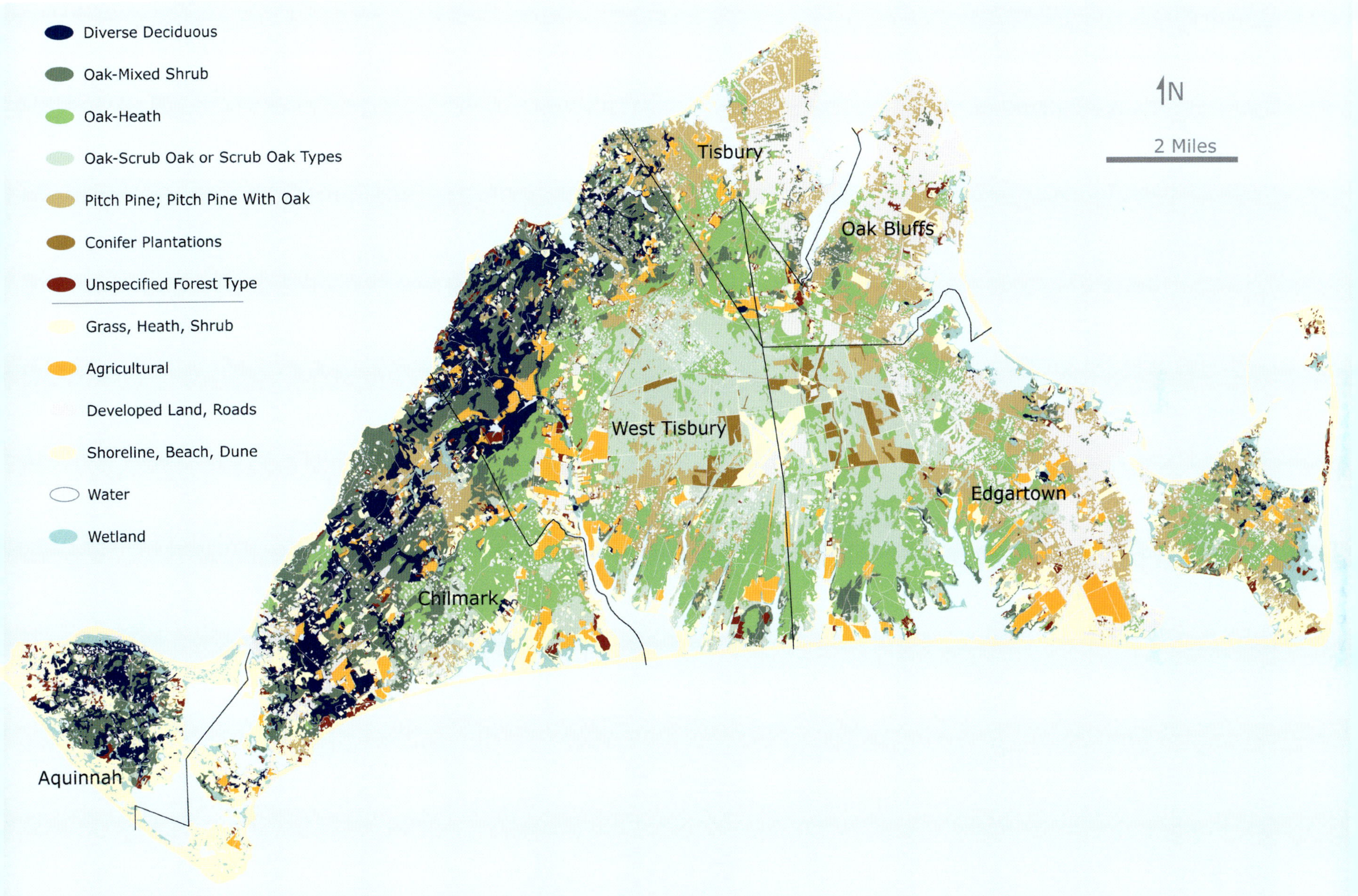

The major types of vegetation and land cover, modified from maps developed in 2002 by The Nature Conservancy and the Natural Resource Assessment Group at the University of Massachusetts. Major changes in farmland, developed lands, and open land vegetation since 2002 have been updated by the Harvard Forest.

along much of the Edgartown–Vineyard Haven Road; in the State Forest along Barnes, Old County, and the West Tisbury–Edgartown Road; or heading to Wasque, Pohogonot, Long Point, and Quansoo. This forest is so common that it is easily ignored, and its regularity of structure and composition strikes some eyes as monotonous. But in its expansive condition, as seen in Wapatequa Woods, the Southern Woodlands, on Long Point, or across the Littlefield Greenways, these open woods are a delight. Under the low oak canopy, light levels are high, the bright bark of the white oaks shines on contorted trunks, and the low shrubs allow easy passage for miles as they brush by your legs.

Scrub Oak Woodland

In sharp contrast to the easily ignored and readily traversed open oak forests, scrub oak woodlands are confrontational in appearance and physical encounter. Even at motoring speed the dense low scrub contrasts with surrounding woods as seen when crossing Deep Bottom on the West Tisbury–Edgartown Road. Meanwhile, the tangle of stiff, sharp branches is generously described as impassible. Broadly, the distribution of scrub oak is tied to dry, sandy soils and concentrated in the center and south of the Island, with notable outliers at spots like the Menemsha Hills Reservation and Litchfield Road on Chappaquiddick. Many species thrive in openings within the scrub, including tree oaks, pitch pine, and planted spruce; low shrubs (huckleberry, blueberries, sheep laurel, bayberry, and bearberry); and many sun-loving herbs such as bracken fern, Pennsylvania sedge, goldenrod, lichens, and little bluestem. Diverse plants proliferate wherever the soil is disturbed along trails and fire lanes.

The local abundance, height, density, and expanse of scrub oak are strongly linked to history. Its strong affinity for ancient woodlands springs from its slow spread onto sites where it has been physically removed by clearance. Under the shade of a continuous forest canopy, it occurs as little more than an incidental shrub. But once the canopy is removed, it expands vigorously. Thus the concentration of scrub oak on the plain is best explained by dry soils, woodland continuity, and repeated cutting and wildfires in the past few centuries. Toward the south shore and on Katama and Chappaquiddick, where it does occur in secondary woods, it expanded from old woodlots or persisted as inconspicuous sprouts in pastures.

While these dense scrub oak thickets slow succession, trees will emerge gradually and weaken the scrub oak. Thus, like other shrubby vegetation across the Island, scrub oak peaked in the late 1800s and early 1900s and has declined ever since as forests have regained an upper hand.

Pitch Pine Forests

Magnificent ancient woodlands of pine and oak with dense heath (huckleberry and blueberry) and scrub oak are scattered across the Great Plain and into East Chop and West Chop. In contrast a sparse understory distinguishes many secondary pine and oak forests that were established by trees spreading onto abandoned fields or planting of pine in the mid-nineteenth to early twentieth century. Pure pitch pine forests are almost exclusively secondary and may range from having little understory, save sedges, to a scattering of diverse shrubs.

Pitch Pine and Oak Woods

The ancient woodland at West Chop Woods provides a superb example of mixed pine and oak where pitch pine dominates and reaches its greatest age. The woods are quite varied, ranging from nearly pure pine to pure oak and even including occasional red maple, beech, and sassafras. Understory diversity is low, with huckleberry predominating and reaching its greatest abundance under open oaks, where it forms a continuous low layer joined by tall and low blueberry, Pennsylvania sedge, occasionally dense wintergreen, and scattered bayberry, bracken fern, lady slipper, and catbrier. While scrub oak is largely absent from West Chop Woods, it

A broad scrub oak bottom in the State Forest slopes up into the surrounding oak forests.

is scattered in the Southern Woodlands and increasing to the south in the State Forest.

The largest pines at West Chop Woods hint at the majesty of old-growth pitch pine and refute its characterization as a scraggly successional species. Trees reach two feet in diameter, and the oldest display thick, platy bark similar to ponderosa pine in the western United States. Many reach thirty feet before sending out robust limbs that form a wide crown outstretching adjacent oaks. White oak is more abundant than black oak; both are often multi-stemmed and range from pole-sized to nearly two feet in diameter.

Scrub Oak and Frost Help Feed an Island's Rare Moths

As we biked out from the youth hostel along the State Forest bike path on the delightful first morning of summer, the oaks and pines flew by us in a blur of green. Then, as we swept sharply up the bottom past the bulbous heath hen statue and rounded the bend, the world abruptly changed. As far as the eye could see, there was a sweep of tobacco-brown foliage. The low expanse of scrub was completely dead. Tiny new leaves and dry flowers hung withered, crisp, and motionless in the stillness of the early sun.

As we biked the plain that morning, that scene was repeated in each bottom and broadly across the flat expanse wherever the tree canopy dropped off and scrub oak dominated. Beyond the fire hazard sign along Barnes Road, the flats were a blanket of brown. So too was the Pennywise area, the open scrub in Pohogonot, and the Littlefield Greenways, as well as stretches near Nats Farm and Old County Road. Wherever scrub oak was fully open to the sky, it hung dead. Yet, the scattered trees that stuck up through this low layer remained green and unscathed. So too was the scrub oak near the coast at Long Point.

It was June 22, 2014—midway through a week of glorious cloudless days in the seventies. Given that warmth, the nature of this new growth's killer was a shock: frost. The past few nights had been still and crisp, in the high thirties around our place up on the moraine. But out on the plain, the dry, cloudless conditions brought widespread frost. This, it turns out, is not unusual for June. And, it is a key factor maintaining the Great Plain's persistent low scrubby state.

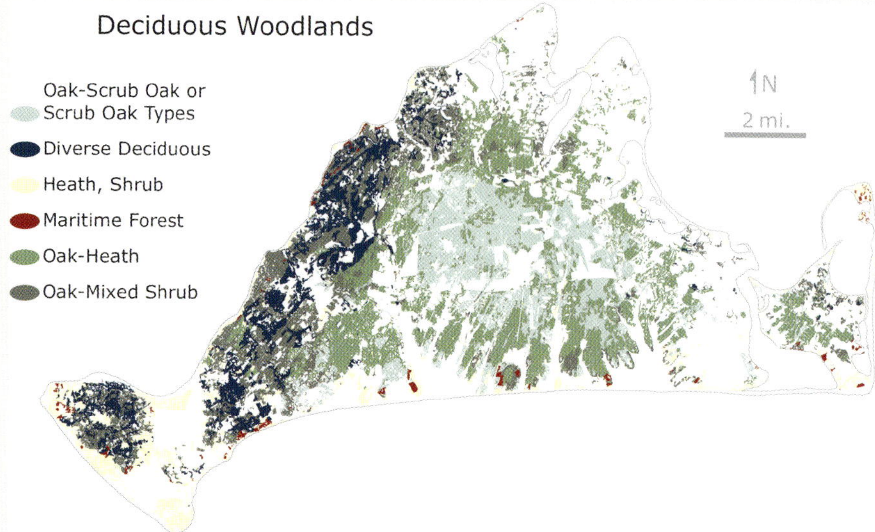

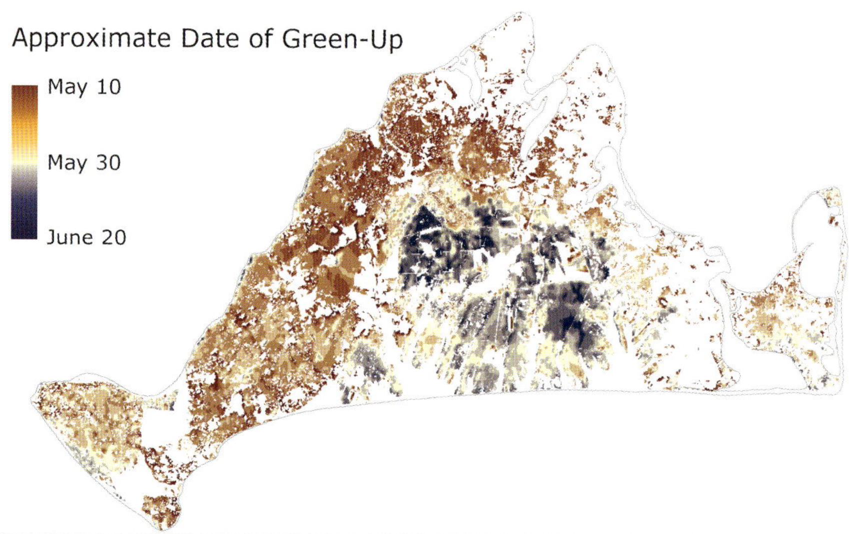

The timing of spring green-up across the Island's deciduous woodlands. Using satellite imagery to detect the greening of leaf-out, Jeremy Fisher at the University of New Hampshire mapped the late arrival of spring to the Great Plain. Large areas of scrub oak on the plain may green-up a full month later than forests on the moraine, which leaf out in early to mid-May.

The extremely cold temperatures result from radiative cooling, which occurs when the day's warmth radiates back into the atmosphere at night, a process accentuated by cloud-free conditions. Temperatures are lowest right above the ground, leading the newly emerged and tender leaves and flowers of scrub oak to freeze and die. The woody twigs survive and, in just a few days, display a new flush of leaves. Nonetheless, the plant is sapped of stored energy and delayed in its yearly growth. Leaves that do persist are deformed and inefficient. As a result, these temperature shocks set the plant back and function as a feedback mechanism to maintain the striking vegetation patterns on the plain. Areas of short scrub oak are preferentially stunted in growth, while the surrounding trees that escape the frost continue to grow taller.

The bottoms and other depressions experience the greatest frequency and severity of frost due to cold air drainage. Growing dense as it cools, the air flows downhill with gravity, pooling in low spots and following the vast system of branching tributaries that drain slowly toward the south shore. This phenomenon is strikingly apparent when driving along the West Tisbury–Edgartown Road; temperatures vary abruptly between the frigid bottom and the warm and wooded surface of the plain. Today, however, this prehistoric drainage system is disrupted by the road, and the many plantations and houses that fill the bottoms. Essentially none of the bottoms stretch unimpeded to the sea anymore. Downslope of these obstructions, frosts are less severe and the vegetation patterns differ from the more natural ones upslope.

Since the structure of the vegetation controls cooling, frost pockets can be created on the plain simply by removing the forest. Once frost sets in, regrowth is hindered and scrub oaks expand and persist. Given this, it is interesting to speculate about the history of the plain. It may once have supported a majestic forest of oak and pine above huckleberry and scrub oak, similar to vegetation that is expanding today. But once the forest was cut, frosts set in, with repeated cutting for fuel and occasional fires removing most trees that did emerge. The result was the thriving scrub and tree oak sproutland described through much of the Island's history.

Throughout this frost scenario, scrub oak is a key player. Walking through the frost-killed expanse it is clear that plants like bracken fern and huckleberry are little affected by the cold, whereas tree oaks and many herbs display some damage. But scrub oak is near its northern range limit and is far and away the most susceptible species. Its tenacious resprouting allows it to persist and produces dense thickets that inhibit other species and reinforce the semipermanence of the low vegetation.

This frost dynamic benefits an unlikely group of organisms—rare moths that thrive by feeding on soft young oak leaves and then mature rapidly under the warm summer sun. Both conditions occur commonly when scrub oak and frost coincide. The leaves of tree oaks emerge early in spring and become leathery long before temperatures warm enough for rapid growth in the moths. But when scrub oak is repeatedly hit by frost, it produces succulent young growth well into June, when the warm days suit the moths well. Hence, a harsh environment and impenetrable shrub combine to feed the Island's rare moths.

The older forests of pine and oak at West Chop, Southern Woodlands, and Ripley's Field have just reached the stage where large trees are dying due to age, defects, ice, and wind. Where trees have fallen, light streams through sizable gaps in the canopy, illuminating the next generation of oak and pine saplings below. Meanwhile, the massive downed trees and uproot mounds provide new habitats that diversify these woods and harken to forest processes and conditions that the Vineyard has not witnessed for centuries.

Diverse Deciduous Forests

The Island's most varied forests occupy the moraine, where glacial deposits and undulating topography create diverse conditions from dry ridges to streamside riparian areas. Clay soils are distributed erratically from hilltop to valley bottom, so that pockets of beetlebung and red maple, which typically sit low in valleys, are often perched on ridges or slopes. Across this varied terrain, oaks reign on sandy, rockier areas; mixed forests of beech, sassafras, oak,

Brown, frost-killed leaves on scrub oaks beneath green open-grown oaks in the Margaret Littlefield Greenways, West Tisbury, June 2014.

hickory, or beetlebung occupy moister sites; and beetlebung and red maple dominate the wet swamplands. The understories parallel the trees: hazel, low blueberry, and huckleberry are associated with the dry oaks, whereas viburnum, sweet pepperbush, and witch hazel increase with moisture. Given that much of this landscape was farmed, most forests are secondary and young, growing rapidly, and changing progressively with succession. Pockets of ancient woodlands do persist in this landscape along with some extremely old trees.

These mesic forests support some notable characteristics. One fascinating and readily recognized feature is the local abundance of holly. Though scattered across the coast, holly reaches great abundance and size on the Vineyard's northwestern shore, especially from Seven Gates and along Mill River to Cape Higgon. In places like Holly Holm, they range to thirty feet tall, a foot in diameter, and more than seventy-five to one hundred years in age. Hickory is locally abundant, giving a feel for pre-European times, when it was more common. On Seven Gates, the slopes above Buttonwood Farm Road, and the upper end of Quammox Road on Chappaquiddick, a few hickories reach a foot and a half in diameter and exceed one hundred and fifty years.

Beetlebung is an unusual addition to many forests, often forming distinctive clones of even-sized trees. Though elsewhere in New England red maple predominates across many soils, it takes a diminished role on the Vineyard. Finally, wherever it occurs in abundance, beech creates a locally barren wood. Its dense shade and competitive root mat eliminate most plants other than beech-drops and beech saplings. Though nothing on the Vineyard approaches the inspiring groves of Naushon Island, beech is undeniably on the rise.

Shrublands

Despite its extensive forests, the Vineyard landscape retains more expansive areas of shrubby growth than most of New England, including Cape Cod. Varying from dense young woods, brushlands, and shrublands to heath, these offer cover for diverse wildlife, along with a seasonally distinctive abundance of flowers, fruits, and seeds. Concentrated on former pastures, many shrublands support diverse herbs, shrubs, vines, and tree saplings that are progressively transitioning to forest. Great examples are seen throughout Wompesket, parts of Waskosim's Rock, and other morainal sites, especially across Chappaquiddick. On the plain, shrublands are dominated by scrub oak, huckleberry, and blueberries.

A range of low woody and grassy maritime vegetation occurs along the coast on former pastures where wind, salt spray, sandy soils, and a lack of trees and seeds slow succession. Maritime shrubland is widespread and variously contains bayberry, shadbush, cherry, viburnums, sumac, and poison ivy, along with vines and invasive catbrier, bittersweet, and honeysuckle. On sites like Moshup Trail, Lobsterville and West Basin, Katama, Squibnocket Ridge, and

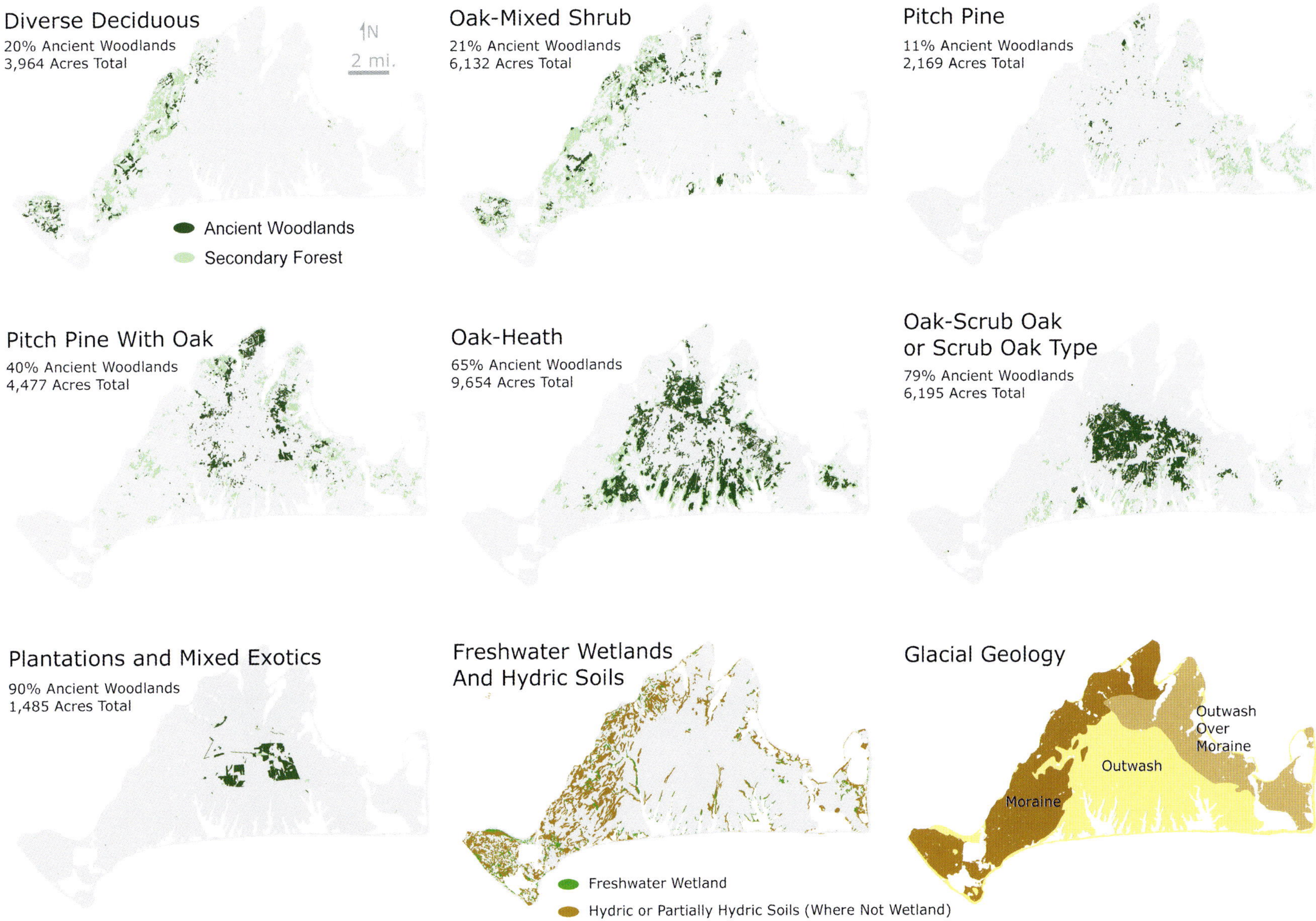

The relationship between forest distribution and history. Diverse deciduous forests and oak forests with a mixed shrub layer are largely concentrated on the moraine, where moister soils support these secondary forests. Pitch pine forests have a northerly and easterly distribution and are composed of historic plantations, successional forests on old fields, and ancient woodlands in and around the State Forest and on West and East Chops. Oak forests and scrub oak are concentrated toward the center and south on the plain and other drier soils, and compose the bulk of the Island's ancient woodlands.

One-hundred-and-eighty-year-old pitch pines in the ancient woods at West Chop Woods.

the west end of Tashmoo Pond, a low and open maritime heath community also includes chokeberry, bearberry, poverty grass, tufted hair grass, Pennsylvania sedge, and little bluestem. On capes and sandy spits like Cape Poge, red cedar woodland may persist for centuries without succeeding to a more forested condition due to chronic disturbance by wind, erosion, salt spray, and human activity. In areas of more active sand, maritime dune communities support coastal grasses and many low sun-loving species listed above.

The Island's Unusual History of Forest Plantations

Tree planting on the Vineyard began early as voyagers and returning sailors imported seeds, rootstocks, and plants from other shores. Edgartown's grand pagoda tree, transported in a pot from China in 1837 by Captain Thomas Milton, is a singularly well-documented example. But countless others, lacking backstories or pedigrees, are scattered around backyards, woods, and streets Island-wide.

The purposeful establishment of forests across the Cape and islands began remarkably early as landowners, especially farmers, planted trees to stabilize soils, create windbreaks, and start a woody crop on barren sandy soils. In the 1840s, after his own experiments transplanting white pine into fields near Walden, Henry Thoreau traveled to Nantucket, where he described Captain Gardiner's successful plantations and then rambled the Cape from Orleans to Truro noting dunes and sandy fields supporting rows of new seedlings. In these early efforts pitch pine was preferred, although Japanese black pine was also planted on Nantucket and in Aquinnah and Chappaquiddick, and white pine was commonly used inland. The source of planted material was largely unrecorded. Seed was collected locally, seedlings were transplanted by the cartload, and many purveyors sold seed collected elsewhere.

In 1846, its first year of publication, the *Vineyard Gazette* fueled interest in forest planting by reprinting articles on the subject by landowners from across the northeast. Brief accounts with local examples from the Cape and islands followed in subsequent years, and by the 1850s the sale of pitch pine seed (origin unspecified)

Large beeches on the slope of Whiting Hill on Seven Gates Farm, West Tisbury.

was regularly advertised. Tree cultivation was advanced in the spirit of progressive agriculture, although the crop's anticipated use was generally unstated. As with twentieth-century efforts in the State Forest, this early tree planting appears grounded more in faith than sound foresight.

Tree farming was readily advanced by the Martha's Vineyard Agricultural Society. In 1862 the fourth annual Cattle Show and Fair included the competition "Forest Trees," judged in three categories that illuminate the range of motivations for tree and forest cultivation.

1. For the best plantation of forest trees of any variety, not less than half an acre, nor less than three years old, with statement of the . . . condition of the land, and the time and mode of planting.
2. For the cultivation of ornamental trees, of any variety, either sown or transplanted, as road side or street trees, or for house shelter and shade . . . their treatment . . . and subsequent culture . . . and their adaptedness to the soil and climate of the Island.
3. For experiments in the planting of pine trees in the shrub plains and the pasture lands of the Island, not less than ⅛ acre, with statement of mode of planting, the kind of pine sown, and amount of seed used. Any further statements throwing light upon the propagation of pine trees on the extensive plains of the Vineyard will be very desirable.

A large American holly in the oak woods on Holly Holm, Chilmark.

Plantation of pitch pine on a former farm field along Willow Tree Bottom near the West Tisbury–Edgartown Road.

The first two categories sought to support the expansion of tree canopies broadly—in towns, around homes, and along streets. With its specific focus on the plains and pastures, the third category reveals something else. On the one hand it indicates that the farm community clearly acknowledged the surplus of farmland and the deteriorating quality of many pastures. By encouraging the planting of pine, they sought to put these low-value lands to better use. The strong interest in planting the "shrub plains" is fascinating testimony of the interest in rehabilitating the Great Plain landscape. Given that oak was naturally more abundant on these sites, the focus on pine is presumably based on its greater value, ease of planting, and availability of seed.

The extent to which these prizes encouraged planting remains uncertain. The persistence of ads selling pine seed in the *Gazette* and the modern abundance of pine forests that appear naturalized from plantings suggest that this campaign was fairly effective.

Nineteenth-century planting was focused on open fields in the outskirts of towns and avoided the heart of the Great Plain. That changed dramatically once the Commonwealth acquired it.

Following the demise of the heath hen and shift in agency and mission from sustaining game to forest production, the state's assessment of the large scrub oak areas under its control shifted from that of distinctive habitat to worthless wasteland. Combining nineteenth-century agricultural and twentieth-century forestry perspectives, state managers committed to convert the ancient woodlands of scrub and tree oak to "productive" forest by planting pines and spruce. Though similar efforts were pursued on public lands statewide, the practice on the Vineyard had a unique aspect. Most planting across the northeast and the Island focused on abandoned farmland where the native vegetation had been eradicated. In the State Forest, non-native conifers were planted directly into thriving native oak scrub—the low "woodlands" mapped by Henry Whiting and noted by Nathaniel Shaler and others. Since that scrubby vegetation had proven resilient to repeated cutting and fire, this planting required a novel approach.

On the Great Plain, the state foresters faced a thick mat of woody roots beneath a dense growth of aggressive scrub. To reach the soil and combat this growth, the crews employed a tracked dozer and large plow to carve a single deep furrow in the ground and flip the upturned soil into a parallel ridge. Behind the plow, the dozer trailed a sled caged in steel mesh. As the strange contraption crawled across the plain, a solitary hapless occupant sitting in the cage inserted seedlings in the dirt exposed along the edge of the furrow and between the sled's runners. In this manner state forestry crews gradually violated the largest expanse of native sandplain vegetation on the globe, perforating it with some four hundred thousand seedlings as they established large blocks of non-native trees.

The scale of these efforts is worth considering. With a professed goal of imparting value to worthless land and providing the Vineyard with a dependable wood supply, more than 1,500 acres were planted with white, red, and Scots pine along with sizable areas of Norway and white spruce. Most of the work occurred in the 1930s, with renewed activity in the 1970s. None of this "improvement" planting gave consideration to native species and ecosystems, the unique qualities of the plain, or the consequences of introducing exotic species.

The plantations' fate has been decidedly mixed, but the expected demand for the wood never materialized. Most of the Scots pine and red pine died of pathogens, including *Diplodia* fungus. Others were windthrown in Hurricane Bob along with pockets of white pine. Many of the remaining white pine and Norway and white spruce plantations are thriving, with densely packed but rapidly growing trees that exert substantial impacts on the native vegetation. Though remarkably resilient to physical damage, scrub oak, huckleberry, and the other native species suffer under the shade of dense conifers. In areas where the conifers have died, the shrubs persist, often mixed with blackberry, poison ivy, and conifer saplings that have colonized due to the weakened state of the native growth. But beneath intact spruce plantations, many native species have disappeared. In the brighter white pine woods, towering trees shade a layer of blueberries, forming a compelling scene reminiscent of central Sweden.

Thriving plantations do not simply dominate the site where they were planted. At sixty to eighty feet tall, the mature conifers disperse seeds widely in and beyond the State Forest. As they spread, these exotic conifers will increasingly compromise native forests and shrub lands.

Largely dominated by native trees, but infiltrated through accident and intention by many introduced species, the Island vegetation is in a novel condition. Though less strongly controlled by human hands than in the past, and increasingly dominated by natural processes, it retains many cultural legacies from history. These conditions are unique, delightful, and challenging. The same can be said of the animals that make use of the rapidly changing landscape.

Wildlife

Besides domestick animals, the quadrupeds which are found on Martha's Vineyard are these which follow: the skunk; the musquash [muskrat]; the mink; four or five species of ground mice; the mole; the rabbit. . . . There are no deer, foxes, nor squirrels.
— Reverend James Freeman, 1807

Even without considering his lengthy discussion of the snakes, birds, fishes, and mollusks, we learn much from Reverend Freeman's account of wildlife at a time when Vineyard farms were flourishing and forests were still falling across the landscape. Compared with today's Island, his is missing species that we are abundantly familiar with, most notably deer, turkey, squirrels, raccoons, otter, and chipmunks. He notes the rabbit, but we now know that this was the native New England cottontail, a species nearly extirpated from the region by the larger, more competitive, Eastern cottontail that was widely imported from Kansas by the thousands and released by Massachusetts agencies and sportsmen's groups in the 1930s. Though he includes the familiar skunk, that species was subsequently eliminated and then reintroduced in some undetermined way. These shifts in mammals and many other groups of species are certainly understandable given the massive alterations that occurred in the land and in human attitudes toward nature since the arrival of Europeans.

When we examine these trends in detail, some general patterns appear, such as notable shifts in the relative abundance of woodland versus grassland birds over time. But it is also clear that each species—deer, heath hen, osprey, brook trout, and each grassland bird—has a separate and unique history. The biology and habitat preferences of every species are distinct. As a consequence, at any given time we are viewing a dynamic scene of many species on distinctly different trajectories. While one species may be rapidly increasing with the recovery of forests or a decline in human persecution, another may be disappearing from a century's loss of openland habitat. Others may be newly arrived and facing an uncertain future, while some species exhibit fairly steady numbers over time. Thus the instantaneous snapshot that Freeman or we derive through observations or annual counts yields no information on the dynamics—past or future—of either the animals or the environments they use. Each species is undergoing different trajectories recognizable only through a long-term perspective.

When we do take a long view, we recognize the challenges

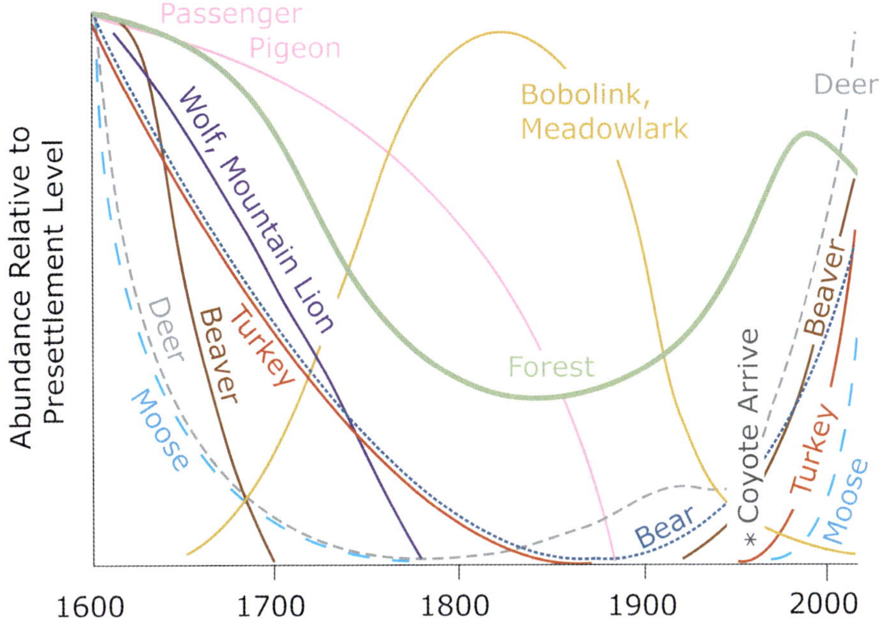

Wildlife across Massachusetts has changed dramatically with changing habitats and human attitudes. While each species exhibits individualistic behavior, general trends appear. Large carnivores (e.g., wolf and cougar) have been eliminated regionally and not returned, whereas the passenger pigeon and heath hen are now extinct. Many native woodland species were driven to low levels (deer, bear) or eliminated completely (beaver, turkey, moose) in the mid-nineteenth century and have returned in great numbers. Species that favor grasslands, shrublands, and young forest increased with the emergence of farms and loss of forest but have been declining since the late nineteenth-century regrowth of forests. An interesting collection of non-native species has arrived through introduction (Eastern cottontail rabbit), long-distance migration (coyote), and shifts in distribution due to changing temperatures and food resources (many birds). Conceptual graph made from data collected by James Cardoza at the Massachusetts Division of Fisheries and Wildlife and the Harvard Forest.

that confront us in managing for and living with wildlife. In past centuries, the Vineyard and the eastern United States have witnessed a remarkable recovery of forests and a shift in human attitudes toward conservation. Much, but certainly not all, of the historically native wildlife has prospered as a result. At the same time, and rather ironically, many species that thrived in the open agrarian landscape of Reverend Freeman or Henry Whiting have plummeted, including the moths, butterflies, birds, and other wildlife of open fields, brushlands, and cutover forests. Today some species flourish to the point of nuisance (deer, skunks, geese), others need special care (piping plovers, many grassland birds, shrubland moths, black snakes, box turtles), and most face a rather uncertain future with changes in the environment and human pressures.

How we begin the narrative on wildlife history depends on our interpretation of the land before European arrival. While some have envisioned a mosaic of Indian fields and open woodlands managed with fire, the paleoecological archaeological records paint a different picture. These reveal a forested land shaped by natural processes—wind, ice, sea level, insects—in which climate was the main driver of change.

In this landscape, woodland animals thrived, though many were regular prey for native inhabitants. Species of shrubby, grassy, and open sites were few and faced challenges. Wetlands on saturated soils, along streams, and fringing the coastal ponds and estuaries provided major breaks in the forest, and were hot spots of activity for these species. The shoreline provided another array of distinctive habitats—beach, dune, and eroding shores, headlands, and cliffs. On the other hand, processes that disturbed and opened terrain on the mainland, such as the flooding and scouring of major rivers and damming by beavers, were missing in small Island watersheds.

Given the broad suite of landscapes, the abundance of wildlife would have been startling. Equally shocking would have been the depredation inflicted by the arriving Europeans. From John Josselyn's 1663 account of coastal New England, we get a sense

Otter walking the eroding beach at Wasque. These mammals are abundant in ponds across the Island and one of the Vineyard's largest native land animals, January 2014.

of a tremendous force of nature confronting a changing cultural landscape: "I have seen a flight of Pidgeons in the spring . . . that to my thinking had neither beginning nor ending . . . and so thick that I could see no sun. . . . But of late, they are much diminished by the English taking them in Nets." He reported similar losses in other species: "the English and the Indians having now destroyed the breed, so that 'tis rare to meet with a wild turkey in the Woods." The arrival of Europeans initiated transformations in the landscape and wildlife populations that continue to reverberate today. Most important were land use and dramatic shifts in habitats. But for many notable species, the early history of persecution and subsequent shift to conservation hundreds of years later are equally important factors.

The predominant change was the clearing of forests and organization of the land into an agriculturally productive system of pastures, hayfields, meadows, cropland, and woodlots. Other activities augmented these changes: the draining of wetlands for farmland, damming of streams for power, and movement of countless stones into miles of walls across the landscape and jetties and breakwaters on the coast. Inadvertent impacts were equally important and enduring, including the gradual disappearance of wood from streams, ponds, and wetlands, and the erosion of farm fields, infilling of ponds, and addition of nutrients to many waters. With diminished forest cover, more water reached the streams, so they ran stronger and erosion increased.

Hunting exerted an intensifying impact regionally. Key species—furbearers, deer, and moose—fueled local trade and were major exports. Native and European hunters and trappers responded efficiently to this huge demand, and the area deprived of valuable species spread rapidly into the hinterlands—well in advance of settlement. John Smith reported that from 1631 to 1636, more than ten thousand beaver skins were shipped from Plymouth to London. From his trading post in Springfield on the Connecticut River, John Pynchon received forty-seven hogshead of beaver skin, accounting for 8,992 animals between 1652 and 1657. Over the next fifteen years, although his trade in beaver declined to 6,480 skins, he recorded the pelts of 718 muskrat, 415 moose, 379 otter, 92 fisher, and 82 marten. When added to animals taken for local use and to eliminate real and perceived threats to humans, livestock, and crops, the impacts were severe.

As the interacting forces of land change and wildlife depredation unfolded, large and fur-bearing mammals, predatory birds, and forest-dwelling species of all types declined. Then, as the agrarian scene emerged, those species that favored grasslands and shrublands flourished. If one did not dwell on the losses (ancient forests, a thriving Native American population, large mammals, and awe-inspiring abundances of wildlife), the cultural landscape of fields, woodlots, and colonial activity created a delightful scene filled with its own inspiring flora and fauna.

Many plants and animals thrived as if they had evolved with

these novel habitats. Certainly, contemporary naturalists took this blossoming of native wildlife in the cultural landscape as a perfectly natural state of affairs, as witnessed by the domesticated descriptions of common bird species by William Peabody in his 1839 *Ornithology of Massachusetts:*

> The Field Sparrow . . . is found in open fields and pastures, where it lives on insects and seeds.
>
> The Cliff Swallow . . . builds under eaves and cornices, where its nest is partially sheltered from rain.
>
> The barn-yard affords [the Whippoorwill] a foraging ground, which it often visits; sometimes it takes its station on the step of the house door.
>
> The Killdeer plover is found in open fields [and] in the presence of horses, cows and sheep. . . . [They] sometimes follow the ploughman, to seize [worms] which he has turned out from the ground.

Indeed, during the height of agriculture and into the period of widespread farm abandonment, many of our favorite and increasingly uncommon species thrived. In addition to those Peabody noted, these included the eastern meadowlark, red-winged blackbird, upland sandpiper, Savanna sparrow, vesper sparrow, grasshopper sparrow, eastern towhee, and many others.

However, as farms were abandoned and forests began returning, some of the forces most destructive to wildlife emerged across the east, including intensive clearcutting and market hunting. These triggered a backlash by early conservationists, beginning with Henry Thoreau and ultimately spurring the formation of conservation organizations, state and federal forest and wildlife agencies, and the adoption of rigorous state and federal regulations.

William Hornaday summarized the state of the situation succinctly in 1913: "Air and water are polluted, rivers and streams serve as sewers and dumping grounds, forests are swept away and

Geese, along Sengekontacket Pond, contribute nutrients to the coastal pond.

the fishes are driven from the streams." But Hornaday's greatest focus was the direct persecution of birds and mammals. He cited a "great increase in the slaughter of songbirds for food" and pointed regionally to unimaginable declines in the passenger pigeon and gray squirrel, the loss of turkey from New England, and the threatened status in Massachusetts of many birds, including the wood duck, red-tailed hawk, great horned owl, upland sandpiper, woodcock, and piping plover. With a call to "remember the heath hen," he identified the "Army of Destruction" responsible for this depredation—overzealous sportsmen, meat hunters, well-gowned women supplied by plume hunters, and predators ranging from cats and dogs to skunks.

Hornaday advocated for legislation that sought to make wildlife the property of the state, ban the sale of wild animals, tighten hunting regulations, protect threatened species, and establish game reserves. In this quest he solicited the support of diverse allies: federal and state governments, sportsmen's groups, the Boone and Crocket Club, Camp Fire Clubs, and Audubon Societies. He also

A withdrawn box turtle in the woods near Great Rock Bight.

argued for the hunting and trapping of predators, including raptors, large mammals, red squirrels (which will eat bird eggs), and cats.

State and federal legislation did emerge, including a model law in Massachusetts that banned the sale of wild meat. With reduced depredation and with improving forest and shrubland habitat, many native species began to rebound. Growing recognition in the early twentieth century that "the most practical wildlife management is accomplished through good land use" promoted the establishment of state and national parks, forests, and reserves. State (and some private) efforts spurred the recovery process by actively reintroducing native species (beaver, deer, quail, many fish, turkey) and studying many others. The greatest driving forces, however, were natural: the resiliency of forests, wetlands, streams, and ponds and the ability of many native plants and animals to repopulate their former range. The consequence has been a progressive rewilding of much of the eastern landscape, including the Vineyard.

Looking over time, we see some major trends in wildlife response. A singular group of species that thrived in the prehistorical landscape has been extirpated regionally (the wolf and cougar, for instance) or driven extinct globally (passenger pigeon and heath hen). These species represent both top predators and iconic symbols of the need for conservation. A second group includes many species driven to low levels or even regionally extirpated that have subsequently rebounded, such as deer, osprey, turkey, beaver (regionally), and squirrel. Then there is a growing group of species that are not native but have appeared in past decades to establish a solid foothold. These include the coyote, which has taken advantage of declines in competitors (wolves) and changes in the landscape, human attitudes, and their own gene pool (through interbreeding with wolves) to expand from the Great Plains and Southwest to much of North America. It also includes the cardinal, mockingbird, turkey vulture, opossum, and others that are expanding northward with warming temperatures and an expanding food base that includes backyard birdfeeders and road kill.

Two final groups are historically fascinating and challenging to conservation and land management. Openland species—including many charismatic birds (woodcock, bobolink, meadowlark, savannah sparrow, and others mentioned above) and equally intriguing but less well-known insects (butterflies and dragonflies) and small mammals—were likely uncommon to rare before the wooded landscape was opened by colonial activity. Almost all show a recent century-long trajectory of decline as their preferred habitats of pasture, field, barnyard, and early successional vegetation are disappearing. Deciding whether, and knowing how, to manage for these species that thrive with certain human activities is a major challenge.

The final group includes native species that remain uncommon. The brook trout, recently rediscovered in Mill Brook and other Island streams, is constrained by dams, water conditions, and many legacies of past land use. Others, like the anadromous herring and shad and catadromous eel, which move between freshwater and saltwater in their lifetime, also suffer from historical overfishing and conditions beyond the land's edge. And a number of species simply remain enigmatic: little studied and possibly never terribly abundant.

By exploring a select few of the important species in greater detail, I seek to fill out this picture of how humans have greatly influenced the dynamics of the Island's native animals.

Heath Hen

Many distant sounds wend their way through the Vineyard landscape—the oscillating roar of South Beach, the low groan of buoys in the Sound as the fog closes in, the yearning honk of a solitary goose as it searches the dawn sky for its flock, and the bright tinkling of pinkletinks (peepers) as darkness descends on a day that promised hope for spring. But one sound is missing that for generations must have been as much a part of the place as any of these are today.

> Here in the gray dawn a strange, weird sound fills the air. It swells and dies upon the ear, but never rises or falls. And becomes intermittent or ceases only when the sun rides up the sky. . . . It is neither whistle nor call; there is no other sound quite like it in nature. One might imagine it the wail of the wind spirit, but no man understands just what it is or how it is made. We know only that it emanates from strutting, dancing heath cocks, and is one of their customary mating sounds. Heard from a distance, borne on the sea wind, it swells to the fullness of a grand undertone, mingling with the ordinary nearer sounds of the rolling plain. Like the trilling of the toads in a million pools, like the morning chorus of a bird song on a thousand hills, it is a vital, virile expression of the fecundity of old Mother Earth. . . . It is a pæan of hope and joy, a forerunner of the pulsating, vigorous life of summer. (Edward Forbush, 1919)

We are without that sound, for the heath hen is extinct.

The bird's last stronghold, if one can call the place where it perished by such a word, was one of the Island's most distinctive habitats—the open scrublands of the Great Plain. There the last heath hens were spied from blinds and documented by a parade of enthusiasts. Here, too, they was prey to hawks and cats, and ravaged by fire and disease. The heath hen spent its final half-century on earth entirely restricted to Martha's Vineyard.

Heath hen display at Felix Neck Wildlife Sanctuary.

It went extinct despite intense efforts to save it. The details of the bird's care, breeding, and ultimate demise have been scoured for insights as conservationists have sought to sustain other dwindling species on earth. But the heath hen's story has another twist that is forcing renewed scrutiny. Today, breakthroughs in genetics are fueling serious efforts to resurrect this extinct bird. These discussions have focused attention on the hen and Island. The Vineyard seems an obvious place to attempt the world's first effort to clone and reintroduce a ghost species. Consequently, questions abound. Is it advantageous or even ethical to bring a species back? What impacts and repercussions might come from the reinsertion of a missing species into a landscape that has moved on? And where might this technology lead . . . to passenger pigeons, mastodons, or more ancient beasts? Of critical importance—who will decide? Which, if any, species, and where?

The questions are posed and the forums begun, initiated in the summer of 2014 at an evening panel that discussed heath hen reintroduction before a full house at the Agricultural Society. Articles in the *Gazette* report subsequent progress: a science team assembled, a funding campaign launched, gene sequences developed, and optimism growing.

All of this is intriguing. But, before reintroduction can be considered, a fundamental question lurks in the background, possibly overlooked in the crescendo of excitement. What exactly was this bird, and what was its role in the landscape? To understand what we are missing and how it might be nurtured if brought back, we need to divert our focus from complex laboratory procedures to the pages of history and early days of wildlife biology. When we do this, we confront a startling realization—woefully little is known about this historic bird. Though it has become an icon—for the Island, for global conservation, and for the movement to resurrect species—the most basic aspects of its distribution, habitat, and food preferences remain the subject of conjecture.

A few facts emerge as we assemble historical sources. Though the heath hen appears to have ranged from Maine to the Carolinas, it was likely never abundant anywhere, and its decline was singularly precipitous. To the north, it was rare and extirpated in New Hampshire around 1792. To the far south its status is vague, with birds perhaps persisting in Kentucky into the 1830s. Its stronghold appears centered on Massachusetts and Connecticut through Long Island and into New Jersey and Pennsylvania. However, even here, documentation is thin, for the bird disappeared well before it was even identified as a distinct subspecies (in 1885) from the prairie chicken of western grasslands. The heath hen was gone from Connecticut and Long Island by the mid-1840s. Writing from Philadelphia in 1857, Elisha Lewis described it as "more nearly extinct than the buffalo." By 1870 it was extirpated from the mainland. At that point, it remained a two-island species until it disappeared from Nashawena, the second largest of the Elizabeth Islands, around 1880. From then until the last male bird wandered into the scrub in 1933, it survived on Martha's Vineyard alone.

The rapid decline meant that there are precious few firsthand descriptions of the bird from the mainland. Most ornithologists wrote secondhand accounts based on reports from hunters and residents. Many simply repeated statements lifted from colleagues. The acknowledged expert Alfred Gross summarized the situation in his 1928 monograph based on Vineyard studies: "Th[ere] is, however, such a dearth of material concerning the Heath Hen during those early times that we know little of the conditions under which it existed . . . and we are unable to determine with any degree of accuracy its relative abundance and distribution."

As is often the case in natural history, the dearth of information led writers to perpetuate a few key and oftentimes questionable anecdotes as they fleshed out popular and even scientific papers. Massachusetts provides two notable examples of the propagation of ecological rumors with little grounding in historical fact. Both served to exaggerate the heath hen's abundance.

The first bolsters the iconic status of the bird, asserting "the Pilgrims' first Thanksgiving dinner featured Heath Hens and not wild turkey." This is part of a larger tale that heath hens allowed the Pilgrims to survive their first few winters. Despite the romance of this tale, there is no support for any specific link between the bird and early colonists other than scattered observations that they were hunted.

The second anecdote comes from the naturalist Thomas Nuttall and has been repeated in nearly every account on the heath hen to assert that it was once extraordinarily abundant. In his 1834 *Manual of Ornithology of the United States,* Nuttall relates that Lieutenant Governor Thomas Winthrop stated, "They were so common on the ancient bushy site of the city of Boston, that laboring people or servants stipulated with their employers not to have the Heath-Hen brought to the table oftener than a few times a week!"

Beyond the fact that Winthrop was describing a scene two

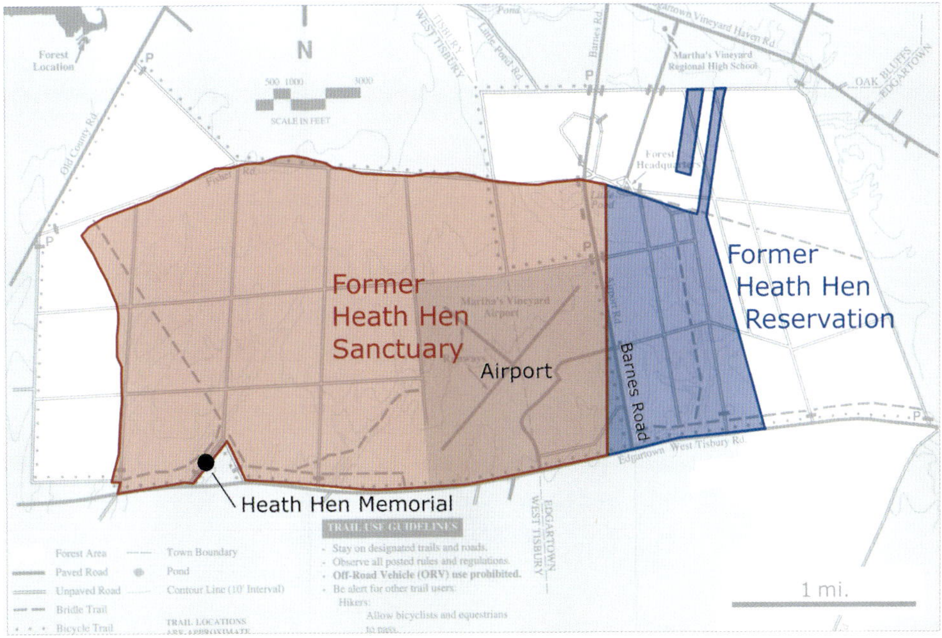

Heath Hen Sanctuary and State Forest, ca. 1920. Overlain on a modern map of the State Forest, the sanctuary lay largely west of Barnes Road and south of the State Forest headquarters. The sanctuary and other lands were subsequently combined into the State Forest, with portions converted into the naval airbase and then the Martha's Vineyard Airport and associated commercial and municipal areas.

centuries old, this assertion is undermined by a revealing fact: identical and equally erroneous descriptions of contracts for indentured servants have been advanced for the Atlantic salmon (which was never abundant in New England) and lobster (which was). No original documents exist to corroborate any of these claims. Both the Thanksgiving and servant contract stories display the apocryphal tendency to inflate the role of extinct species, the American chestnut providing another such example.

With the rapid contraction of the heath hen's range, the only truly detailed accounts of its habitat and behavior come from a single landscape—Martha's Vineyard. This is worrisome, for the Island is unique and presumably represents only a fraction of the conditions under which the species existed. But this is all that we have.

The most valuable eyewitness report by a knowledgeable expert came from William Brewster, curator of birds at Harvard's Museum of Comparative Zoology. In 1885 and 1890, Brewster traveled to the Vineyard for firsthand observations. He captured the wide-ranging qualities of the bird before its population collapse and the frantic attempts to concentrate its rescue on the state reservation.

> Throughout Martha's Vineyard, the heath hen . . . is well known to almost every one. . . . Its range extends, practically, over the entire wooded portion of the island, but the bird is not found regularly or at all numerously outside an area of about forty square miles. This area comprises most of the elevated central portions of the island, although it also touches the sea at not a few points on the north and south shores. In places it rolls into great rounded hills and long, irregular ridges, over which are scattered stretches of second-growth woods, often miles in extent, and composed chiefly of scarlet, black, white and post oaks, from fifteen to forty feet in height. . . . At all seasons the heath hens live almost exclusively in the oak woods, where the acorns furnish them

> abundant food, although, like our ruffed grouse, they occasionally at early morning and just after sunset venture out a little way in the open. . . . They also wander to some extent over the scrub-oak plains, especially when blue-berries are ripe and abundant. . . .
>
> Only one person of the many whom I questioned on the subject had ever seen a heath hen's nest. It was in oak woods, among sprouts at the base of a large stump, and contained either twelve or thirteen eggs. . . .
>
> The estimates of the average number of birds per mile varied from three to five, giving from 120 to 200 birds for the total number. . . . If these breed freely and at all successfully, there should be a total of fully 500, young and old together, at the beginning of the present autumn. When one considers the limited area to which these birds are confined, it is evident that within this area they must be reasonably abundant.
>
> I was assured that with the aid of a good dog it was not at all difficult to start twenty-five or thirty in a day, and on one occasion eight were killed by two guns.

This is the only comprehensive description of the heath hen in its native habitat. Among Brewster's most striking reports is that the bird lives "almost exclusively in oak woods" and only wanders over scrub oak plains "to some extent." Rather than being an openland specialist, the bird nests in woodlands and is only seen "occasionally" on the edges of open areas. These descriptions stand in striking contrast to the notion commonly advanced that the bird was a specialist of sandplain grassland and scrub. This critical insight for restoration provides a sad commentary on historic efforts to save the bird, as these focused on open areas.

From Brewster's day on, the bird's history is harried and well documented. With its population dropping and extinction feared, land was acquired by the state to provide protection and a focal point for extraordinary efforts at restoration. In disregard of Brewster's observations, and with a nod to the western prairie chicken, game managers cleared scrub oak areas, tilled small fields, and planted grains for feed. To facilitate research, they built blinds and invited experts to conduct studies. These activities seemed more geared toward centralizing and easing protection and research than on aligning the bird with its native habitat. On more than one occasion, prairie chickens were introduced to augment the population and possibly breed. Despite these efforts, the population faltered and eventually collapsed.

In the end it was fire, disease, and predation—by hawks, skunks, cats, and people—that eliminated this residual population. Inbreeding, incorrect habitat, and reduced genetic diversity likely played an additional role.

But as we look back at the bird's sketchy history, the factors driving its trajectory toward extinction are clear. The bird was never widely abundant. Though originally distributed across a sizable geography, it was concentrated in unusual habitat—dry open woods and adjoining brushy sandplains—that is scattered. Throughout history, it was overhunted. Like other species, this occasionally showy and easily hunted bird declined precipitously with the predatory pressure of colonial history. As individual populations then blinked out across its range, only the Vineyard remained. In contrast to deer, beaver, turkey, and many other animals, the heath hen did not sustain multiple populations that fostered recovery when persecution ceased.

Beyond its haunting call, one defining characteristic is its elusive history. This alone should give pause to any speedy resurrection. Facts should be sought and caution should prevail. After all—and this observation is truly jolting—despite the fact that the heath hen took its last stand on the Vineyard, we are not absolutely certain that it was native here. The Island's great historian Charles Banks, though no naturalist, was convinced it was not. And the man who studied it authoritatively had doubts. In 1928 Alfred Gross wrote, "We cannot be positive . . . whether the Heath Hen was native to Martha's Vineyard or introduced from the mainland after the coming of the white man."

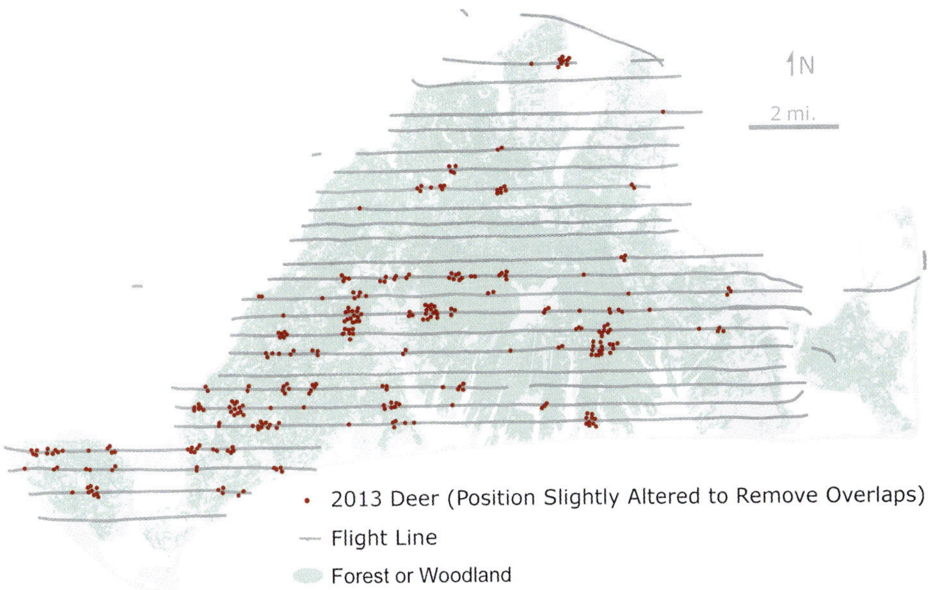

Using thermal imaging and high-resolution photography from a plane in January 2013, Thomas Millette of Mount Holyoke College mapped the Island's deer (each dot is one deer), revealing clustered groupings in the southern and western regions and forests on the moraine. Most of Chappaquiddick and the Squibnocket region were not flown. Mapping in 2014 confirmed an Island-wide estimate of forty to fifty deer per square mile. Note that the dots represent a sampling, not a complete mapping of the total population on the Island.

Northern White-Tailed Deer

Henry Thoreau would have been staggered by the map of deer on the Vineyard recently produced by Thomas Millette at Mount Holyoke College. Hundreds of dots fill the sheet. Clusters of three or four are frequent, but what catches the eye are the concentrations of a dozen or more. Each red dot represents a warm-blooded animal, for Millette's approach ingeniously uses an airplane equipped with two cameras to identify each individual. One camera captures a slightly fuzzy infrared image that discriminates the warm bodies of animals (or running cars and heated buildings) against the frigid snow-covered ground of midwinter. The second camera takes high-resolution photographs, so Millette can separate out deer from dogs, cows, and sheep. The resulting map, compiled through systematic flights, improves tremendously on hunting records, observation, or the counting of piles of deer pellets that are commonly used to estimate deer populations. Millette's map confirms what wildlife officials and health experts have speculated for years. With nearly fifty deer per square mile, the Island rivals Nantucket and some Boston suburbs in having the highest density in the state.

In Thoreau's day, deer were effectively extirpated from the state. Incredibly, his journal relates that the largest wild animal around Concord was a muskrat. He also notes with wistful envy that a friend's grandmother once saw a deer in town when she was a child. In contrast, deer were plentiful through prehistory, when open oak forests provided abundant acorns supplemented by other browse. The population had significant predators—wolves, cougar, bear, and especially people. By stalking, snaring, and driving the animals into the water, where they were taken easily by canoe, native people were deer's major predator for ten thousand years. In fact, the heavy use of deer raises the intriguing question as to whether and how the Island's Indian population regulated hunting pressure. Were they able to maintain a thriving deer population continuously over thousands of years? Or, were deer occasionally extirpated and forced to recolonize by swimming from the mainland, as they have been known to historically?

The sustainable relationship that persisted for millennia was shattered with European arrival. Venison and hides were valuable commodities, and both natives and settlers became market hunters. By 1638, a decline was noted around Plymouth. The first deer law was instituted in Portsmouth, Rhode Island, in 1646, and the position of deer reeve was established in many Massachusetts towns by the mid-1700s. While forest cutting and clearing created good habitat for this species that thrives in forest edges, it also made them more vulnerable. Noted as widespread on the Vineyard in the 1600s, deer were eliminated early in Island history. By the 1800s, deer were extirpated statewide, save for remnant populations in the Berkshires and at the base of the Cape.

Three factors aided deer's regional and overall recovery: the expansion of forests, which provided cover; hunting regulations and a shift toward wildlife conservation; and a high birthrate that allows deer to increase by at least 30 percent annually. In 1897 Massachusetts enacted a five-year closed hunting season. In 1905 the Division of Fish and Game estimated the statewide population at five thousand, and by 1910 a six-day hunting season was established in western Massachusetts. In his otherwise pessimistic 1913 volume, *Our Vanishing Wildlife,* William T. Hornaday reported one glimmer of hope—deer were increasing throughout New England.

Until recently, deer management in Massachusetts and elsewhere encouraged population growth, given the animal's singular popularity for viewing and hunting. The approach worked spectacularly. By focusing hunting on male animals, statewide numbers increased over the past forty years to around 85,000 to 90,000. In the process, deer have gone from rare and magical to common and a nuisance. A 2009 *Vineyard Gazette* article complained that the "Vineyard is a virtual Tokyo of even-toed ungulates." More broadly, nature writer John McPhee railed, "In my opinion they are rats with antlers, roaches with split hooves, denizens of the dark primeval suburbs."

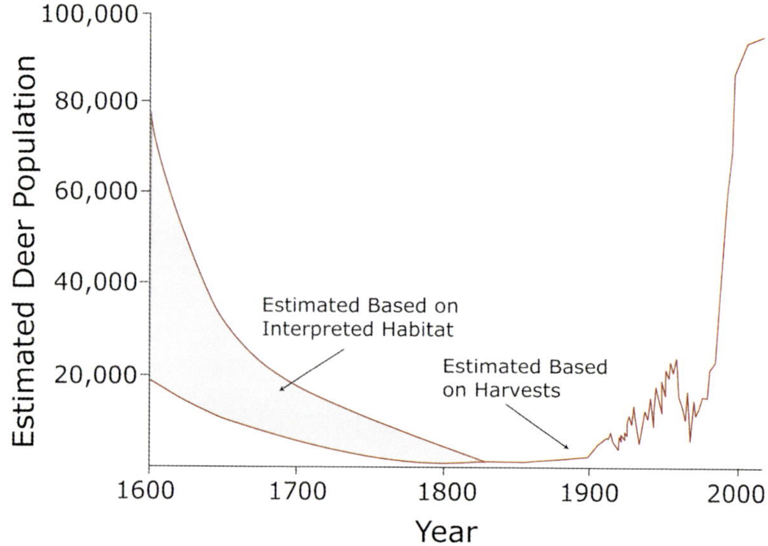

Historical changes in white-tailed deer in Massachusetts. Two scenarios bracket the early estimates. The population disappeared from all but the Berkshires and the base of Cape Cod in the nineteenth century, increased with strict hunting regulations and increased forest cover in the twentieth century, and then grew steeply. Currently the population is estimated at 85,000–95,000 statewide and is roughly stable due to hunting. Before 1966 reporting was voluntary and likely underestimated. Recent data from the Massachusetts Division of Fisheries and Wildlife.

Now recovered to levels that equal, or in many places surpass, pre-European densities, deer have become an environmental, health, and safety issue. Nationwide, a few hundred people die annually in deer-car crashes, with more than two thousand accidents in Massachusetts every year; huge sums are lost in crop and horticultural damage or paid in fencing and deer repellent; the composition of forests is altered; and human health is threatened by diseases carried by ticks supported in part by large deer herds.

With local support in many areas shifting to favor deer control, the Massachusetts Division of Fisheries and Wildlife is working with towns and conservation organizations to reduce the population using the means employed for millennia—hunting. Indeed, the state has adopted a strategy once advanced by Aldo Leopold, based on

Deer along a pond on Buttonwood Farm Road, West Tisbury.

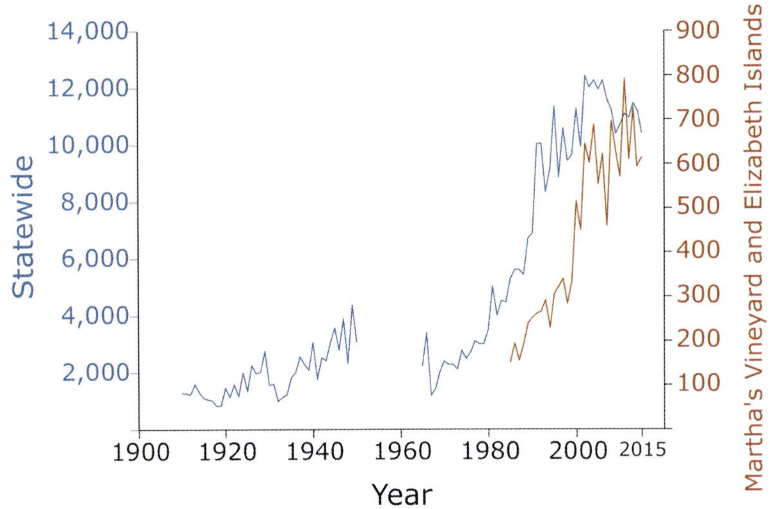

Statewide and in Dukes County (Vineyard and Elizabeth Islands) the number of deer harvested each year has increased steadily recently due to local and statewide interest in controlling the population and limiting car accidents, Lyme disease, and damage to native vegetation, farms, and backyards. Data from the Massachusetts Division of Fisheries and Wildlife.

hunting antlerless deer (does and young bucks). Leopold's original recommendation unfortunately coincided with the release of the movie *Bambi,* which doomed its initial application. But decades later, it is working in Massachusetts and on the Vineyard. Despite an aging and declining population of hunters, there has been a consistent increase in hunting success. The number of males taken on the Island has risen consistently, but a great increase in females substantially increased the total kill since 2000.

These increases have been aided by other regulations. Deer hunting season now extends from late October to the end of the year and through a succession of seasons for archery (six weeks), shotgun (two weeks), and black powder (two to three weeks). Statewide, hunters can take two antlered deer and as many antlerless animals as they receive permits for by lottery. On the Vineyard and Nantucket, up to 2,700 permits are available, greatly exceeding the demand. Each deer killed must be reported to either a computer-based system or checking station. The checking stations provide more than regulatory oversight. The sex, weight, health, and condition of each animal are also recorded. From these data comes one surprising bit of news. Despite their high density, the Vineyard deer population is in excellent health. One agency staffer described the meat as "marbled with fat" due to the abundance of food, especially acorns. Given its diverse habitats, the many masting species, abundant browse (including gardens), and relatively mild and generally snow-free winters, the Island is superb range for deer.

One growing restraint on hunting is accessible land, a big challenge on an island with a highly fragmented ownership. The density of houses is one problem, as hunting is prohibited within five hundred feet of buildings without permission. Another restraint is the increased posting of properties against hunting (or access altogether). Historically, many conservation groups and owners established wildlife preserves that prohibited hunting to protect animals. However, with increasing concern regarding habitat damage and human health, attitudes are changing.

Deer-browsed white cedars at the base of the fire tower on Indian Hill, West Tisbury.

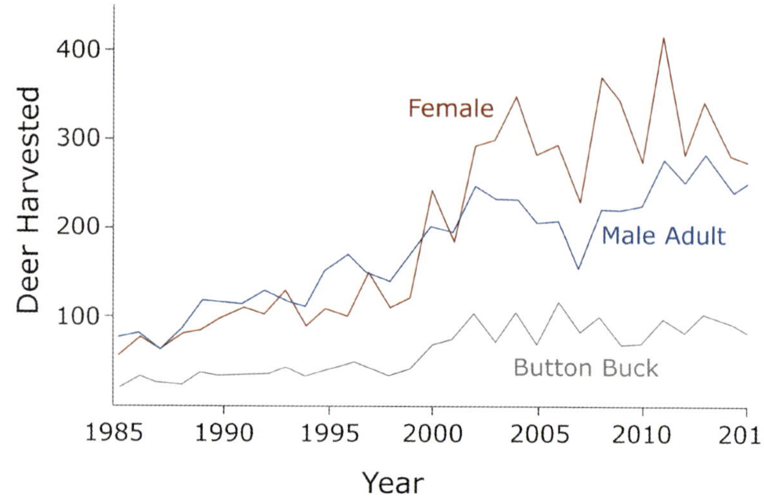

To control deer herds, wildlife agencies encourage the hunting of does, a shift from the historical focus on males. On the Vineyard, female mortality has surpassed that of males since 2000. Button bucks are males less than a year old with small nubs instead of protruding antlers. Data from the Massachusetts Division of Fisheries and Wildlife.

For many hunters, the State Forest is a central hub for hunting that is augmented through access on some town, Land Bank, and conservation lands. With a need to remove 35 to 40 percent of the population annually to keep it stable, hunting requires public and private support. According to David Stainbrook, the state deer and moose project leader, deer numbers have stabilized on the Island in well-hunted areas but are still climbing in areas with little hunting.

Concerns about human health have stimulated the effort to quantify and reduce the Vineyard deer population. Deer serve as a major host for deer ticks, which can harbor illnesses including anaplasmosis, babesiosis, Ehrlichiosis, Lyme disease, Rocky Mountain spotted fever, and even tularemia, which ranks fourth behind smallpox, the plague, and anthrax on the Centers for Disease Control and Prevention's list of infectious diseases. Alarmed in 2000 at the death of a Chilmark resident from tularemia, and rising cases of babesiosis and Lyme disease, the boards of health advanced a major public information initiative and commissioned Millette's aerial survey. The resulting maps have sparked calls to reduce the population by up to 80 percent.

Deer and other large mammals, however, are only a reproductive host for ticks; they do not carry and transmit the infectious disease. That job is left to small rodents, especially white-footed mice, shrews, and chipmunks, which harbor the disease and transmit it via ticks. These animals thrive in fragmented landscapes and backyards with few or no predators such as fox, owls, weasels, and bobcats, few of which occur on the Vineyard. To fully control ticks and associated diseases will require a combined effort to reduce the deer population and maintain the Island's biodiversity, including a suite of small, moderate, and large predators. The nature of these wildlife-based conservation challenges and opportunities would have amazed Henry Thoreau.

Grassland Birds

In 1846, while his U.S. Coastal Survey crew was surveying the Vineyard, young assistant Samuel Gilbert recorded in his journal that Henry Whiting and Robert McCleary, the senior members of the team, took a deserved break gunning and "had very good luck, some fine Plover, etc." A decade later the *Vineyard Gazette* noted that a group led by Dr. Daniel Fisher shot eight plover in a day's hunt on the Great Plain. In like fashion, on his two trips to the Island, Daniel Webster enjoyed shooting plover by day and conversing with Fisher and others on summer evenings in Edgartown. This mid-nineteenth-century period, when the Island was widely cleared for agriculture and heavily cut for firewood, witnessed the peak of open habitat and species like golden plover that flourished in it.*

By the late 1870s and 1880s, ornithologists began to note a decline in the numbers of migrating plover and other breeding birds. One that dropped steeply due to habitat change and hunting was the charismatic upland sandpiper. So common in 1850 that it was seen in the hundreds and commercially harvested in the Northeast, the upland sandpiper was locally abundant on large Vineyard fields. By the 1930s this bird, which is extreme in its requirement for large (one hundred acres or more) expanses of grassy uplands, was on its way to becoming one of the rarest grassland birds in the Northeast. In Massachusetts it was reduced to a few prime locations—the broad outwash plain at Agawam, expansive fields in Hadley and Newburyport, grassy moors on Nantucket, and farmed portions of the Vineyard's Great Plain. By the 1950s it was restricted to the Island's largest grassland—the plain surrounding Katama Farm. Now, though individuals are regularly spotted around Katama, there are no breeding pairs. Statewide, by the 1990s only twenty-five to thirty-five breeding pairs were nesting at seven sites and the species was classified as endangered.

The upland sandpiper is an extreme example of a sizable group of birds that frequent grassy expanses. Others include readily identified species such as the whippoorwill, meadowlark, and bobolink, as well as those recognized by more discerning birders such as the savannah, vesper, and grasshopper sparrow. All were once described

* The golden plover should not be confused with the piping plover that inhabits the shoreline and remains a major focus of conservation today.

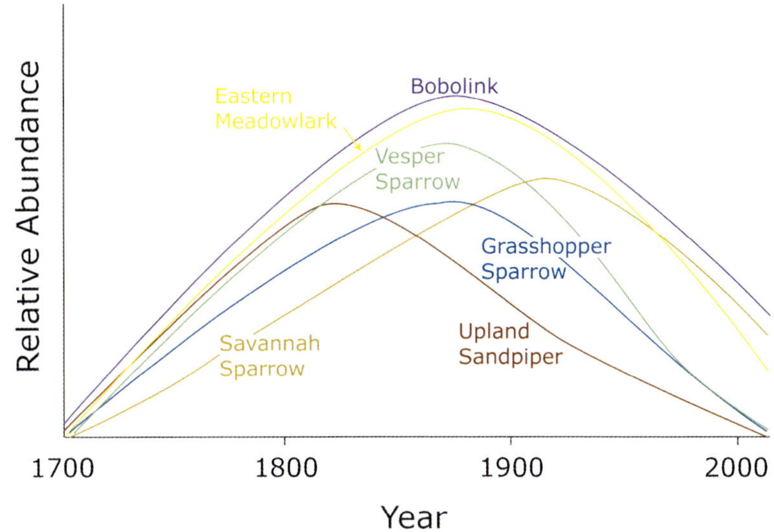

Grassland birds through time. Deforestation and agriculture favored many species that prefer open habitats. Individual species peaked in abundance at different times according to their preference for specific habitats. Conceptual graph based on data from Massachusetts Division of Fisheries and Wildlife, Jones and Vickery (1997), and Harvard Forest Archive.

as abundant to thriving in agrarian landscapes. The grasshopper sparrow was "spectacularly abundant" on the Cape, Nantucket, and Vineyard in the 1870s, where it bred "in dry fields and sandplains." The vesper sparrow was noted by Forbush (1907) as "next to the song sparrow, the most abundant ground sparrow in Massachusetts . . . wherever there are open fields and upland pastures." Turning to the "most abundant" bobolink, Forbush remarks that its migrating numbers were so great that it was frequently shot as an agricultural pest on southern rice fields.

These species began plummeting in the late nineteenth or early twentieth century. In cases such as the sandpiper and bobwhite, initial declines were attributed to overhunting and even rapacious killing for regional markets. But even as hunting was reduced through state and federal regulations, these and many other species continued a downhill slide. Birds persisted in fewer and fewer locations, and those that did nest had less success. The ultimate cause was habitat loss, as abandoned farmland gradually reverted to shrubs and then forest, and fragmentation as properties were subdivided and developed. On the Vineyard, the loss of farms across the moraine focused these birds toward the south shore and sites like Quansoo, Long Point, and especially Katama. Even here development and increased woody cover reduced and fragmented grasslands from the 1960s on.

With remaining farmland deteriorating in habitat quality due to intensive hay, corn, and crop production, birds shifted to a uniquely modern and vulnerable cultural landscape. The best habitat for the constellation of grassland specialists is found today at airports, military air bases, and large landfills. Statewide, the premier site is Westover Air Base in the Connecticut Valley, where 1,600 acres of grass is kept at low to medium height to maintain the habitat for many birds, including more than 150 upland sandpipers, 75 percent of the state's population. In these settings, management details are critical. In the 1980s, Westover shifted to later summer mowing and saw the population of grasshopper sparrows increase from 50 to 168 nesting pairs.

Similar facilities dominate the list of the top fifteen sites for grassland birds: Logan Airport, Otis Air Base, Worcester Airport, Hanscom Air Base, Martha's Vineyard Airport, and Katama Air Field. Only on the Elizabeth Islands, particularly Nashaweena, which is heavily grazed, do agricultural and seminatural landscapes appear on this list. With limited populations centered on a handful of sites, openland birds remain vulnerable. Decisions to close defense facilities or cover large landfills and agricultural fields with solar arrays, as has occurred recently in many Vineyard towns, reduce the remaining habitat.

These grassland species share one final characteristic: in most cases we know nothing of their early history or abundance. The first comprehensive description of the upland sandpiper in Massachusetts came in the 1850s when Henry Thoreau recorded it as commonly breeding on the open plains of Eastham, Wellfleet, and Truro. The history of other species is similarly sketchy. Either due to low abundance, limited ornithological knowledge, or their minor value to the colonists, most are not described until the mid-nineteenth-century fluorescence of natural history. Most are interpreted as rare to absent in prehistory. For example, the upland sandpiper is considered a prairie species that spread eastward from western and Midwestern grasslands as farmland increased.

Each species differs in habitat requirements, distributions, and trajectories of change. Whereas upland sandpipers require immense fields of mixed short and tall grasses, vesper sparrows prefer sparse short grass, and bobolinks and meadowlarks are fond of mature fields of five to ten acres with a mixture of grasses and herbs.

Their future is highly uncertain. Major questions revolve around how much effort should be expended to support species of uncertain origins that persist on highly artificial landscapes. If there is a hope, it is captured by the words of Bob Askins, noted ecologist at Connecticut College: "Although appropriate management of airfields and protected natural areas is important for the future of grassland birds in the Northeast, farmland remains the greatest potential source of habitat for many of these species." We shall examine this option on the Vineyard in detail as we turn to the Island's conservation future.

Survival of the Salters

These days much of the focus on coastal streams centers on the charismatic migrating fish that historically returned in the millions from the open ocean to turn freshwater into a frenzy of activity. Herring, shad, and eel amazed visitors with their spectacular abundance and supplied coastal communities with income and a seasonally reliable source of food. Through the twentieth century, many of these species declined as a consequence of impacts suffered in both fresh and salt water. Overfishing, pollution, damming, and impeded access to spawning grounds were the primary culprits. Despite solid efforts to reverse these conditions, the future of many of these fish remains an open question.

Largely lost in the concern over commercially abundant species is a lesser-known fish with an unusually adaptable lifestyle that has allowed it to persist literally in the shadows, an unsuspected survivor. The coastal brook trout, or salter, as it is known when it chooses to venture out into the open ocean and back, has somehow survived in small populations, having weathered four centuries of abuse to land, waters, estuaries, and sea. Recent studies show that, although present in as few as seventeen of seventy-five coastal streams in Massachusetts, salters may persist in a dozen tiny brooks across the Vineyard. From these tenuous outposts, it is poised to rebound and claim a greater role in waters that over recent history have been stocked with non-native fish, ironically, to improve local fishing.

Coastal brook trout are a local race of the native species that once dominated northeastern streams and are themselves the passion of fly fishermen all across the north. While native to freshwater, the species displays the unusual flexibility to adapt to the salty water of estuaries and even the open ocean, where it may spend two months or more feeding before returning to reproduce in its native waters. Access to the salt environment conveys major advantages. Whereas Island streams are nutrient poor and low in productivity, coastal waters are teeming with food. Once established in these salty environments, the salters grow rapidly in size and change from their dark cryptic appearance of the narrow streams to shimmering silver with pearl-white belly.

Early on, the distinctive species earned a wide reputation on the Cape and islands, with fish reportedly reaching two to three pounds in size. The Vineyard stood out in some circles, as indicated by this lively description of Mill Brook by Dr. Jerome Smith in 1833: "In No place, however, do we remember to have seen [salters] in such abundance as in Dukes County, upon Martha's Vineyard. . . . It was

Like many small streams in the up-Island landscape, the upper reaches of Tiasquam River continue to support native brook trout.

here in the month of November last, and of course in their spawning time; while returning home from a ramble among the heaths and hills of Chilmark and Tisbury, that crossing the principal brook of the island, our attention was attracted towards the agitated state of the water, and never do we recollect so fully to have realized the expression of its being 'alive with fish.'"

But human impacts on fresh and coastal waters were profound, and by the early 1900s, it was thought that most of the streams had lost their native trout. Forests had been cleared, and streams had warmed and were polluted. Fallen branches and logs that once structured the flowing water were gone, erosion muddied beds where the fish once spawned, and dams restricted access throughout the stream system and to the sea. One of the greatest threats to the survival and integrity of remaining native populations was the introduction—by landowners, anglers, and state agencies—of many non-native fish along with new strains of native ones. In an effort to expand fishing opportunities, the Division of Fisheries and Wildlife stocked a wide array of species, including non-native rainbow trout, brown trout, tiger trout (a brown and brook trout hybrid), and hatchery-raised brook trout. The latter come from western Massachusetts and have been bred for rapid growth under hatchery environments. Stocking has occurred on the Vineyard since the nineteenth century, with consistent efforts by the state commencing in the 1950s. Since the 1960s, West Tisbury's Mill Brook has annually received 150 to 300 fish comprising a mixture of the three trout species. Stocking is directed to Old Mill Pond as well as to Seth's, Duarte, and Upper Lagoon Ponds. While fishery managers, including Steve Hurley, the state's southeast fisheries manager, view the stocking programs as a buffer to protect wild fish from overexploitation, the possibility for interbreeding or competition from stocked individuals poses a threat to the native species.

Given the history of impacts, the results of the first systematic surveys of the Island's brooks in 1988 were surprising. Native brook trout populations persisted in Old Mill, Mill (Chilmark), Blackwater, Fulling Mill, Witch, Paint Mill, and Roaring Brooks and the Tiasquam River. Although genetic studies have not yet been conducted, research on the Cape where stocking rates were much greater (up to 91,000 fish added annually to the Mashpee River) shows no evidence of hybridization with salters. Steve Hurley is optimistic that native populations on the Vineyard remain intact.

A few factors likely facilitated this surprising survival act. Most important is salter biology itself. Able to reproduce in freshwater, they likely sought out the remaining coolest spots in the headwaters of these heavily impacted watersheds. Here they persisted all these years as one of the many small and inconspicuous species that are found in most networks of tiny brooks. The otherwise damaging compartmentalization of the streams by dams also helped to keep the gene pool intact. Hatchery fish were largely stocked in downstream ponds and thereby remained isolated by intervening dams and warm waters from the native populations in the upper reaches. The final saving grace for the natives is the fact that hatchery fish are bred and stocked to be caught. Unloaded in locations and on days that are well advertised to the fishing public (these days by social media), the vast majority are hooked in days. Few persist to reproduce.

Though persistent and providing an optimistic sign for the future of Island streams, the status of native trout remains precarious. The populations are tiny. Resampling has failed to yield fish in some streams. It is also quite possible that each stream supports a strain that is genetically differentiated due to isolation, local adaptation, and small population size. Much care must be taken in their conservation. But, if the habitats are guarded, native fish will likely persist in sites that have become wilder and cleaner in recent decades. And, if the downstream obstructions are removed, and the fish are able to move readily through the full course of their modest streams, thriving salter populations may reappear in Vineyard waters.

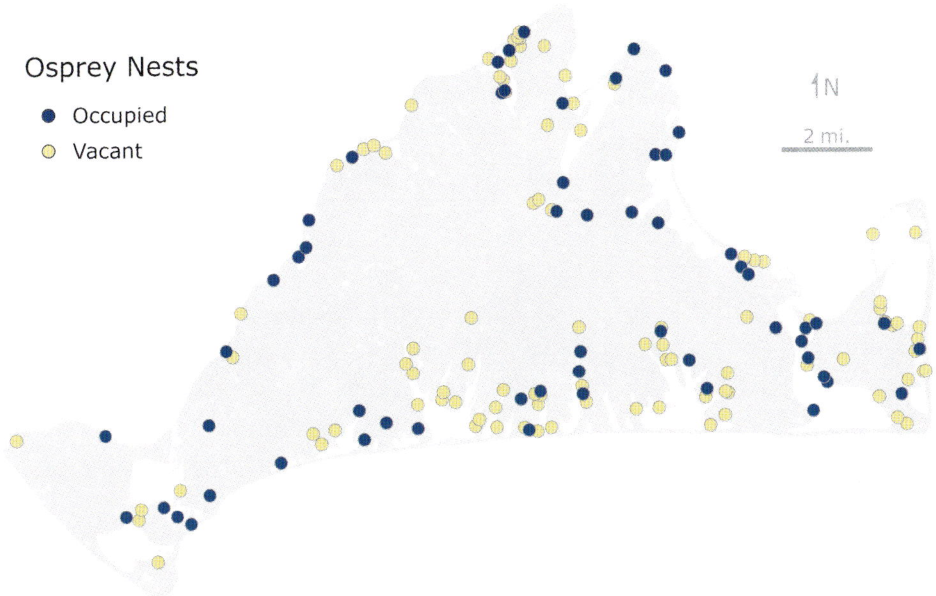

Occupied osprey nests (2013) have a wide distribution but tend to occur adjoining large bodies of fresh and brackish water. Data from OspreyWatch.org.

Fish Hawk

The sight is simultaneously incongruous and a marvelous testimony to the osprey's return to prominence from near extirpation. In the center of Goodale's monstrous gravel pit, some fifty feet below the surface of the Great Plain and with a bleak view of little more than the layers of sand and gravel that wall in the seventy-five-acre hole, sits a lone telephone pole sporting a huge nest. Inside are two fledglings. Amid the industrial commotion and riotous noise of excavators, conveyors, and a parade of dump trucks, a family of ospreys is thriving. On weekends they accommodate the pits' conversion into the most active rifle and gun range on the Island.

A couple of years back, the Goodales noticed two pairs of birds fighting to nest atop one of their cranes. Perplexed but hopeful, they erected a telephone pole topped with a nine-foot-square platform. In short order, and like others across the Island, one pair took to the artificial site. These days, when they aren't caring for their young, they spend their time flying up and out of the cavernous hole seemingly oblivious to the commotion as they head the half mile to the closest water at Lagoon Pond.

Unlike many hawks that were persecuted for their perceived depredations on poultry, livestock, and game animals, ospreys feed almost solely on fish, so were largely tolerated by colonial farmers and hunters. According to Thomas Nuttall, the birds' arrival each spring toward the end of March was greeted enthusiastically by local fishermen as it coincided with the appearance of shoals of shad and herring. Characterizing them as gregarious, he recounted that Gardiners Island and the tip of Long Island supported the birds in extraordinary numbers—up to three hundred nests in the early 1800s. In the early twentieth century, it was common for twenty nests to be seen within any half-mile stretch on the east coast.

Though accurate numbers are elusive, most reports suggest that the birds did well on the Vineyard. Ten pairs or so were variously described through the first half of the twentieth century, with birds

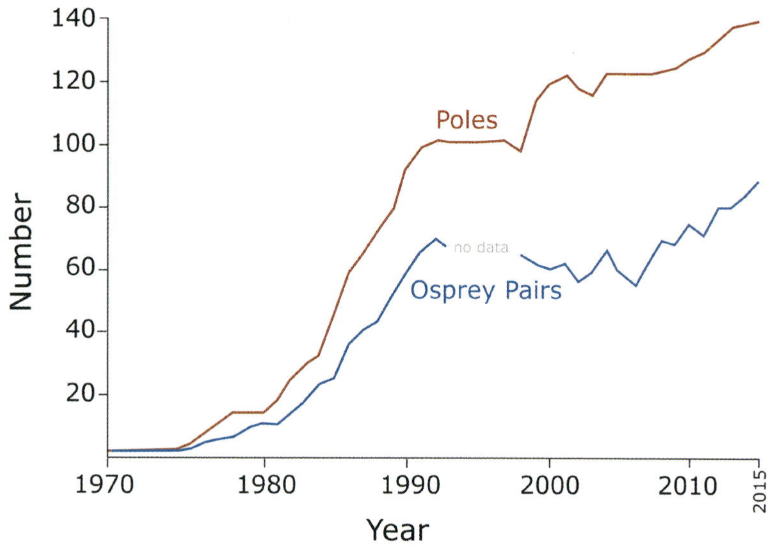

Recovery of the osprey population on the Vineyard. Beginning in 1970 with pairs located at Deep Bottom and Mink Meadow, osprey numbers rose dramatically through the early 1990s and have held fairly steady for the past two decades. The data were collected by Gus Ben David and Rob Beauregard.

largely nesting along the shore in the tops of old pitch pines. Then in the 1950s through the 1970s, the cascading effects of DDT caused a collapse in the populations of most predatory birds, especially osprey. Sprayed to control mosquitoes in salt marshes and shallow water, the poison was taken up by insects, retained in the fatty tissues of fish that consumed these, and then concentrated up the food chain. In raptors, the consequences were catastrophic—a thinning of eggshells led to nest failure. Standing atop a trophic pyramid, ospreys were decimated. From Boston to New York, active nests dropped more than 90 percent to less than one hundred. In 1970, merely two nests were known on the Vineyard—one on a pole in Deep Bottom and a second in a dead pine at Mink Meadows.

The recovery of the population is attributed globally to the reduction of DDT and locally to the activism of individuals like Gus Ben David on the Island. In the early 1970s, as DDT regulations began having their desired effect, a few nests began producing young. Yet a problem emerged. Despite the fledging and return of more birds, few new nests appeared and the population did not expand. Chance conversations revealed that electric crews had been struggling with birds and impeding their attempts to erect nests on poles and transformers. In an effort to attract the birds from these contested sites, the first artificial platforms were erected atop adjacent telephone poles. The simple solution worked. Subsequent platforms were installed speculatively on spots with access to productive waters and distance from competing nests.

The spectacular results stand as one of the great achievements in environmental regulation and species recovery. Over the next two decades, the Vineyard osprey population doubled about every five years. By the early 1990s, there were over seventy pairs, a number that according to data compiled by Rob Bierregaard and others held steady for twenty years. At this point, nearly 130 nest poles have been established, with fifty or more breeding pairs producing a maximum of about 120 chicks annually. The numbers fluctuate. Factors like a cold wet spring or reduced food in the nesting season lead to fewer young. There is also considerable variation in site success. While more than 130 sites have been used recently, only about 100 have yielded chicks. Some thirty-five nests have produced chicks faithfully each year for a decade. Of these, just a dozen are responsible for more than half of the young. Some sites and some pairs are better performers than others.

An explanation for the lengthy population plateau and persistence of empty platforms remains elusive. The health and fledging success of young birds suggests that food was not limiting. Territoriality appears to control the geographical spacing among some sites, and great horned owls, northern goshawks, and predators kill chicks. But a lack of truly long-term data precludes conclusive determination of whether the population reached a saturation limit or how it compares to numbers centuries ago. However, the advent of satellite GPS technology has enabled

Osprey on a nesting pole in the bottom of Goodale's gravel pit, Oak Bluffs, July 2014.

Bierregaard and others to track bird movements in great detail over their entire range. The eye-opening data have filled many gaps in the understanding of osprey mortality and behavior.

With transponders strapped to their backs, osprey send hourly (or more frequent) signals for up to a few years' duration. The distances traveled and rates of movement are stunning. Females head south in mid-August and travel to separate winter quarters on lakes and wetlands stretching from Colombia to Argentina. Males and juveniles leave about three weeks later and overwinter in Venezuela or elsewhere across northern South America. Each bird follows somewhat different routes to the wintering ground each year, and juveniles remain a full eighteen months before returning home.

In their travels individual birds may fly twenty to thirty miles per hour and average 130 miles per day, although they have been recorded as flying straight for up to fifty hours and one thousand miles. Given that osprey can live around twenty-five years, an individual bird can log over 150,000 miles simply in migrational flights. In our consideration of population sizes on the Vineyard, these transcontinental data fill important holes. The long flights, stopovers, and winter habitat expose the birds to multiple dangers. One of the most significant is being shot by fish farmers along the Caribbean and coast of South America. While food may limit reproduction up north, mortality in transit may be a more important factor controlling population growth.

Given the osprey's propensity to move its nest site occasionally, it remains unclear whether the gravel pit will become a permanent extension of the bird's historical habitat. Nonetheless, the successful fledging and departure of two chicks from that nest represents yet another unusual chapter in the ongoing saga of the recovery of this unique bird.

■

Skunks and Cats

Attempts to build populations of the federally endangered piping plover have highlighted the heavy toll that predators exert on the Island's ground-nesting birds. Along with crows, two of the most notable culprits are skunks and cats. Though once extirpated, skunks now thrive on the Vineyard. The major controls on their populations are automobiles, outbreaks of distemper, predators like red-tail hawks and great horned owls, informal removal, and animal control officers, who remove up to 750 animals annually. Skunks naturally prey on a diverse array of insects, eggs and young birds, amphibians, mammals, reptiles, and garbage. The most effective means of controlling the population is by reducing refuse around buildings and beaches.

Cats, including both feral animals and roaming domestic pets, have been repeatedly identified as a top threat to wildlife in the United States. For birds alone, the annual depredation is estimated at 1.4 to 3.7 billion annually. While few people subscribe

Dozens of vehicles congregate at the eastern tip of Norton Point, across from Wasque, August 2009. Although a long-standing Island tradition and significant revenue source for conservation organizations, the presence of vehicles on the barrier beaches is controversial due to the environmental impacts, including damage to organisms and their habitat and the distribution of food waste that attracts skunks and other scavengers that then predate on bird nests.

to Hornaday's suggestion of an open hunting season on unleashed cats, attitudes toward wild-roaming felines vary widely. The simple solution of restraining domestic cats indoors remains an inexplicably challenging step for many owners to take. In places where cats roam freely and especially in locations supporting uncommon species such as plovers, many conservationists argue for trapping and removal. On the Vineyard, Luanne Johnson and Liz Baldwin at Biodiversity Works, an Island-centered ecological nonprofit that studies and helps sustain a broad range of species, provide one of the few voices articulating the threat of cats, dogs, and other extensions of human activity on native diversity. The issue is growing and has yet to come to a head or resolution.

Coyotes Around the Corner

So, in this trajectory of change and shifting species, where does the future lie for wildlife and how will we cope with it? Though the possibilities are myriad, one likely challenge has begun to wash up on Island shores: coyotes. Their closest source—Woods Hole or Naushon Island—lies three and a half miles from the Vineyard, which doesn't seem insurmountable for a species that has covered much of North America in the past century and adjusted to urban as well as wilderness frontiers. A few have washed up dead, at least one has roamed the Island in recent years, and yet no breeding population has established. The Vineyard has had time to contemplate what coyotes might bring and how we might respond.

Naushon Island provides some clues as to the consequences, for coyotes transformed this 4,700-acre island ecologically and culturally. Before coyotes' arrival, sheep were casually distributed around the Island and deer were numerous and scrawny. The appearance of a pair of coyotes in 1986 was marked by the death of many sheep, leading to their relocation to the inhabited eastern end of the island. The abrupt removal of the sheep from an island already greatly undergrazed from a historical perspective initiated a further explosion of woody growth—catbrier, huckleberry, bittersweet, many invasive plants, and other shrubs and trees. In the battle to recover the open fields, the grazing animal of choice is now cattle, which can handle the coyotes.

The ecological transformation of the deer population was equally profound. As documented by Gon Leon, a longtime island resident, the population in the early 1980s was about five hundred (seventy per square mile), with the weight of hunted animals averaging eighty to ninety pounds for bucks. As the coyote population grew, the animals turned to pack hunting of the deer, and eventually reduced the numbers to about forty by 1999. Since then the deer population has increased to about one hundred more-robust animals. The average buck now weighs in at about 140 pounds, and the deer exhibit a newfound wariness.

The range of human-animal conflicts is evident in the chemical products available in Island stores.

Coyotes can exert a transformative influence on other species, the land, and people; consequently, their potential arrival to the Vineyard has generated considerable debate. The benefits of a top carnivore could be great, including a reduction and increased health in the deer herd and a decline in rodents and other small mammals. Studies suggest that this would be accompanied by a reduction in tick populations and Lyme disease. The populations of geese would also decline, accompanied by reduced nitrogen loading to coastal ponds. Cat owners would likely restrain their free-roaming pets or lose them.

Yet, the downsides are also great. Farmers would be severely challenged in grazing sheep freely, and enhanced vigilance would be needed to manage poultry. Ground-nesting native birds would likely suffer, as would many on beaches, as coyotes scavenge shorelines regularly. Thus, while a sentiment of protecting wildlife prevails in many circles, a vocal group of farmers, naturalists, and homeowners would likely seek to eliminate any arriving coyotes. The fear is that once established they would be as impossible to extirpate as they have proven to be on Naushon.

Many of the shifts in wildlife witnessed through history have been gradual and exerted only modest repercussions broadly. In contrast, coyotes would incite great human response and, if successful in forming a resident population, cascading layers of ecological change. It is the potential for planned management and serendipitous changes that we will need to consider as we look forward to the future of the Island.

Planning the Island

By the time strong county-wide or regional government with sufficient powers to deal effectively with the developer is created, Martha probably will have been raped. This was certainly the case in Long Island, Staten Island, along the New Jersey coast, and Cape Cod, where attempts to protect the environment and improve the year-round economy were found to be largely ineffective on a town by town basis.
— Metcalf and Eddy Comprehensive Plan, December 1971

Preserving the Island's character is vitally important for the visitor-based economy. The natural environment, the scenic beauty and the simple Island way of life are the engines of our economy and must be carefully protected.

Virtually all interviewees felt that the MVC was the single most important reason that the Vineyard has managed development so well in the past generation.
— Mark London, *Looking at the Commission*, 2003

In late 2009, the Martha's Vineyard Commission (MVC) released a comprehensive and visionary document—the *Island Plan*—that provided a portrait of the Island and blueprint for a sustainable future for its one hundred square miles. Embracing eleven overarching goals and 207 specific strategies, this "living plan" proposes to focus residential and commercial development, preserve traditional settlement patterns, safeguard critical natural areas, create a more balanced economy, increase the local production of food and energy, develop affordable housing, and prepare the Island for climate change—an ambitious plan by any measure. Implicit throughout the document is recognition that it's not solely the job of the Commission to implement these strategies. That will depend on Island-wide collaboration among municipal governments, private institutions, and regional agencies that are empowered to operate across all six towns. These key institutions include the Commission itself, with its strong regulatory, research, and planning role; the Martha's Vineyard Land Bank, which can purchase and manage land and easements to conserve ecosystems, farms, and scenic landscapes and provide access to trails and shorelines; and the Vineyard Transit Authority, which moves more than one million passengers efficiently across the Island.

Given New England's age-old love of local control, how did this mode of Island-wide thinking and action come to be? New England is unique nationally in its partitioning of land into individual townships, each with local autonomy known as "independent home rule" that outstrips the authority of intentionally weak county governments. For more than three centuries, the Vineyard has presented an extreme example of home rule. Each town is geographically distinct; collectively, they are more physically and culturally varied than any other group of neighboring towns in the entire Northeast. So the question looms: how did six towns come together to forge the will, institutions, and resource base for collective oversight, to conserve 40 percent of their land area, and to assemble a bold path toward a sustainable future?

One simple answer—that such thinking comes naturally to an island—collapses under even casual scrutiny. Though the desire to maintain Island character is a strong sentiment across the Vineyard, these are fiercely distinctive communities filled with residents loyal to their piece of the Island. It took the threat of rape and environmental cancer, an initial and unexpectedly giant step toward planning, and the looming potential of federal control of the Island in the 1970s to galvanize the six towns to think and act with many (but far from all) qualities of a single entity. The road traveled has been uneven and remains rocky. But, just like the landscape that we have been exploring, regional planning and action on Martha's Vineyard now has a lengthy history and much to show for itself. Like the woods, fields, and harbors, it has its own legacies—of accomplishments—and momentum and institutions that carry it and its new plan forward today.

In the late 1950s, with a growing postwar population newly focused on summer leisure, the emerging interstate highway system fueled a conflagration of tourism and second-home development along the shores of New Jersey and Long Island, up through Connecticut and Rhode Island, and out onto Cape Cod. The Vineyard did not escape, and the pace, quality, and scale of building began to shift. Outside developers began buying large tracts of land from farmers and old families. Longtime summer residents got into the act, too, as, for example, brothers and developers Alvin and Moses Strock, who accumulated a couple thousand acres. Summer days in the down-Island towns of Vineyard Haven, Oak Bluffs, and Edgartown became increasingly disrupted by throngs of visitors, cars, and diesel-spewing tour buses. New summer homes sprouted in woods, fields, and beachfront settings, and in the mid-1960s a fifty-eight-house complex—the Strock brothers' Waterview development on Sengekontacket Pond—jolted the Island with its aggressive transformation of a perennially quiet coastal scene.

Growth brought throngs of day-trippers, packed beaches,

Line in Oak Bluffs for the Rhode Island passenger ferry.

traffic jams, and environmental damage to formerly quiet towns. Dunes began to exhibit signs of trampling, trash proliferated along roadsides and beaches, and a major widening of the Edgartown to Tisbury road shocked residents with its jarring highway breadth, wide shoulders, and invitation to speed. The late 1960s brought a proposal for jet service and major expansion of the airport—all of this on a sleepy and largely rural scene with its many dirt roads, cozy houses, farmscapes, and history-laden villages.

A new reality confronted the Island, that this delightfully varied landscape could be subsumed in the homogenizing coastal blight that residents and visitors thankfully left behind as they boarded the ferry. A survey took the pulse of the Island and confirmed that 90 percent of residents felt that the Vineyard was in jeopardy. The fear and concern spread from dinner table conversations to open distress expressed by locals and notables alike. In a telegram to the

Massachusetts House and Means Committee in May 1969, James Cagney wrote, "For more than 30 years I have watched Martha's Vineyard go downhill as a place of natural wonder and haven. Now they are talking of runways for jets. Is there no end of destruction of all that is natural and worthwhile?"

Taking stock, many recognized the cause of the Island's vulnerability. The place, with its ages-old administrative institutions, was completely unequipped to handle the growing desire of people to share in its beauty and leisurely pace of life. Oversight of new construction and commercial growth was casual and outmoded. Local regulations varied and were applied with the idiosyncratic flavor of the personalities of the individual town authorities. Coordination was lacking across town boundaries, most glaringly on issues that affected everyone, like transportation and the environment.

Town representatives did meet collectively in an All-Island Selectmen's Association, there was a Dukes County Commission (DCC), and by 1966, the Dukes County Planning and Economic Development Commission (DCPEDC) was established with a small staff that played an advisory function to the county and towns. But the volunteers who served on town boards were outgunned in knowledge and legal firepower by the shrewd investors and outside consultants lured by the availability of developable land. Towns lacked the regulatory structures to cope with new challenges.

A few voices called for professional help and collective action. In 1965, the DCC contracted with the Boston engineering firm Metcalf and Eddy to create a master plan to "preserve and enhance the character of the island." But funding proved inadequate and the effort stagnated. Then, in 1969, spurred by interim chair Anne Hale's charge "It is time for us to do something about planning and stop living by crisis," the DCPEDC secured $40,000 in federal and county support that allowed the firm to get solidly to work. Over the next two years Metcalf and Eddy scoured the Island's landscape and records.* Along the way, they produced six reports covering land use and open space; population; the economy; housing; major assets and problems; and land-use planning and oversight.

Sunset picnickers at Menemsha Beach, Chilmark.

The engineering firm's work was unexpectedly detailed and comprehensive. And, as its staff became passionately engaged in the subject of their study, the writing and assessment of the growing problems became increasingly and then shockingly frank. While extolling the Island's unique qualities, the reports employed stark terms, egregious examples, and raw imagery to depict the vulnerabilities of the regulatory system, the looming environmental crises, and the clear threat of the deterioration of the Vineyard's natural and cultural characteristics. Major concerns included:

- Indiscriminate building and destruction of woods, heathlands, and farmland

* Though county-wide in scope, Metcalf and Eddy's study focused on the large, diverse, and rapidly growing Vineyard.

Planning Timeline on Martha's Vineyard

1930s U.S. Fish & Wildlife National Wildlife Refuge proposed for Vineyard's South Beach.

1956 National Park Service recommends shoreline acquisition for Cape Cod, Martha's Vineyard, and Nantucket.

1961 Cape Cod National Seashore legislation signed by President Kennedy; federal land acquisition totals $42.9 million by 1983.

1966 Senator Ted Kennedy endorses the federal acquisition of the Elizabeth Islands.

1967 Strock Enterprises/Island Properties builds fifty-eight houses on sixty acres at Waterview Farm on Sengekontacket Pond.

1969 Poll: 90 percent of Islanders fear over-development. Metcalf & Eddy (M&E) study launched.

1970 Interior Department *Islands of America* promotes the Island Trust (federal-state-local) approach to land conservation.

1971 Metcalf and Eddy *Draft Comprehensive Plan Summary* delivered to DCPEDC, which unanimously votes it "not accepted."

1972 Senator Kennedy introduces Island Trust legislation to a divided audience.

1974 Strock Enterprises Waterview II–IV (850 lots, 507 acres) approved by Oak Bluffs.
Governor Sargent signs legislation to establish Martha's Vineyard (Land & Water) Commission.
Vineyard Gazette publishes the suppressed strategy section of the M&E Plan.

1976 MVC designates Waterview and Sengekontacket Pond area as its third District of Critical Planning Concern.
Interior Department rejects the Island Trust Bill, which dies in the House.

1977 Massachusetts Supreme Judicial Court upholds MVC's power to regulate the Waterview plan.

1978 Edgartown withdraws from MVC (675–554); Oak Bluffs votes to remain (426–356).

1979 McDonald's defeated from establishing on Island.

1980 Tisbury withdraws from MVC; Edgartown suffers intense development.

1984 Edgartown and Tisbury rejoin MVC.

1986 Legislature establishes MV Land Bank Commission; funded with 2 percent real estate tax.

1988 Harris poll: residents and summer people agree that development is overrunning the Island.

1990 MVC & MVLB first joint project: Island Open Space plan, released in 1991.

1997 Martha's Vineyard Conservation Partnership formed of conservation groups and MVLB.

1999 Henderson-Smith-The Nature Conservancy study "Choices for the Future" predicts housing buildout by 2005.

2005 Housing Bank legislation drafted to assist affordable housing.

2008 MVC proposes Island-wide waste management plan.

2009 Island Plan released after five years of effort.

2013 Harris poll: 80 percent support MVC preserving Island character, natural resources, and open space.

2014 MVC releases study predicting buildout by 2089; 40 percent of Island land is conserved.
Island-wide Districts of Critical Planning Concern and regulations established to reduce fertilizer use and water pollution.

DCPEDC—Dukes County Planning & Economic Development Commission
M&E—Metcalf and Eddy
MVLB—MV Land Bank
MVC—MV Commission

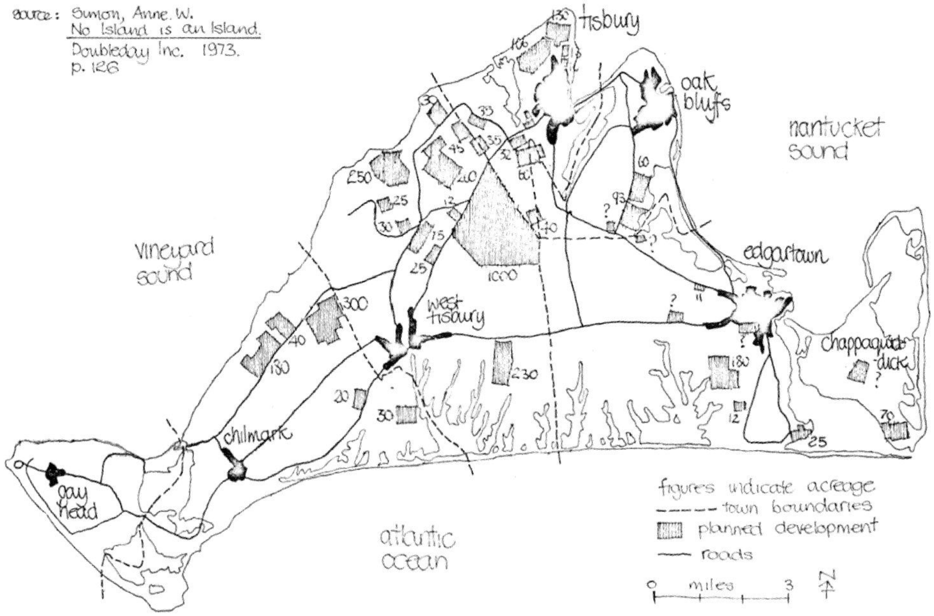

The development challenges facing the Island that spurred the Metcalf and Eddy study are captured in this map from Anne Simon's 1973 book *No Island Is an Island*. Darkened polygons represent contemporary development threats totaling more than two thousand acres. Reproduced from Anne Simon, *No Island Is an Island*. Garden City, N.Y.: 1973. Drawn by Raphael Palacios; original research for map done by Susan Whiting.

- Loss of habitat
- Shoreline development, erosion, and coastal flooding
- Breakdown of established land-use patterns and distinct communities
- Open dumps with illegal burning
- Open disposal of raw sewage
- High coliform counts in ponds and coastal waters
- Groundwater contamination
- Burgeoning traffic with air pollution, noise, and gridlock
- A growing dominance of short-term tourists, including day-trippers
- A widening gap between residential income and the cost of living

In juxtaposition to these cascading forces, Metcalf and Eddy noted the meager examples of regional institutions—the hospital, high school, airport, courthouse, and jail—and the ill-equipped state of local governments to cope with existing pressures, let alone anticipated growth. No town was adequately armed with the appropriate portfolio of zoning, planning, conservation, and subdivision regulations; housing, building, and sanitary codes; or professional advice.

In the absence of immediate action and shift to regional planning and regulation, Metcalf and Eddy painted a grim, sobering, and prescient prognosis for the future.

> The County will become more and more tourist-oriented and dependent upon sources of capital from outside . . . and more influenced by the desires of large off-island investors. . . . The amount of gross dollars guaranteed by tourist trade will continue rising, and the percentage retained by year-round residents of the County will continue dropping.
>
> Urban spread or indiscriminate land use will be the order of the day. . . . Roads and parking facilities in the centers of

Edgartown, Vineyard Haven and Oak Bluffs by 1980 will be jammed and by 1990 will be beyond the level of tolerability. . . . Noise will be the rule. . . . Bicyclists and pedestrians will take their lives into their own hands.

The loss of the distinctive villages, rural communities, natural landscape, and quality of life that characterized the Islands was the foremost concern.

[How do you measure] the value of animal, bird, and fish life, heath, grass, the natural habitat, and the music generated by the relentless ebb and flow of the tide, the smell of clean air, and the quiet. . . . It is on these latter values that the forecasts presented . . . will have their greatest impact.

The time for action was unequivocally at hand. And so, the authors issued the proclamation that made their report so toxic in its day and notable in ours.

The next five years are critical. If a definite and well-ordered program of preventive and prescriptive medicine is not undertaken almost immediately or within the next two years, by 1975 the Vineyard undoubtedly will have contracted environmental terminal cancer.

Federal strategies for the Cape and Islands had been floated episodically in the early and mid-twentieth century with little traction and scant local awareness. Back in the 1930s, the U.S. National Wildlife Refuge had proposed acquiring the entire south shore of the Vineyard. In 1961, Congress established the Cape Cod National Seashore, a unit of the National Park Service intended "as an experiment to conserve a fragile and precious resource that overlays six established communities" and extending across the outer Cape to Provincetown. The authorizing legislation for this unusual national park established a citizen's advisory committee and federal funding for land acquisition. First introduced by Senators Leverett Saltonstall and John Kennedy, the final legislation was signed by then President Kennedy. Eventually, $43 million worth of federal land was purchased.

Turf for a new Island lawn traveling the ferry from Woods Hole on an eighteen-wheel truck.

With the Cape example firmly in place, a federal role emerged as one serious remedy to the threats enumerated by Metcalf and Eddy. Such a top-down solution offered special appeal to many minds due to the inadequate nature of local oversight. In an eyebrow-raising *Vineyard Gazette* article in early 1967, Senator Ted Kennedy responded to state senator Edmund Dinis's suggestion that the federal government should utilize eminent domain to acquire the Elizabeth Islands, simply stating, "This proposal is an excellent idea." As Metcalf and Eddy was completing its first reports in 1970, the Department of the Interior released *Islands of America,* an overview of the country's invaluable and vulnerable islands. Packed

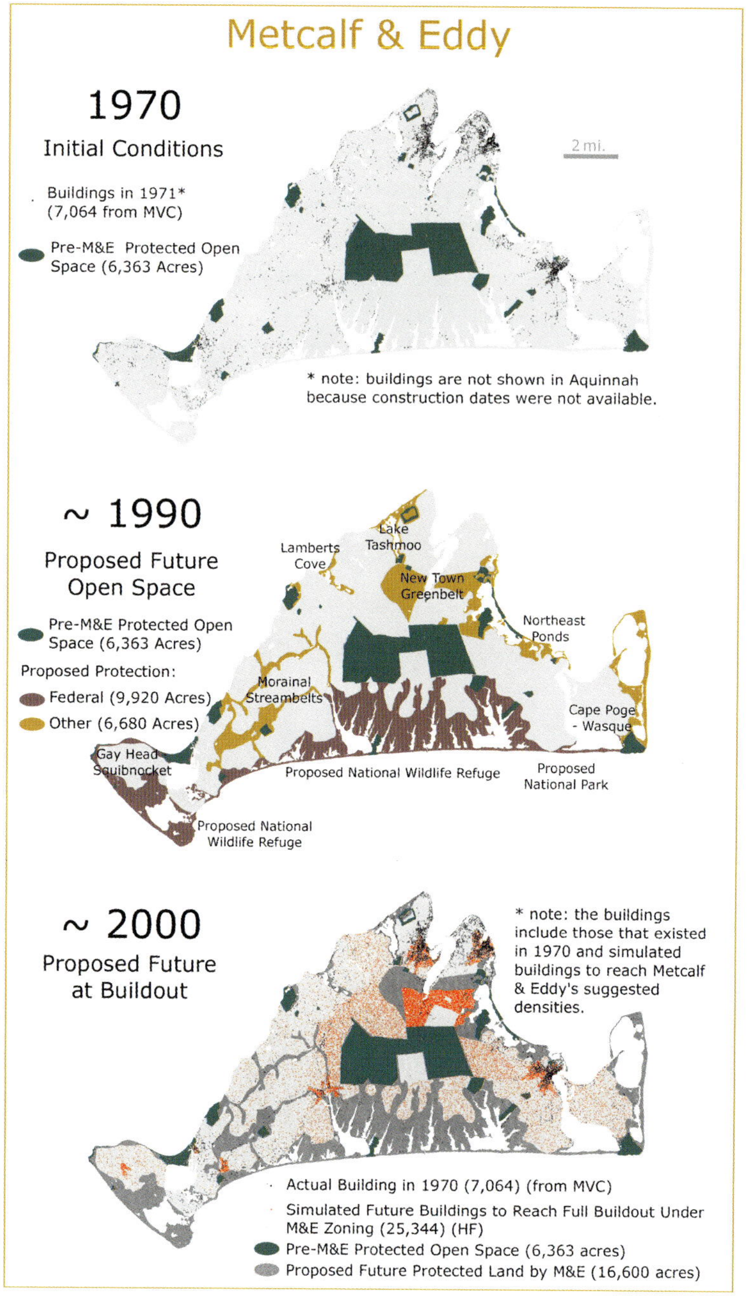
Metcalf & Eddy
1970
Initial Conditions
Buildings in 1971*
(7,064 from MVC)
Pre-M&E Protected Open Space (6,363 Acres)
2 mi.
* note: buildings are not shown in Aquinnah because construction dates were not available.
~ 1990
Proposed Future Open Space
Pre-M&E Protected Open Space (6,363 Acres)
Proposed Protection:
Federal (9,920 Acres)
Other (6,680 Acres)
Lake Tashmoo
Lamberts Cove
New Town Greenbelt
Northeast Ponds
Morainal Streambelts
Cape Poge - Wasque
Gay Head Squibnocket
Proposed National Wildlife Refuge
Proposed National Park
Proposed National Wildlife Refuge
~ 2000
Proposed Future at Buildout
* note: the buildings include those that existed in 1970 and simulated buildings to reach Metcalf & Eddy's suggested densities.
Actual Building in 1970 (7,064) (from MVC)
Simulated Future Buildings to Reach Full Buildout Under M&E Zoning (25,344) (HF)
Pre-M&E Protected Open Space (6,363 acres)
Proposed Future Protected Land by M&E (16,600 acres)

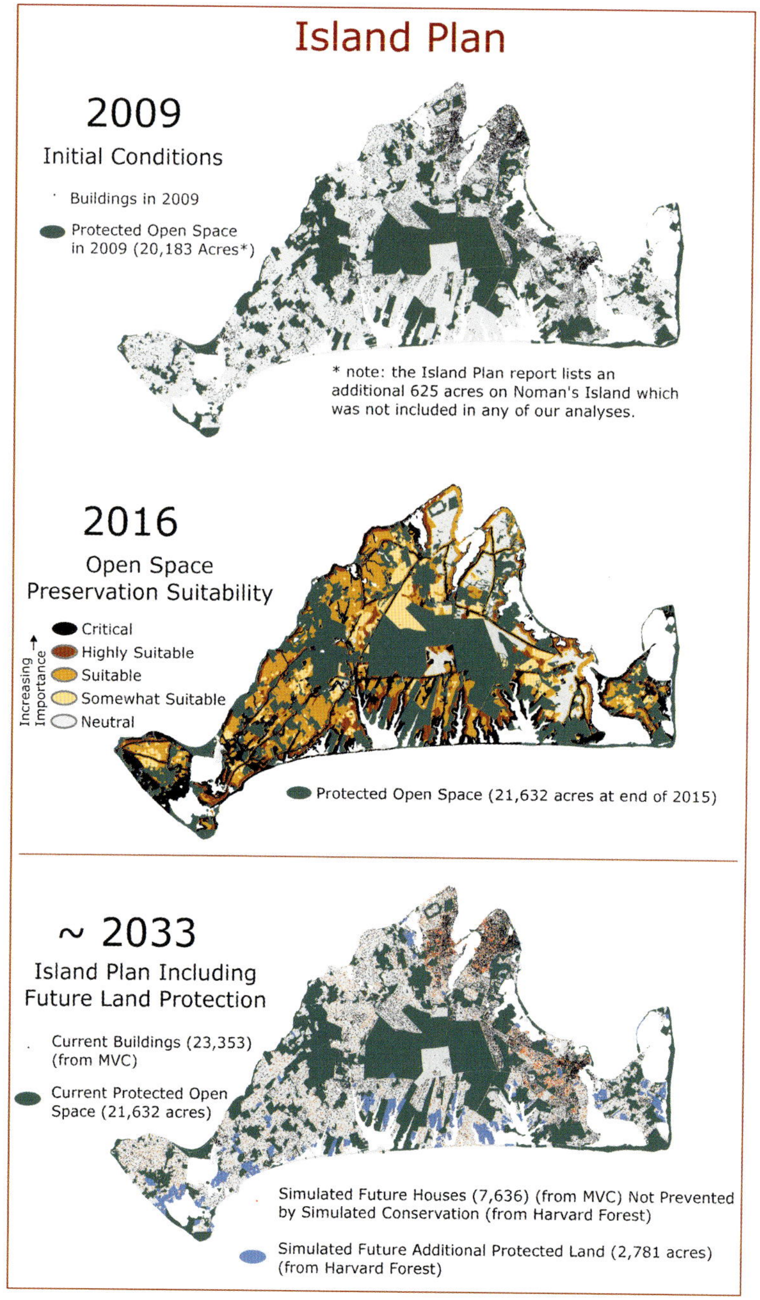
Island Plan
2009
Initial Conditions
Buildings in 2009
Protected Open Space in 2009 (20,183 Acres*)
* note: the Island Plan report lists an additional 625 acres on Noman's Island which was not included in any of our analyses.
2016
Open Space Preservation Suitability
Increasing Importance
Critical
Highly Suitable
Suitable
Somewhat Suitable
Neutral
Protected Open Space (21,632 acres at end of 2015)
~ 2033
Island Plan Including Future Land Protection
Current Buildings (23,353) (from MVC)
Current Protected Open Space (21,632 acres)
Simulated Future Houses (7,636) (from MVC) Not Prevented by Simulated Conservation (from Harvard Forest)
Simulated Future Additional Protected Land (2,781 acres) (from Harvard Forest)

with spectacular photographs, the glossy book included a proposal for a study of the Vineyard, Nantucket, and adjacent islands for possible protection in a new entity, a National Island Trust.

In late 1971, the Harvard University Research Bureau responded to Senator Kennedy with a lengthy memo that recommended strategic federal purchases of beaches on the Vineyard and Nantucket. It read, "At this point in time, market forces are such that much land on the islands is ripe for development. On both Nantucket and Martha's Vineyard, new large-scale tracts are, generally speaking, completely out of character with the patterns of growth established on the islands over the centuries. . . . Only immediate action can guarantee that future generations will find Cape Cod's offshore islands almost as this generation has found them."

Island politics, the steady release of interim findings, and emerging recognition of the potential for top-down solutions fueled growing interest in the anticipated release of the final Metcalf and Eddy summary document. Each of the firm's topical reports had been reprinted by an attentive *Vineyard Gazette* and generated favorable, albeit sober appreciation among readers and town boards. But when the draft Comprehensive Plan Summary was delivered to the DCPEDC in June 1971 with its prognosis for environmental cancer, commission members were dismayed and voted unanimously to "not accept" it due to "basic philosophical differences and disagreements."

When the *Gazette* sought further information regarding this decision, it was denied access to the rejected document. Rumors spread, including word that the report was killed due to its frank portrayal of the situation and the far-reaching proposed solutions. Such talk intensified as months passed and the Commission and authors met behind closed doors. Finally, a compromise was forged. On December 10, 1971, an abridged version of the final summary was released with a letter from Chairman Dean Swift indicating that the DCPEDC had expunged the section titled "Strategy" that addressed the Island's future. In some ways, this was of little importance. The released version was adequately inflammatory given its stark language and clear leaning toward federal versus local remedies. The offending section remained censored for three years, until Henry Beetle Hough published it in a special edition of the *Vineyard Gazette* with the assistance of its original author, who by that time was a vice president at Metcalf and Eddy.

Though struck of its most colorful language and forceful proposals, the 1971 summary contained sweeping recommendations along with prescient projections. Many of its strategies were ultimately adopted or paralleled through actions of the DCPEDC, various towns, and years later, by the Martha's Vineyard Commission and Martha's Vineyard Land Bank.

(*opposite*)
Planning the Island, 1970s and 2000s. Comparison of the condition of the Island, proposed plan, and scenario for the future as presented in the 1973 Metcalf and Eddy study and the 2009 *Island Plan*.

Metcalf and Eddy: Top: When the Metcalf and Eddy (M&E) project began, less than 10 percent of the Island was protected from development, largely in the State Forest, and the year-round population was 6,117. *Middle:* To conserve open space and concentrate development, M&E proposed federal purchase of the south coast, establishment of large greenbelts around a new seventh "town" south of Lagoon Pond, and the permanent protection of watercourses and shorelines by conservation entities. *Bottom:* M&E anticipated a future of densely populated towns surrounded by moderately to heavily populated suburban and exurban areas and sparsely populated rural regions.

Island Plan: Top: When the *Island Plan* (IP) was released in 2009, three times as much conservation land included some extensively connected landscapes. *Middle:* The IP identified priorities for permanent conservation based on ecological, environmental, and recreational characteristics. *Bottom:* Future conditions were mapped in the IP by projecting new housing on all developable properties until "buildout," which was estimated to occur in 2086.

Recognizing the important role of land protection in shaping the future of the Island, we revise the IP future projections by (i) adding conservation land protected since publication of the IP through 2015 or missing from the original IP (a total of 1,317 acres) and (ii) simulating future land protection at a pace and pattern similar to the past twenty years. The resulting buildout occurs around 2033, with approximately 43 percent of the Island in conservation (an increase of ~2,900 acres over today or the IP) and ~24,824 houses (an increase of 7,636). We estimate that the additional conservation land reduced the maximum number of houses projected by the IP by about 1,200.

Water views for the Major's Cove development along Sengekontacket Pond, Oak Bluffs.

The firm's conclusions were grounded in the belief that the scale of the problems confronting the Islands outstripped the ability of the individual towns to cope with them. Consequently, the overriding recommendation was for "a publicly created organization with both county-wide jurisdiction and adequate powers and financing to effectively deal with the problems." This institution would have the expertise, authority, and resources to advance the Future Land Use Plan proposed by Metcalf and Eddy. That proposal contained three far-reaching elements:

1. An Open Space Plan for land acquisition to protect natural resources.
2. A new (seventh) community surrounded by greenbelts that, in conjunction with the State Forest, Sengekontacket and Lagoon Ponds, and Katama Bay, would reinforce the identities of existing villages.
3. An Island-wide housing plan that matched the physical and cultural landscape.

Beyond this sweeping regulatory and land-use vision, the plan called for regional solutions to transportation (on and to the Island), housing, and economic development, all reinforced by parallel capabilities in each town.

Though the uncensored text focuses on regional solutions, the plan's maps reveal a major federal role. Most boldly, the map of future land cover depicts the Island's entire south shore in the hands of the National Park Service and Fish and Wildlife Service in the U.S. Department of the Interior.

While the DCPEDC struggled with the draft summary, Senator Kennedy met with a small group of people on the Island and determined that the time for federal action had come. Warned by the analysis from the Harvard University Research Bureau that the Islands presented a much more complicated setting for federal involvement than had the Cape, Kennedy began favoring the so-called Island Trust concept proposed by the Interior Department. Under this model, the heavy federal hand would be moderated by the involvement of local representatives on a commission making recommendations to the secretary of the Interior concerning federal acquisitions.

Despite federal interest since the mid-1960s, coastal communities professed shock on May 26, 1972, when Senator Kennedy abruptly submitted his Island Trust bill to Congress, complete with its federal-regional-local solution to the Vineyard's emerging crisis. Though presented by the senator and his point man K. Dun Gifford as "a work in progress" and "open to major revision" by the towns with public input, the bill's unexpected appearance and sweeping scope of federal intrusion had a stunning effect. To many, it represented a wholesale usurpation of local control. Most glaringly, it established

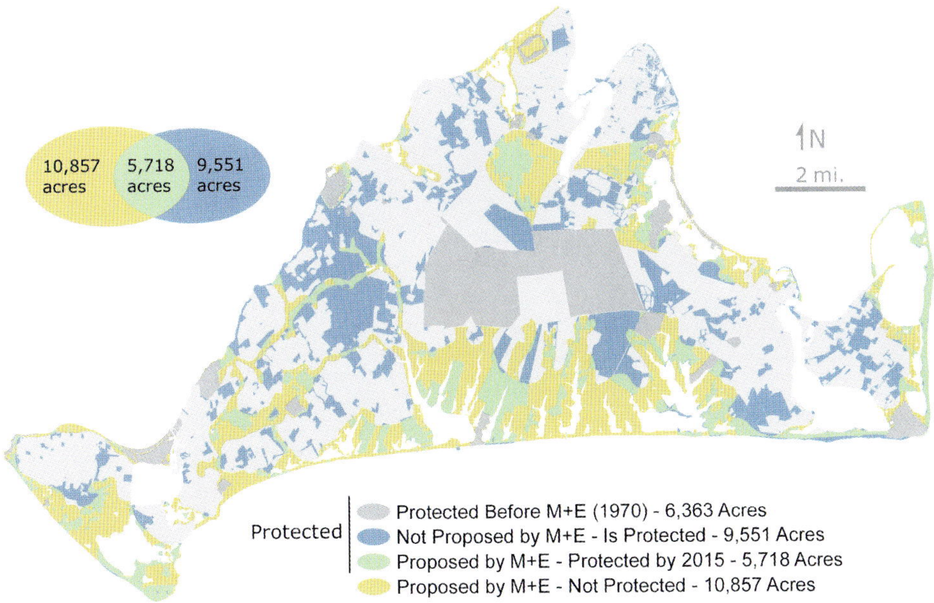

Conservation lands as envisioned by Metcalf and Eddy (M&E) in comparison to 2015 reality. Since the 1970s the pace of land protection has been strikingly similar to that recommended by Metcalf and Eddy in terms of total additional area (15,269 acres protected versus 16,575 acres proposed), but the pattern and method of land protection differ. In contrast to the large federal role M&E envisioned, private organizations, the Land Bank, towns, and the Commonwealth of Massachusetts have advanced land protection. While there is considerable overlap between M&E priorities and subsequent conservation acquisitions (5,718 acres), there is a lack of congruence over nearly twice as much land (9,551 to 10,857 acres). The modern conservation landscape has a much greater inland focus than M&E.

three broad land-use zones: *Forever Wild* public lands in which development was prohibited and existing private ownerships would extinguish within twenty-five years; *Scenic Preservation* lands where owners could maintain but not expand existing houses or initiate new construction; and *Town and County Planned* lands under local jurisdiction and ordinances approved by the Interior secretary. The federal government preferred to acquire land through purchase from willing landowners, but it would invoke eminent domain in situations where that was not possible.

The response on the Island was immediate, passionate, and utterly divided. The *Vineyard Gazette* led off with Henry Beetle Hough's editorial "The Kennedy Bill must be enacted." That same day, the All-Island Selectmen's Association asked Kennedy to withdraw the legislation. A new group, the "Island Action Committee," was organized by Bob Carroll, owner of the Harbor View Hotel, and selectman Shirley Frisch to defeat the bill. Though Carroll was head of the local Democratic Committee and a resolute Kennedy ally in the weeks following the Chappaquiddick incident in 1969, he fought the bill furiously, claiming that it would "ruin the economy . . . turn the Vineyard into a summer playground . . . and remove rights without compensation."

Voices up-Island were split, forging long-running divisions. West Tisbury voted in favor, with shopkeeper James Alley spearheading the Committee to Amend and Support the Bill. In contrast, Chilmark selectman and jack-of-many-trades Herbie Hancock enlivened a subsequent congressional subcommittee meeting by stating that Congress sought to "vote these islands into a miniature socialistic land set-up" in which "the land would become forever worthless to all but raccoons and conservationists." Hancock spoke for many with his strident opposition to federal intrusion. "Home rule," he said, "though not perfect, is far better than puppet rule, which didn't work in occupied France and won't work here." In contrast, the bill

The proliferation of houses across the old up-Island farm landscape is evident in this view from Squibnocket Ridge to the north toward Nashaquitsa Pond, Chilmark.

was strongly supported by the Martha's Vineyard Garden Club, the Vineyard Open Land Foundation, and the Vineyard Conservation Society.

Other divisions emerged. A non-binding referendum following Kennedy's announcement showed that while a majority of Island residents opposed the bill, it was favored overwhelmingly by seasonal homeowners. Meanwhile, the single-town island of Nantucket immediately embraced the legislation.

Though the packed congressional subcommittee hearing held in Edgartown in July 1973 revealed animated disagreement, the bill's biggest weakness came from a lack of unanimity in the state's congressional delegation. Senator Edward Brooke, a seasonal homeowner for three decades in Oak Bluffs, announced his opposition. In the ensuing months, Massachusetts house members split, while Governor Francis Sargent signaled strong interest in forging a state and local solution. Kennedy remained convinced that the federal role was critical but, true to his word, reworked the approach to enhance a local role. Over the next three years the bill was amended and resubmitted five times, until the major forces came together. The final version was jointly submitted by Senators Brooke and Kennedy, with a parallel House version filed by Representative Gary Studds. It received the backing of the governor and a consensus committee of Islanders that included members of Island Action, Amend and Support, the All-Island selectmen, DCPEDC, and even the high school student council. The unifying sentiment was the immediate need "to prevent irreversible damage to the Island and ultimately its economy."

Yet compromise doomed the bill. The last version conceded so much authority to local boards that the Interior Department abandoned its support. Fearing that Island representation on the Island Trust committee might place federal funds and agencies under local direction, the Interior secretary raised questions concerning the bill's constitutional legality. The bill died in a subcommittee hearing when the deputy assistant secretary for the Interior rejected any further involvement with "a problem of local land use that ought to be solved locally." The legislation, he asserted, was designed "to perpetuate the exclusive life-styles of the inhabitants of these areas."

Through the period of bitter fighting, a prediction by James Reston in the *New York Times* came true. The bill itself fueled real estate speculation and development. Consequently, during the period of inaction, the need for land protection and regulatory solutions became even more evident. In 1973, Anne Simon, a noted environmental writer and longtime seasonal homeowner, published *No Island Is an Island,* railing against the growing tide of destruction while extolling the Island's distinctive communities and natural features. The book featured a map highlighting two thousand scattered acres slated for housing developments and lamented that the threshold of one housing start a day had been reached. The book's power and veracity were apparent in the positive reviews by Senator Kennedy in the *New York Times* and by Governor Sargent in

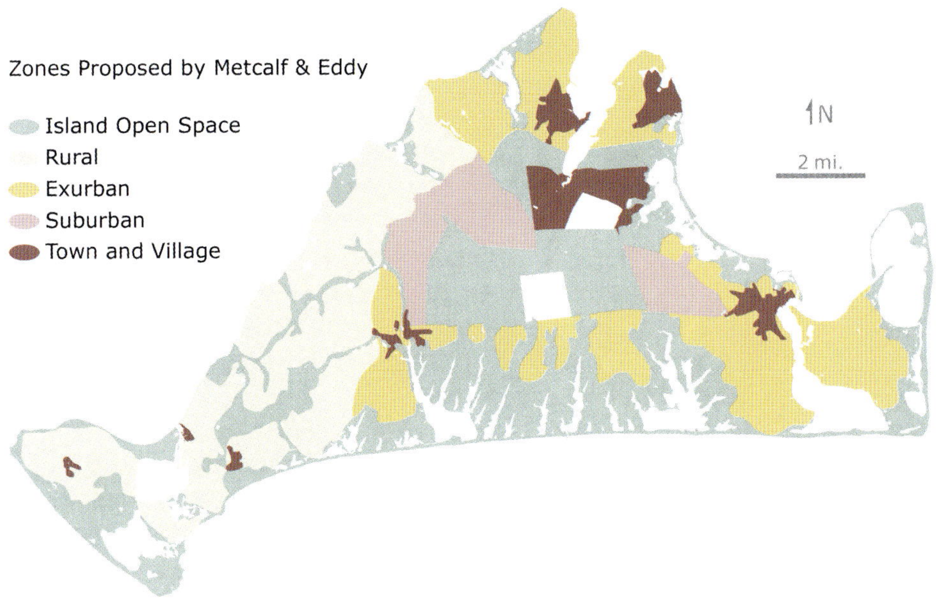

Development as envisioned by Metcalf and Eddy in comparison to modern reality. Metcalf and Eddy envisioned a faster rate of growth than occurred and proposed a pattern that concentrated housing in dense town and village centers while retaining large lightly populated conservation regions. Today, their proposed Island Open Space conservation zones have much higher populations, their rural and exurban areas have developed about as envisioned, and the suburban, residential, and existing and proposed town centers have fewer people.

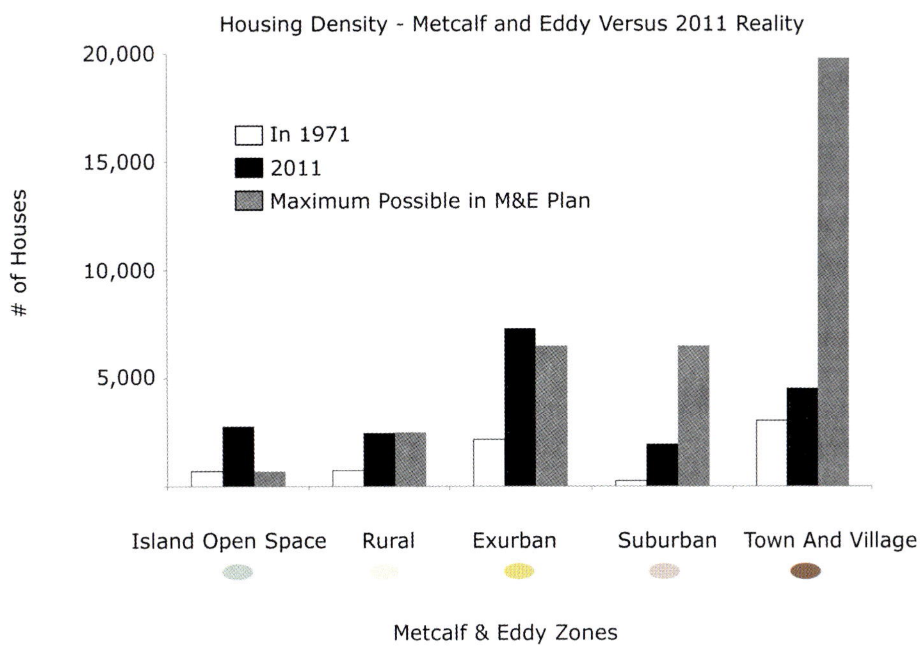

Anne Simon, 1914–1996

With their compelling documentation and powerful argument that survival on land and sea depend on functioning oceans, *The Thin Edge* and *Neptune's Revenge* brought Anne Simon comparisons to Rachel Carson in the *Boston Globe* and the *New York Times*. But the plain language and passionate plea for regional and federal solutions to rampant development and the destruction of nature in her *No Island Is an Island* fanned the conflagration of conservation debate ignited by Metcalf and Eddy and the Kennedy bill. Anne Simon excelled in scholarship, literary success, doggedness, and the ability to trigger divergent passions. *No Island* was praised by Senator Kennedy and Massachusetts Governor Francis Sargent as well as those gathered at Stan Hart's home following a book signing at his Red Cat bookstore, including William Styron, Jules Feiffer, David McCullough, Vance Packard, Henry Hough, Carly Simon, and Anne's sister Barbara Tuchman. Yet, remarks like "[Home rule] is a lot of crap" and "Islanders are too close to the clinking dollars of short-term growth to recognize the rumble of long-term disaster" led her Menemsha gardens to be trampled and her positions to be vilified in local conversations and national press.

With the remove of time from the acrimonious battles of the day, two emergent qualities of Anne's work shine brightly in *No Island*. First, the Island serves as a microcosm for the entire country. What ails the Vineyard is a cancer spreading nationwide. Solutions are desperately needed, and those forged to save her beloved "national treasure" may offer lessons for those grappling with environmental destruction more broadly. Then there is Simon's ability to convey the intriguing diversity and rich qualities of the Island itself. Detractors and some reviewers criticized her for caring more about nature than people. Yet, no writer has exceeded her ability to capture the Vineyard's cultural variation, its blend with the Island geography, and the attraction that both instill in diverse people. While the book shines insightful light on conservation's early days, its depiction of Island qualities and needs resonates today.

the *Boston Globe*. Kennedy used the opportunity to paint the Island's battle as emblematic of larger issues confronting the nation that demanded coordinated federal, state, and local efforts:

> There is a compelling and urgent message in this book, a message which we cannot ignore and from which we cannot escape. It is that unless we squarely face the issue of land use, we cannot expect to raise the quality of life in this country to the high levels we all desire. . . .
>
> Large tract subdivisions, waterside condominiums, sewage spillovers, traffic jams, dune buggies—these symptoms of unplanned and dramatic suburban growth are newcomers to Martha's Vineyard. And they are, right now, irreversibly destroying once and for all the unique and quiet beauty of that island. . . .
>
> Existing state and local laws are not adequate to the task. . . . Legislation at both Federal and State levels can help. But only long, hard, patient, and cooperative work among all interested parties can bring success.

Governor Sargent traveled to the Island and agreed: "There is not one inch of available land on the Island that is not being eyed by developers today." He promised to help pass new state legislation and succeeded on his second attempt. Though frequently cast as a weak compromise to Kennedy's federal bill, the state legislation was forged through close consultation with the senator and the Massachusetts federal delegation. It was clearly designed to be complementary to rather than a substitute for Kennedy's legislation, which by then the governor also supported. In essence, each bill supplied half of the regional authority promoted by Metcalf and

The Vineyard is far from self-sufficient. Transmission lines cutting along the utility corridor on West Chop bring power in from Woods Hole below Vineyard Sound.

Eddy. As reported by the *Boston Globe,* "Governor Sargent and Senator Kennedy both acknowledged the need for partnership and dual legislation to accomplish this goal [of saving Martha's Vineyard]."

While the federal bill advanced the funding and authority for land protection, the state legislation created a completely new form of regulatory body—the Martha's Vineyard Land and Water Commission—equipped with a budget, staff, planning capacity, and regulatory authority over what were termed "Developments of Regional Impact" and "Districts of Critical Planning Concern." This new commission would replace the DCPEDC, assume its staff, and operate with true regulatory authority and enforcement capacity, which the DCPEDC lacked. In the last versions of the Kennedy bill, this new commission was the authority to receive funding from federal and other sources to acquire and hold land. When the federal bill failed, the land protection half of the regional authority was left unfulfilled.

The first significant and concrete step toward realizing the Metcalf and Eddy goal was taken in November 1974 when what is now known as the Martha's Vineyard Commission held its inaugural meeting. The unique scope of this novel planning and regulatory body was put to the test and ratified almost immediately. With a self-protecting eye on all the looming legislation, the Strock family's Island Properties had announced the next phase of its Waterview development in 1973—a proposal for a staggering 824 houses on about 500 acres along Sengekontacket Pond. Though it dwarfed any contemporary real estate venture and was widely condemned, the project was formally approved by Oak Bluffs in June 1974, five months ahead of the formation of the MVC. One of the MVC's first acts was to designate the Sengekontacket area a District of Critical Planning Concern and then review the Strock project as a Development of Regional Interest.

These steps immediately raised the question as to whether MVC authority superseded that of the town and allowed for retroactive review of a previously approved plan. Oak Bluffs and attorneys representing Island Properties argued no. But in a major validation of MVC authority and regional regulatory action, the Supreme Judicial Court of Massachusetts upheld the Commission's broad powers. Its decision said, "The Act establishing the Martha's Vineyard Commission was intended to respond . . . to the threat of destruction of the ecological and other special values of the Island through steeply increasing development of its land and water resources which would not be adequately contained or regulated . . . by the individual towns."

The near-term result was Island Properties' declaration of bankruptcy and piecemeal sale of its holdings. The long-term legacy was a powerful regional agency and the eventual conservation of many Strock parcels that had once filled Anne Simon's map of future development. The first sale was the five hundred acres along Sengekontacket Pond, which was purchased by a group promoting the creation of the Farm Neck Golf Course. That project included the "limited" development of 50 houses, donation of 85 acres to Massachusetts Audubon's Felix Neck Wildlife Sanctuary,

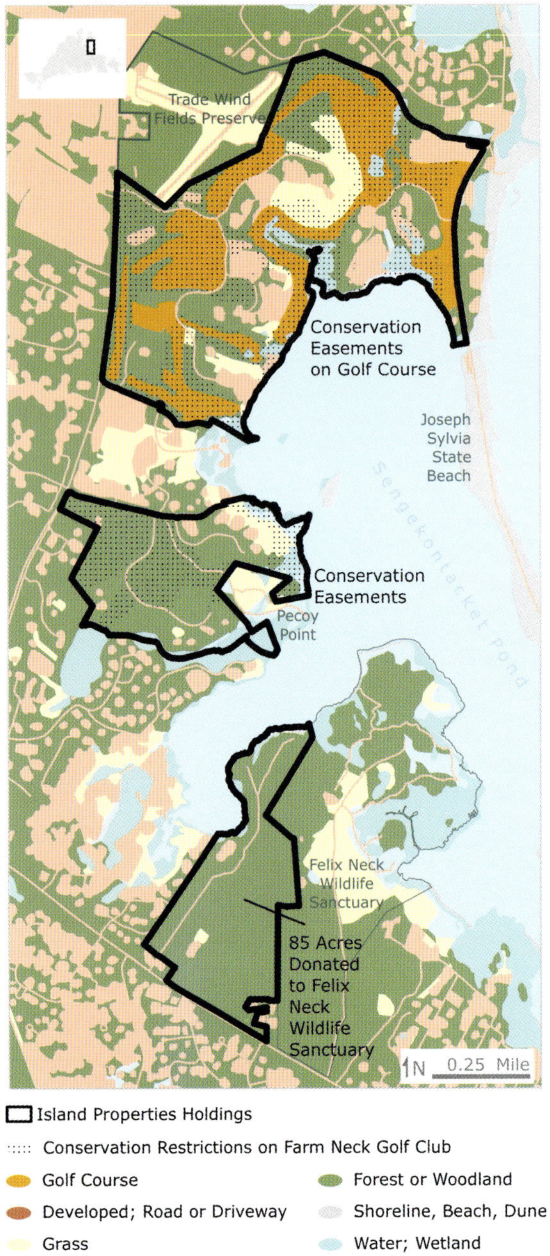

Farm Neck Golf Club—an ambivalent conservation success. Upon the bankruptcy of Island Properties and its plan to build 824 houses on about 500 acres, a group purchased the land and developed Farm Neck Golf Club and approximately 50 homes. With 85 acres donated to the Felix Neck Wildlife Sanctuary and conservation easements established on much of the golf course, Farm Neck became an unusual example of a high-end golf course development lauded by conservationists.

and establishment of a conservation easement on 230 acres of golf course. Although not preserving the intact landscape, this outcome was championed as a conservation success when compared to Island Properties' plan. Surprisingly ringing endorsements appeared in *Vineyard Gazette* editorials coupled with support by leading open space advocates including Mary Wakeman, Maitland Edey, and Henry Beetle Hough.

The tumultuous battles over the Island's future also spurred the strengthening of town regulations. Over the next decade, communities up- and down-Island formed boards, passed zoning and other regulations, hired professional staff, and produced town-wide master plans. But the public authority to conserve the landscape by acquiring land and easements as envisioned by the Kennedy bill and Metcalf and Eddy came more slowly and in the form of a regional rather than federal entity. In 1983, the state of Massachusetts passed legislation authorizing a land bank for Nantucket. Three years later, it followed suit on Martha's Vineyard. Like its regulatory counterpart, the Martha's Vineyard Land Bank Commission operates Island-wide, but it has a unique funding base derived from a 2 percent fee on most real estate transactions and the ability to assume debt. Ingeniously, the Land Bank is structured to balance equally its investment in regional and town priorities. Guided by town advisory boards and regional commissioners, the Land Bank has emerged as the leader in a century-long effort that has now conserved 40 percent of the Vineyard. When combined with state, town, and private efforts, the Land Bank and Martha's Vineyard Commission have accomplished many goals articulated by Metcalf and Eddy.

The *Island Plan*—2009

It is striking how much and how little has changed in thirty-some years. By 2006, when the MVC publicly launched its new Island planning effort, much that Metcalf and Eddy had predicted and much that it had recommended had come to pass. Building had

Though it never became the seventh Island town envisioned by Metcalf and Eddy, the region embracing the high school, YMCA, ice rink, skate park, and Island elderly housing comprises a busy center of activity.

skyrocketed and many fine farm, woodland, and coastal landscapes had been transformed by development projects that often clashed with Island values and views. But over the same period the MVC had established three Island-wide Districts of Critical Planning Concern—the Coastal, Island Road, and Special Places districts—that form the backbone of protection from inappropriate development of the Island's most sensitive areas. It worked with the towns to create several dozen other DCPCs to protect focused areas, and lessened the impact of many other projects. Meanwhile, efforts by the Land Bank and conservation groups had tripled the acreage of permanently conserved land.

Traffic continued to evoke extreme consternation. Summer automobile jams and road noise had reached extreme levels, and major roads had been straightened, widened, and lined with grotesquely larger poles and cables. On the other hand, bike paths had proliferated, the Vineyard Transit Authority had been established, and bus ridership increased to 270,000 in 2000 and over 1.1 million by 2013. The MVC accomplished all of this in its unheralded role as the coordinator of the public transportation committee.

Down-Island towns had sprawled as feared, but the six towns remained distinct, aided by substantial new conservation greenways (Southern Woodlands, Wapatequa Woods, Littlefield Greenways, Caroline Tuthill Preserve, and the Pohogonot tract). Though Metcalf and Eddy's peculiar concept of a seventh "town" never advanced, the region between the State Forest and Lagoon Pond that they targeted emerged as a bustling center of community, but not commercial, activity. The high school, YMCA, skate park, ice rink, and ball field lie at the center of this dense area along with elderly housing, community services, the Island's only funeral home, churches, and a Masonic Hall.

Some issues identified in the 1970s proved more intractable, particularly affordable housing and nitrogen loading to groundwater. Other concerns had escalated, especially sea level rise. In addition, the cancer of housing continued to spread. Since 2000, the islands of Martha's Vineyard and Nantucket have endured the fastest growth rates in the state. The MVC gathered concerned Islanders together in 2006 to discuss the need and scope for a new plan, and invigorated the group with a time series of maps that projected the history of development and staggering potential for an additional twelve thousand houses. Clearly, this was an unsustainable trajectory of growth and consumptive behavior. Without renewed efforts at land-use planning and conservation, the special qualities of the Vineyard remained in jeopardy.

But the planning process that was adopted to address these long-standing challenges this time around underscored the fundamental change that had occurred on the Island since the 1970s. The Vineyard was now fully equipped with the appropriate people, tools, and resources to tackle significant planning, regulatory, and conservation efforts. The project undertaken by Metcalf and Eddy and the efforts advanced by Senator Kennedy had been one-off, top-down efforts reliant on external powers. Both tapped outside

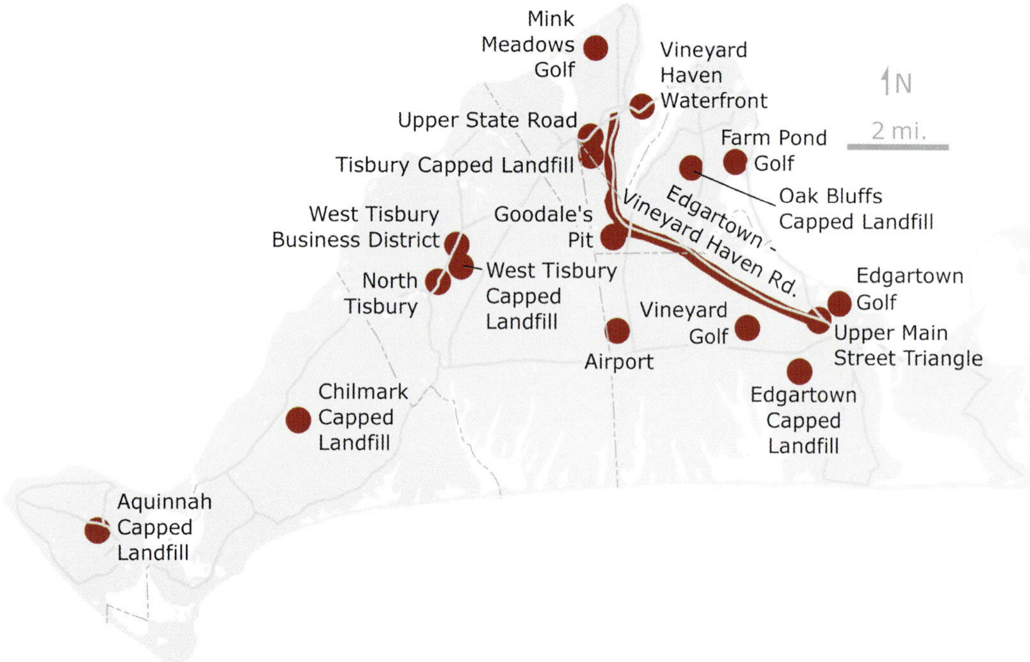

Opportunity Areas. The Island Plan's Work Group on the Built Environment highlighted what they termed Opportunity Areas for activities including redevelopment of low-density commercial, warehousing, or industrial sites (e.g., Upper State Road in Tisbury, Upper Main Street in Edgartown, West Tisbury Business District, and Airport Business Park), Revitalization Areas with focused infill, Disturbed Areas (e.g., gravel pits), and Landfills. Projects have advanced in these areas, including solar projects at Goodale's gravel pit and many capped landfills.

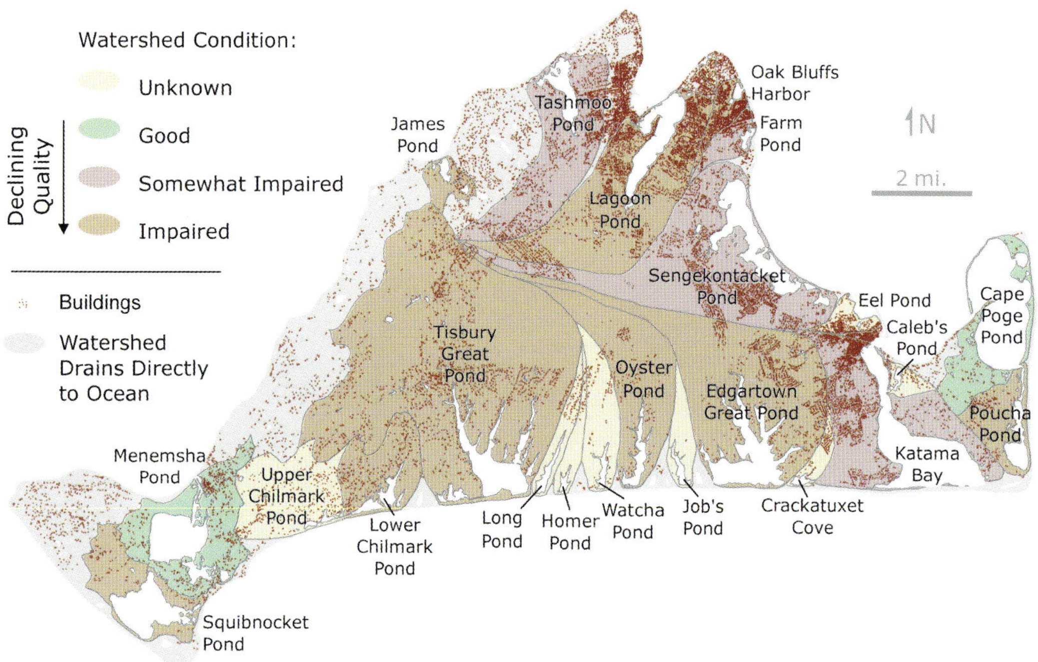

Compromised water quality in coastal ponds and groundwater is a legacy of historic development and land management that has led to environmental degradation and poses huge remediation costs. Controlling nutrient inputs of pollutants such as nitrogen is the only solution, but adequate action and full recovery lie in the distant future. Data from the Martha's Vineyard Commission.

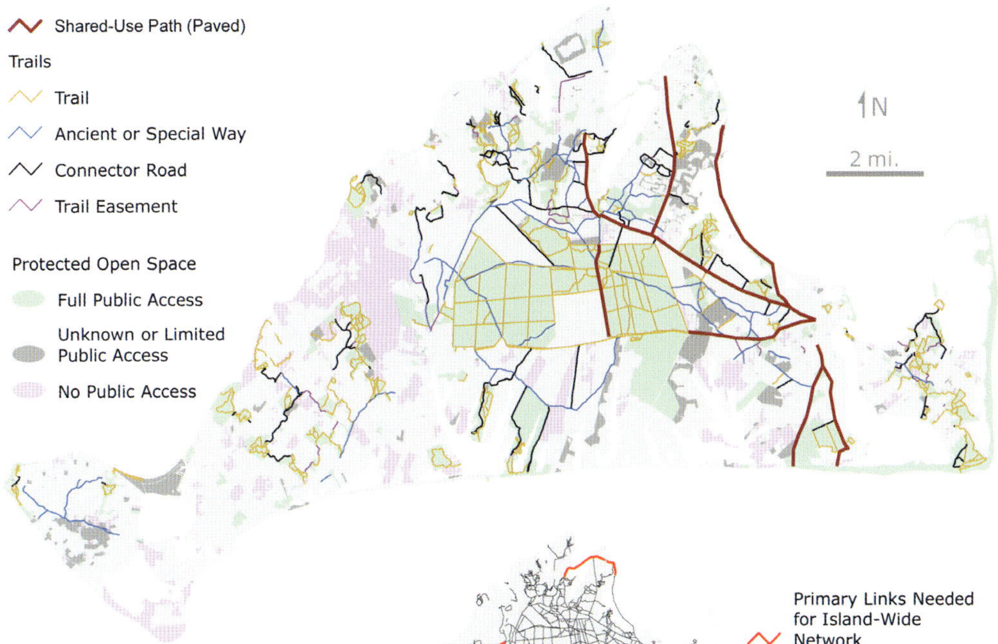

Although Chappaquiddick and sizable sections of the Vineyard support continuous trail systems, an Island-wide network remains a distant goal. While trails easements cross some private property, many large conservation lands contain trails (not shown) that are inaccessible to the public. Data from Martha's Vineyard Commission.

consultants, with the federal government perceived as the ultimate source of solutions and funds to accomplish future objectives. Though each effort was informed by town boards and the public, their major elements originated with outsiders.

The *Island Plan* is strikingly different in origins, design, and execution. In its capacity as the Island's regional planning agency, the Martha's Vineyard Commission played the central role. But the effort was led by a steering committee of sixteen residents, and the goals and strategies were identified by eight working groups comprising 50 to 160 local residents. Throughout, the process benefited from regular input by more than 550 local volunteers, called network planning advisers. This combined effort led to twelve supporting studies and plans, six surveys of the population, sixteen televised public forums, and extensive newspaper coverage. Opportunities for engagement in open forums ranged from agricultural fair booths to a comprehensive website. Most important, Islanders recognized that they now had the private, municipal, and regional capacity to begin to implement the plan once it was completed.

The list of challenges that the *Island Plan* identified could have been lifted directly from Metcalf and Eddy: unsustainable growth, a loss of young families, disparity between local incomes and the cost of living, and water pollution. The plan's major goals were also strikingly resonant with those embraced earlier.

Solar arrays and wind turbines are producing energy in backyards, on farms, and across the grassy expanses of many capped landfills. The siting of these structures must be done carefully so as not to destroy valuable habitat or infringe on scenic or other values. *Turbines:* Allen Farm (*left-hand corner*) and Morning Glory Farm (*right-hand corner*). *Solar (clockwise from top):* Edgartown town land, Packer's Farm, West Tisbury landfill, Tisbury landfill, Vagelos property in Chilmark, Whippoorwill Farm in West Tisbury.

Oyster and other shellfish operations such as this one in Katama Bay seek to provide local produce and employment while also removing excess nitrogen from the ponds.

- Conserve key ecological regions for biodiversity, scenic character, and recreation.
- Direct development to town and village areas, not sensitive areas.
- Forge an Island-wide perspective but protect the distinct character of the six towns.
- Stimulate a balanced local economy that is more self-reliant.
- Sustain the year-round community through affordable housing and living.
- Address climate change through conservation, renewable energy, and adaptation.
- Maximize local food and energy production; recycle waste.
- Restore salt ponds and bays.
- Maintain economic, cultural, and ethnic diversity.

The resulting *Island Plan* is an impressive package of information and vision—two hundred pages of facts, maps, aspirations, and strategies for addressing the major goals. The authors correctly cast it as more than a simple plan and more a "portrait," "guide and inspiration," and "living document" that can grow with the Island. At the same time, it has many specific strategies for implementation. Each of eight working groups identified fifteen to thirty-eight detailed activities or structures that were lacking, in need of resources and champions, or critical to expand.

Since its completion, the plan's implementation is proceeding unevenly. As I cross the Vineyard speaking to people concerned with the Island's future, few recall the broad sweep of the plan, let alone its numerous recommendations. Many doubt that it is being implemented. Yet a closer look shows that the plan has inspired many actions and informs the work of the regional commissions, local boards, and committees on a regular basis. The concrete steps toward implementation that its promoters hail are substantive even if less than glamorous.

- Protection of nearly seven hundred additional acres that link existing open space, increase access, and reduce the future number of houses.
- Adoption of Island-wide regulations for fertilizer use to reduce nitrogen loading into groundwater and the great ponds.
- Adoption by all six towns of the MVC's Wind Energy Plan, which guides turbine development on the land and ocean.
- Revision of the MVC's checklist for Developments of Regional Interest to increase the vigilance of reviews in ecologically sensitive areas while relaxing it elsewhere.

Through this period, the MVC advanced new projects that made subtle but key improvements to the Island: installation of a roundabout to relieve horrendous congestion at a key intersection; improving pedestrian movement through the summer crush of Oak Bluffs; and inserting a trail system for Beach Road in Vineyard Haven in the design of the new drawbridge across Lagoon Pond. Mark London, who retired in 2015 after thirteen years directing

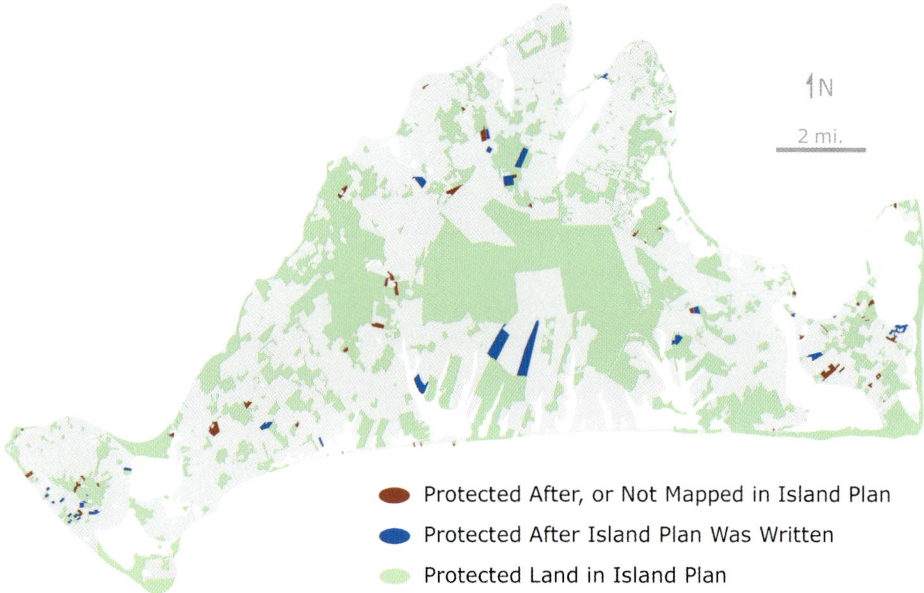

Land protection continues on the Vineyard. This 2015 map displays more than 1,300 acres of conservation land than the 2009 Island Plan, including more than 700 acres conserved since the Plan. Many new small parcels extend or connect previously protected blocks, expand access, or protect scenic landscapes. Data from the Martha's Vineyard Commission and the Harvard Forest.

the commission, strongly believes that the success of centralized planning comes as much in unrecognized improvements and minor tweaks within major decisions as through signature projects and grand planning exercises.

Future implementation of the plan's regionally coherent vision will depend on diverse actors—the MVC, local boards and the state, and nonprofit organizations. In particular, the towns can do much by improving bylaws to direct development from sensitive areas, requiring open space protection with each development, and adopting green building and energy standards for construction. But above all, the regional vision and retention of Island character depend on the passionate engagement of every individual who lives on and uses the Vineyard. The legacy and momentum of Island-wide action are there to seize and advance.

A Conservation Future

Our woodlots, of course have a history, and we may often recover it for a hundred years back, though we do not. . . . Yet if we attended more to the history of our lots we should manage them more wisely.
— Henry D. Thoreau, October 1860

Massachusetts offers an unusual and urgent opportunity for . . . conservation. Following widespread agricultural decline in the 19th century, the landscape reforested naturally and . . . has more natural vegetation today than at nearly any time in the last three centuries. . . . We urge the people and Commonwealth of Massachusetts to launch a bold, comprehensive initiative to conserve these precious [lands] and the ecological and social values they possess. . . . Seldom does history provide us with second chances. Seldom does an investment in the infrastructure that supports both nature and human activity offer the promise to yield so much.
— *Wildlands and Woodlands: A Vision for the Forests of Massachusetts,* 2005

The Vineyard landscape is distinct in many ways—most notably in land values, pace of development toward full buildout, the assemblages of plants and animals, and past success in land protection—but it typifies many qualities of Massachusetts and the greater New England region, including their conservation challenges. They share the history of agricultural and woodlot land use, the ongoing growth of their forests, the tension among farmed, open, and wooded lands, the relentless sprawl of development, the fragmentation of the land by many small private landownerships, and the looming threats from climate change, insect outbreaks, and other stresses. Nevertheless, the Vineyard has put itself into a particularly strong position to address the looming challenges due to its expansive breadth of conserved lands, its forward-looking planning efforts and knowledge base about the landscape, and the capacity for ongoing land protection and stewardship.

The single most important step in mitigating future environmental impacts is to conserve our natural and food-producing landscapes intact. Before we turn to the question of *how* the land should be managed, we must adhere to the bumper sticker designed by Islander Tweed Roosevelt—*"Martha's Vineyard—Save What's Left"*—so that we have a future to manage. To actively keep nature and farmlands intact is a powerful commitment that every citizen, landowner, and Vineyard advocate should make to mitigate future environmental change and support human well-being. The Vineyard landscape faces many environmental challenges. But whether we focus on climate change or other forces of nature or man, the most resilient landscape capable of both buffering and recovering from diverse impacts is that of nature itself.

Overshadowed in our focus on climate change is the fact that the foremost threat to our environment—on the Vineyard and beyond—remains the direct destruction of nature and conversion of forests and farms to other damaging land uses. Every further reduction in the size of our natural areas and farms, or perforation of these by houses, driveways, and other hard infrastructure, undermines

Bumper sticker on a car parked at Larsen's Fish Market in Menemsha shares the conservation sentiment of its designer Tweed Roosevelt and many Vineyarders.

the ability of nature, and us, to cope with the changing world. We should do everything humanly possible to maximize the extent to which forests remain forests, farms continue as farms, and streams, wetlands, and ponds stay free from modification.

With all of its stellar attributes, the Island Plan contains one glaring flaw. The future condition that it projects in decadal maps that lead to complete buildout shows development as the only possible future for open space that is not currently conserved. That future ignores one of the Vineyard's great strengths and most critical options—land protection. The nearly one thousand acres conserved since the plan was completed have already brightened the Island's future by enlarging existing areas of permanent habitat, forging critical connections among others, securing more food and wood-producing capacity, expanding the protection of Island aquifers and watersheds, guaranteeing public access to more landscape, and reducing the maximum potential number of houses by more than two hundred. Continuing the current pace of land protection would further reduce the consequences of buildout. Yet a future of steady conservation is not the Island's inevitable destiny. The pace of land protection could slow. Or, a determined effort to invest in the Vineyard's natural infrastructure could increase it and leave the Island even more intact than pessimistic predictions suggest. The Island bears substantial costs in coping with the degradation of its waters, importing more food and resources and exporting more

Land conservation pyramid. (i) The foremost priority, forming the foundation of the pyramid, is keeping the land intact. Today's good management will not matter if the land is developed in the future. (ii) Think before managing. Do as little as necessary; every impact leaves reverberating legacies. Consider doing nothing. (iii) Consider large ecological processes and the broader context of neighbors, surrounding landscapes, and the entire Island. (iv) Reference the past when developing a plan for the future. What shaped the land's history? How can I work with that? (v) Assess your work. Did it achieve the intended results? Adjust accordingly.

Continually work with neighbors and Island organizations to protect adjoining lands and expand the sphere of well-managed land. Adapted from David Kittredge, University of Massachusetts and the Harvard Forest.

waste products, and mitigating the damage from rising oceans. The more we conserve the remaining land, the lower these future costs will be.

As land protection efforts advance, we should also heed the Island Plan and target select properties for un-development. The Island discussion of un-development began in a 1990s *Gazette* interview with James Lengyel, executive director of the Land Bank. The approach, which involves strategic removal of buildings and

Grassy sheep pasture with oaks and a wind turbine on the Allen Farm, Chilmark.

The Bathtub and Norway Spruce: What Are We Managing For?

How should we value and manage habitats that arise as a consequence of human activity? If they add diversity, should they be appreciated and therefore perpetuated, as has been suggested here for agriculture, or should they be managed against because they conflict with the prevailing patterns of nature? The sight of a bathtub sitting in the middle of the State Forest alongside a robust plantation of non-native spruce provides focus for the question and the divergent opinions it generates.

According to longtime superintendent John Varkonda, the bathtub, along with a windmill and sizable pit adjoining it, were part of an ill-fated scheme by state wildlife managers in the 1960s that might have transformed the State Forest. Concerned by what seemed to them the monotonous qualities of the Great Plain, and apparently unimpressed by its subtle landscape patterns or diverse array of uncommon moths and plants, the wildlife professionals concocted a plan to enhance the property's diversity and apparent value. Their simple solution? Create ponds to attract ducks, aquatic plants, and amphibians and serve as oases of biodiversity in the dry and sandy environment. The bathtub and associated infrastructure are the remaining legacies of the first trial in this program of landscape enhancement. Given that the groundwater lies fifty feet below ground, it was necessary to line a dug pit with clay to retain water and then install a windmill-driven pump to fill this from the underlying aquifer. The water ran through piping into the bathtub at the base of the windmill and then into the pit/pond. The scheme evidently worked for some initial period, but the clay liner ultimately failed, allowing the water to drain and leaving the pit, derelict windmill, and bathtub sitting incongruously alongside a dusty fire road.

While maintaining duck ponds on a dry sandplain may seem ludicrous, consider the adjoining spruce plantation. Who would have guessed that a boreal tree would prove to be among the most successful non-native species planted in the State Forest? Indeed, spruce were planted in the bottoms with the hope that they would tolerate the cold air that accumulates in these "frost" bottoms. Surprisingly, it turned out that they grow superbly throughout the coastal landscape, where they form dense single-species stands that shade out all other species and look completely incongruous alongside the native oaks and pines. Moreover, the shade-tolerant spruce naturalize and have spread readily to become one of the Island's most aggressive invasive trees.

Given these facts, as Glenn Motzkin and I developed our 1999 report on the conservation management of the State Forest, we proposed that all plantations be removed in order to allow native species to flourish and the natural vegetation pattern to reemerge. While this seemed like an intuitive recommendation, we were surprised when some staunch conservationists arose in protest. With their great skills in natural history, these folks had identified the spruce as providing a unique habitat with boreal affinities that, not surprisingly, supported species that were otherwise uncommon on the Vineyard. According to Allan Keith, these included breeding pairs of winter wren, purple finch, red-breasted nuthatch, and red crossbill; the only known nesting of golden-crowned kinglet; and more saw-whet owls than any comparably sized area. As one local conservationist stressed to a *Vineyard Gazette* reporter, "It's a rare thing when a nonnative bird has sought out what it feels as home," adding: "We need to examine the vitality of that habitat." This intriguing perspective opened our eyes and, along with other complicating factors, eventually diverted the state's restoration efforts.

So, how is conservation management to deal with such issues? Do we accept historical land use and legacies of past management as necessary parts of the landscape to maintain? Do we strive to increase biodiversity for its own sake and add elements like ponds and spruce to sandplains? Or do we prioritize native species, natural landscape patterns, and conditions that prevailed centuries ago? In the words of Dick DeGraff, emeritus professor at the University of Massachusetts and one of the great minds in wildlife management, there is no such thing as the "right" management. Every management activity and habitat will be good for some species. In words from a different realm, "if you build it, they will come." So, as we advance our thinking about the future, we should have a very clear set of goals and sense of exactly what we are striving to achieve and enhance. Is it a duck pond, a spruce plantation, a farm that yields much-needed produce along with habitat, or a marvelously diverse and yet desolate scrub oak barren?

Remnant chimneys from two summer homes on Cape Poge, Chappaquiddick, September 2012.

infrastructure to reverse the fragmentation and perforation of conserved landscapes, has been widely promoted by TNC's Tom Chase and employed at Duarte Pond by the Land Bank and on Chappaquiddick by Sheriff's Meadow Foundation. The most notable example also centers on Chappaquiddick, at Hickory Cove, where a total of five buildings have been removed in a project coordinated by the Land Bank. With the removal of the final structures on land previously owned by the Self family, a magnificent landscape on Cape Poge Bay of ponds, marshes, upland forests, and meadows is now unencumbered or visually interrupted by built structures.

We must also look beyond designated conservation lands and strengthen the ecological value of adjoining and supporting lands. Backyards, bundles of neighborhood woods, lawns, fields, and even town centers, cemeteries, and the roadway margins, have value as semi-natural habitat that can be maintained with the same care that we apply to our conserved areas. Everything we do to reinforce nature in our own backyards and that we often take for granted in our daily lives augments our conservation landscape.

As we then turn from securing the land to caring for it, Henry Thoreau's words can guide our deliberations and actions. Thoreau recognized that history is embedded in every landscape. As he pondered the land's future and evaluated the land-use practices shaping his landscape, he admonished his neighbors to learn from and manage with the history of the land. Knowing what processes have shaped a given woodlot, field, wetland, or shore in the past, we can better anticipate how it might change in the future. We can also better identify what approaches to take if we seek to sustain the present condition, replicate it elsewhere, or shape it in different directions.

On the Vineyard, as in Thoreau's Concord, important lessons emerge from such a historical perspective. Most important, the fields and meadows, heathlands and shrub lands, bottoms and scrub oak plains, and majestically contorted oaks all arose from a recent history of intensive use. These landscapes support birds, butterflies, sights, and sounds that we cherish today. By acknowledging the actions that produced these scenes and ecosystems—the felling of trees, grazing of pastures, mowing of fields, planting of crops, and a myriad of related environmental changes—we can come to value farms, actively worked woodlots, and timberlands for more than the local products they yield. Each can be embraced for its beauty, its resonance with Island history, and its ecological values. These culturally contingent ecosystems have value, and deserve our care.

History has also taught us that we must anticipate, cope with, and frequently accept change and surprises. Views will disappear, cherished lands will erode into the sea, oaks and pines may die, our favorite bird, wildflower, and butterfly populations may wax or wane, and new species will undoubtedly arrive. That the Vineyard landscape will inevitably continue to change is due to many factors: every inch of the Island is recovering from past events and will continue this rebound far into the future; hurricanes, nor'easters,

For hundreds of years the expansive grassy plain at Katama was kept open by sheep and cattle grazing. With rampant development and tree planting since the 1960s, open areas have shrunk to less than a quarter of the plain in a heterogeneous mixture of farmland at Katama Farm (*middle right*), grassland on Herring Creek Farm (*middle extreme left*, including the Bamford Reserve of TNC), shrubland on the Katama airfield managed by TNC with fire (*center*), and beach, grassy dune, and shrubland along South Beach below Atlantic Drive. Shrub cover is expanding across the airfield, pitch pines are filling the shrublands surrounding the houses at lower right, and all of the developed residential areas are becoming thickly wooded.

erosion, ice damage, insects, pathogens, and more will keep coming to these shores; the climate will change and ocean will rise; and lands will be managed by their owners in diverse ways. Some of these changes are predictable, such as sea level rise or the successional dynamics of our unmanaged fields. But much of it—the arrival of the next 1635-caliber hurricane, a new insect pest, a breeding pair of coyotes, or proposal to develop a pristine landscape—will surprise us. Some we can prepare for, some we can regulate, or vote on, and others we will need to cope with. But, by expecting and remaining open to change, we can better anticipate, accept, and accommodate it.

Finally, as we turn to address change, we should learn from our past successes and failures. These tell us that success more often accompanies efforts that work with and accommodate nature rather than struggle against it. Past efforts to confront or alter the course of natural processes have more frequently failed than succeeded. Often they exacerbate a perceived problem that we had hoped to eliminate. Good examples of this include coastal armament that diverts the energy of the ocean from one place to another and interrupts the natural replenishment of eroding areas with much-needed sand. Another is our attempt to "improve" or "restore" nature following disturbances like hurricanes, coastal erosion, or insect outbreaks. In most cases, such efforts to assist nature are misinformed and cause additional damage.

Overall, we could benefit from less hubris. Except in our actions to conserve it from ourselves, nature doesn't need us to care for it. It is rarely improved by our actions, except in our own view and for our own very local benefit. We would be well advised to heed this before acting. The hillslopes of dead trees that spanned Middle Ridge in Chilmark and West Tisbury and the collapsing cliffs of oaks and pines at Wasque have been extremely well treated by being left alone. Neither requires our pity or our assistance. Despite their initially shocking appearance, both comprise thriving ecosystems and embrace magnificent examples of natural processes of recovery and adjustment that will yield new and different ecosystems. The forests of dead oaks are rapidly regrowing, yet with a bit more beech. The Wasque landscape stands as an ever-dynamic and awe-inspiring enigma for all who wander its shores and those who seek to predict its future. Who could or should devise a better, alternative path for either?

The approach taken here to conservation management seeks to balance pragmatism, human needs and desires, and

With the sheep herd now preyed on by coyotes, the trustees on Naushon Island are using cattle to maintain old pastures and recover areas overgrown with catbrier, vines, and shrubs. Vast inland parts of the island remain heavily forested with ancient woods comprised of old beech and oak, May 2014.

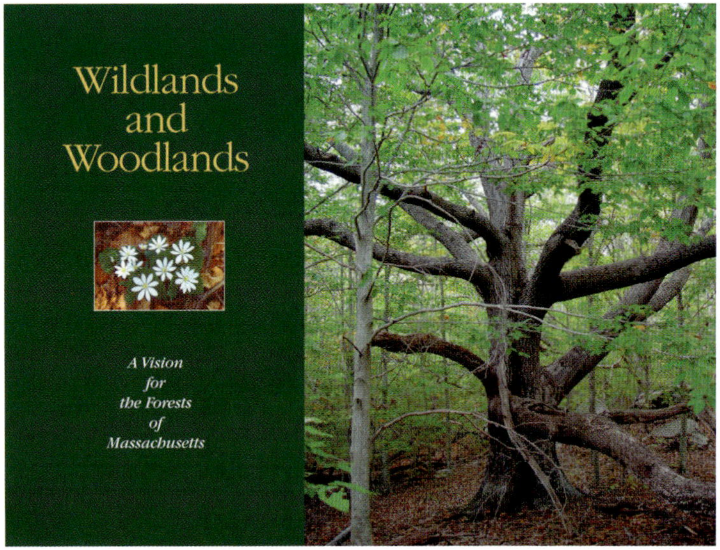

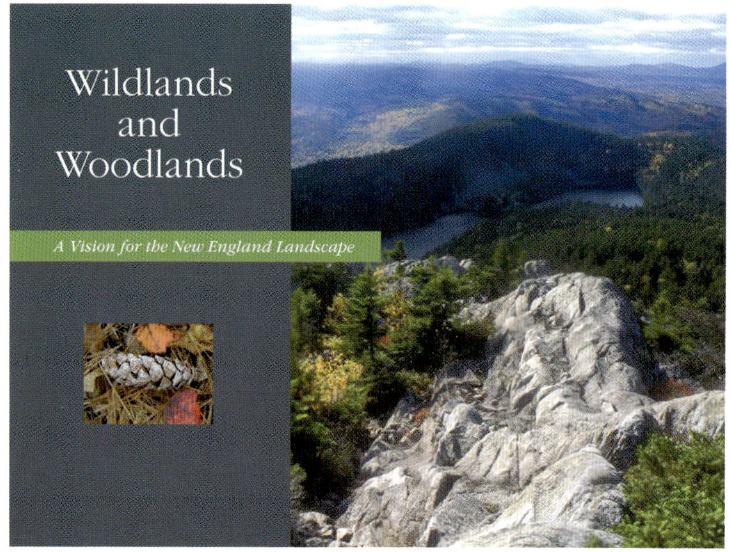

Wildlands and Woodlands visions for Massachusetts and New England. *Left:* Written by Harvard Forest scientists to aid and challenge conservationists and policymakers in addressing the needs of nature and society, the 2005 vision proposed that 50 percent of Massachusetts be conserved permanently in forests through conservation easements and purchase from willing landowners. *Right:* In 2010 academics from across New England called for the conservation of 70 percent of the region in forest, 10 percent in agriculture, and the balance in a mixture of these and developed areas. While requiring a near doubling of the pace of land protection, the effort is buoyed by the fact that New England has accomplished this feat three times in the past century. Cover images reproduced with permission of the Harvard Forest, Harvard University, Petersham, Massachusetts.

A New England Food Vision. In 2014 the Food Solutions New England consortium produced a regional vision for food production and farm expansion emphasizing local production, improved diets, and healthy food for all. By expanding farmland from two million to six million acres (to 1950s levels), the region could produce up to 50 percent of its needs. The consortium envisions an increase in pastureland for grass-fed beef, lamb, dairy, and poultry while advancing to the forest protection goals of *Wildlands and Woodlands.* Developed space could increase from 10 to 12 percent and accommodate projected growth through clustered and compact development and redevelopment of villages, town centers, and cities.

A New England Food Vision complements the 2010 report *Agricultural Self-Sufficiency* written for the *Island Plan* by Jo-Ann Taylor and Mark London of the Martha's Vineyard Commission. Cover image reproduced with permission of Brian Donahue, Brandeis University, and the Harvard Forest.

nature's impulses. It acknowledges the constraints imposed by laws and regulations, ownership and institutional differences, and conservation priorities. It is guided by keen attention to the ecological history of the land and especially these questions: What processes have shaped this landscape? How have such conditions been achieved in the past? And, what future directions might this area take? As such, it draws heavily from local tradition and the knowledgeable people who have worked the land. But it is rooted in science and experience-based projections of what may come to pass under different scenarios of management and future conditions. And it demands regular observations and the collection of data to test the ideas and study the consequences of the approaches employed.

To balance human needs and natural conditions in conserving the Vineyard, we can think in terms of three broad management approaches presented in two recent papers—*Wildlands and Woodlands: A Vision for the New England Landscape* (2010) and *A New England Food Vision (2015)*. This integrated approach combines hands-off management of a limited portion of the land (wildlands) with active management of the bulk of the land through forestry (woodlands) or agriculture (farmlands) undertaken by people who produce crops, livestock, timber, cordwood, and other products for a living. Combining active and passive management encourages a diverse mosaic of conditions and enables conservation lands to address multiple societal needs—recreation, scenic beauty, food, water, wood resources, biodiversity, and local employment.

The Argument for More Working Farmland and Woodland

Widespread management for agriculture or wood production on the Vineyard's limited land area may seem counterintuitive or trivial when stacked up against the Island's economic reality and high rates of consumption. Shouldn't the Vineyard's landscape simply be set aside for all to enjoy and as backdrop to the tourism that dominates the economy? Shouldn't the Island's food and fiber come from mainland sources where it can be produced at grander scales and lower cost? After all, the Vineyard has never been self-supporting; it

Over the past two decades Jim Athearn has brought back the old pasture on his family's Bethaven Farm in Chilmark.

has been importing spruce timber from Maine and much more from elsewhere for centuries. Yet, local production can actually make a meaningful dent in the Island's food and wood needs. And, there is a strong environmental argument for the Vineyard to increase its sustainable production of resources, put forth in the booklet *The Illusion of Preservation.*

There is a strong inclination on the Island to stand back and simply let our forests age and fields go brushy, or to manage them and the rest of our land neatly with power equipment for tidiness and aesthetics. But the reality is that the Island consumes immense quantities of resources. Some 96 percent of Island food is produced elsewhere. For timber and other wood products it is pretty much 100 percent. Yet producing these resources locally has many benefits: they can be obtained under the highest environmental oversight and in ways that are compatible with local interests and healthy for local populations; they support local people and enterprises; and they yield important lessons for consumers and onlookers of all ages. These lessons include a simple message of environmental responsibility: all of us depend upon the earth's limited and environmentally precious resources for life. Land is finite. What we consume here translates somewhere directly into impacts on the land and water. Our consuming habits have real consequences that we should consciously embrace and participate in.

This is where the *illusion* comes into play. When we lock up our own conservation lands solely for passive enjoyment and focus resource production on distant places, the environmental impacts and social consequences of sustaining our lives and lifestyle are removed from our minds and oversight. Awareness of this reality behind the *illusion of preservation* provides three arguments for shifting more production of our chickens, beef, vegetables, fruit, gravel, wood, mulch, and other products into our own backyards. First, it involves us more fully in understanding our own lives. Wood comes from trees that must be cut, gravel (and many resources) comes from immense pits in the earth, and meat comes from the killing of animals. These are fundamental truths that we should embrace and participate in at some level. Recognition of how and where we obtain the stuff of our lives can make those lives more complete.

Second, when we oversee the source of our food and other products, it is likely that we will apply greater care—environmentally, ethically, and socially—in their production than will others in distant places, because these consequences and the quality of the products affect us directly. We likely have real concerns seeing the Island mined for gravel, chickens and cattle slaughtered, trees felled in the State Forest, or manure applied to neighboring fields. But as these activities take place, we and local officials will pay close attention and ensure that they are undertaken in as sound fashion as possible. Local production can therefore translate into better stewardship for the earth as a whole. Conversely, when resources are extracted on a larger, "more efficient" scale distant from consumers'

Once a broad pasture fenced at the north end and surrounded by Long and Middle Point Cove and the ocean, Long Point is managed as a grassy shrubland by The Trustees of Reservations using fire and mowing.

eyes, there may be overwhelming pressure to cut corners and accept externalities—for instance, the crowding of animals, excessive chemical applications, or the dead zone in the Gulf of Mexico resulting from massive nutrient runoff along the reach of the Mississippi River. Intensification at great scale lacks sufficient cultural means to inspire and enforce good stewardship. Much better that we "own" and are responsible for the environmental impacts of what we consume.

Finally, and this is undoubtedly more aspiration than proven assertion, it is possible that as we witness and better appreciate where our resources come from and how they are produced, we may come to alter and reduce our patterns of consumption. As we tour our gravel pits, slaughterhouses, harvested lands, mulch-production facilities, and farmyards and understand the varied consequences of our resource use, we may choose to live differently. We may reduce our intake, increase the reuse and recycling of materials, compost

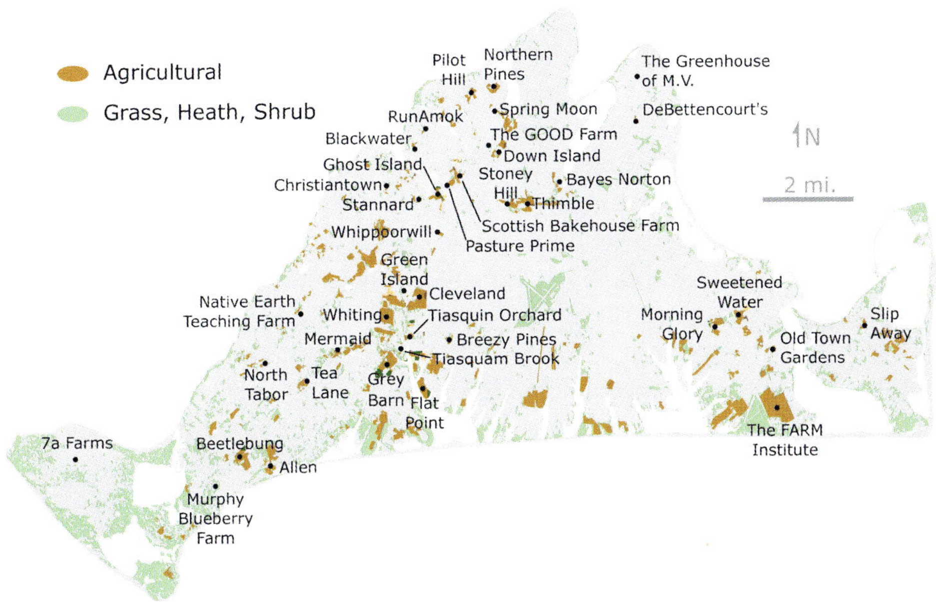

Map of the active farms and openland habitats of grassland, heathland, and shrubland. There is great potential to expand agricultural production by increasing the intensity of use on some existing farmlands and expanding farming onto conservation land managed through mowing, brush cutting, or fire. If advanced carefully the latter would meet conservation habitat goals with less expense and consumption of fossil fuels. Information from Island Grown Initiative, The Farm Institute, Martha's Vineyard Commission, and local farmers.

unused products from farm and table, and reduce the streams of waste that leave the Island in dumpsters (nearly invisibly to all of us) to unseen places. And we may come to appreciate locally produced products for their known healthful qualities, association with land that is well cared for, and the real and known people who provide them. The consequences of local production *can* be a powerful positive feedback loop in which sustainable local production tapping expansive conserved lands yields multiple local benefits, an improved quality of life for consumers and producers, and lessened global environmental impacts.

Efforts to increase the proportion of locally produced and consumed food and resources can be meaningful on the Vineyard and more broadly across New England and beyond. As part of the Island planning process, Jo-Ann Taylor and Mark London explored this issue in a revealing though completely unheralded study—*Agricultural Self-Sufficiency on Martha's Vineyard.* Following their analysis, they advanced three specific strategies for the Island to produce more of its own food: increasing the land area in food production; improving the productivity of those lands; and relying more heavily on foods that can be grown locally. The most ambitious realistic estimate showed the Vineyard increasing production from the current 4 percent to 50 percent of local residential demand. A collaborative effort by agricultural and food experts headed by Brian Donahue at Brandeis University came to a similar conclusion for the six-state region in *A New England Food Vision.*

Taylor and London confirmed that the single most important step toward increasing Island food production is the protection of existing and potential agricultural land and doubling or tripling the amount of actively farmed land, which was about 935 acres in 2005. The additional land would come from former fields that remain open, including many conservation lands, forests on prime agricultural soils that occupy about 10 percent of available land, and backyard gardens on parcels greater than one acre. Many inefficiencies were identified in the current agricultural system.

Pigs rooting around a coastal oak forest on Everett Whiting's land west of Black Point Pond, Chilmark.

Cattle, sheep, and hogs are currently exported for slaughter in western Massachusetts or southern Rhode Island, which involves sizable transportation, energy, and personnel costs and unnecessary stress on the animals. By developing an Island-based meat-processing capacity with coordinated transportation, meat production could increase substantially. A good model exists in the mobile chicken processor developed in recent years by the Island Grown Initiative, which increased the local production of chickens from three hundred to nearly ten thousand annually.

Similar arguments can be made for wood products, especially landscaping material, firewood, and custom timber. Though some Vineyarders do harvest local firewood, much still comes from the mainland at great expense in transportation, fossil fuel consumption, and climate impacts. Each load of imported wood also increases the potential of transporting yet another insect pest or fungal pathogen.

Spreading farming and wood harvesting onto conservation lands helps return conservation to its roots. In the days of Henry Thoreau, Gifford Pinchot, Nathaniel Shaler (a founder of The Trustees of Reservations), and Aldo Leopold, conservation meant a balance of strict preservation and the sustainable use of resources. Fortuitously, new approaches in farming and timber harvesting lend themselves to playing a more active and positive role in conservation. Agricultural practices in New England are becoming more diversified, organic, environmentally sound, and focused on the health of land and people and less intensive, extractive, and dependent on fossil fuels and industrial chemicals. Animals—cows, sheep, pigs, goats, alpacas, poultry, and more—are moving outside (again) to be grass-fed on large pastures and native hay, supplemented by imported grain. Organic nutrients are prized and being carefully cycled back onto the land through composting, manuring, and the rotation of animals, grains, and vegetables.

In Vineyard locales like Mermaid, Slip Away, and Allen Farms, and the geographically spread operations of the Athearn family that stretch from Morning Glory to Uncle Leonard's and Bethaven Farms, we can see that farm landscapes can again form diverse mosaics of crops, pasture, hay, and orchards complemented by new infrastructure such as glasshouses, farm-based dairies, compost-tea breweries, and wind turbines. Such varied operations harken back to the nineteenth century in their small scale and diversity. As these farmers effectively fit varied practices to the land, they also generate varied habitats for insect pollinators and an array of openland birds, amphibians, reptiles, and mammals. Meanwhile, wood harvesting practices offer increased flexibility in their scale, treatments, products, and resulting landscape condition. While it remains possible to harvest entire swaths of forest to create early successional landscapes, nimble equipment also allows highly selective operations to produce savannas or thinned woods. We have many options to manage more of the Island's land base to benefit society in a sustainable fashion.

Farmland

Farmland is an appropriate and yet perhaps surprising place to begin discussing the Vineyard's conservation landscape. Appropriate because agriculture has been a defining Island activity throughout written history and surprising because even its farmer practitioners seldom view their actions as conservation management. We protect and value farms for their aesthetic, economic, and nutritional benefits, but they are seldom recognized for their important habitat value. Farmers are seldom included in discussions of conservation management, and indeed few of them think of their work as complementing that of the staff at the Land Bank, Sheriff's Meadow Foundation, or State Forest. And yet, farmers may be our most important land stewards, and their actions are critical to the retention of many species and ecological conditions. As we've discussed, many of our most distinctive species thrive because of the Vineyard's history of agriculture. The single most important threat to dwindling populations of birds, butterflies, and plants is the loss of open space associated with farm abandonment. Now, in a time of farm revitalization and growing interest in local food production, we have an opportunity to embrace farming as part of an integrated approach to conservation management. Following years of treating openland habitats as though they were natural, legacies of Indian practices, or in need of highly specialized conservation care, we can recognize and manage these old farm landscapes for what they are. The argument for expanding environmentally sensitive farming as a conservation strategy is based on ecological history.

The Vineyard is a cultural landscape shaped by centuries of human use. Though forest is the land's predominant natural condition, open expanses exist as compelling legacies of a more active agricultural era. Many factors lend appeal to these treeless areas—familiarity, memories, and photos; expansive views and open skies; variation in plants and wildlife; and their simple contrast with dark green woods and the deep blue sea. As grasslands, heathlands, and shrublands shrink along with vistas and many species, there is a growing desire for their retention. Openlands, of many sorts, are a major focus for conservation and have been widely protected. Yet without active management, they change rapidly. They want to become forest: meadows will soon become woodlands. Constructing

Slip Away Farm, Chappaquiddick, produces vegetables, chickens, eggs, and meat on property owned by the Martha's Vineyard Land Bank and Martha's Vineyard Preservation Trust. The old Wampanoag Chappy schoolhouse (*left*) was moved and refurbished to serve as a farm stand adjacent the original house.

a tall building or rooftop platform may maintain a valuable water view, but it does nothing to support the disappearing habitat and character of the land.

Over the years, landowners, conservationists, and property managers have struggled and, in most cases, progressively lost the battle to retain the expanse and quality of meadows, fields, and heathlands. This is largely because they have overlooked or been unable to replicate the human activity that gave us these lands. Today, many areas called "sandplain grasslands" are managed with fire with the stated purpose of restoring the pre-European vegetation and native heritage of the Island. Elsewhere we expend huge resources on machinery, personnel, and fossil fuels to mow or brush-hog the annual growth of grassy or shrubby areas and

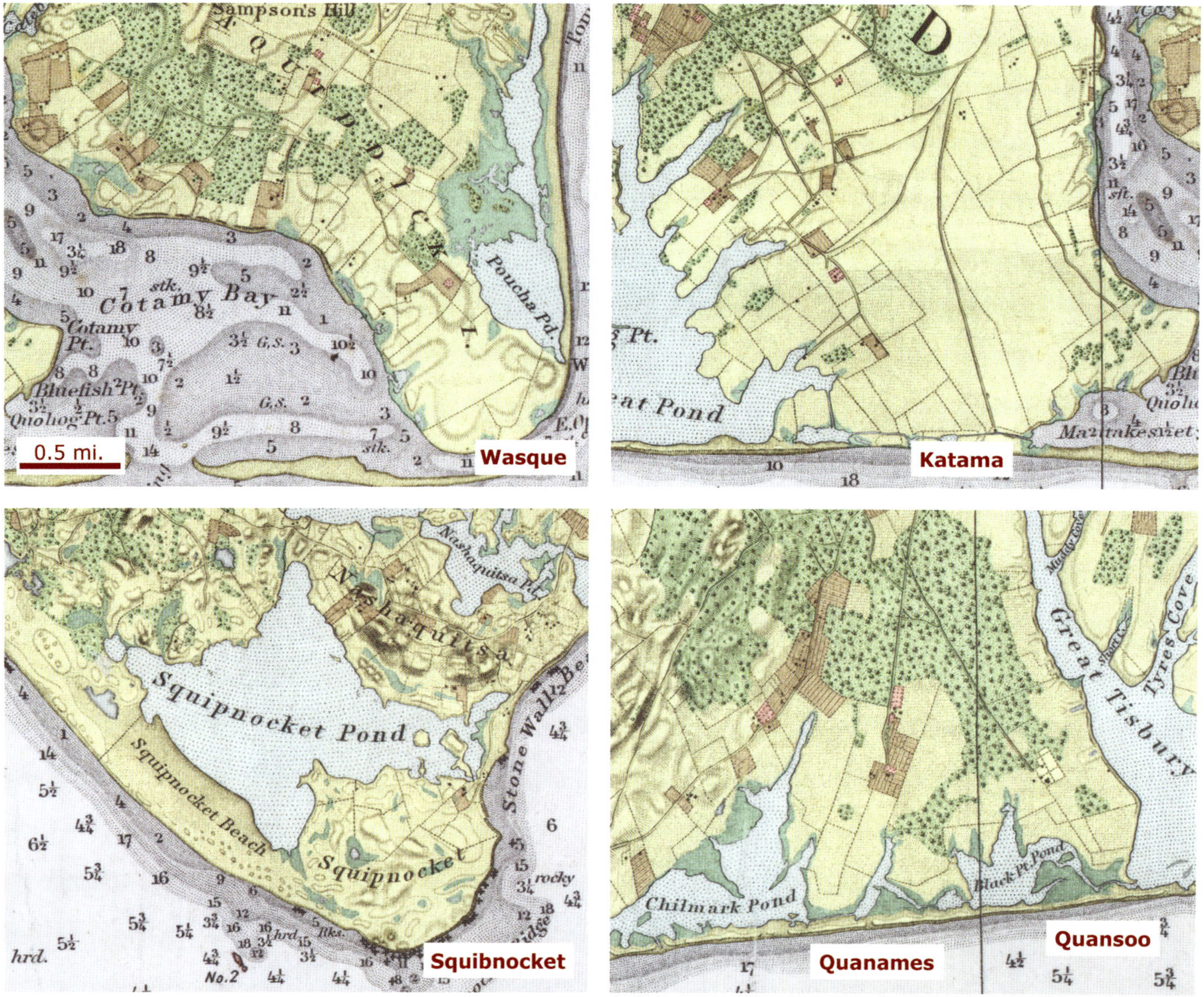

A historical argument for conservation management through farming. The fact that all of the large open conservation areas are former farmlands supports the argument that agriculture could provide effective management of this habitat. Henry Whiting's map ca. 1850 depicts Wasque, Squibnocket, and the bulk of Katama separated into expansive pastures by fences (indicated by thin broken lines) of stone (Squibnocket) or wood rails over ditches and ridges (Wasque and Squibnocket). North and west of Black Point Pond at Quansoo is an established pattern of pastures and large crop and hayfields. Map details from NOAA, 1860, Chart No. 12, *From Muskeget Channel to Buzzard's Bay and Entrance to Vineyard Sound Mass.*

Fire on the Vineyard

While paleoecological records document a great increase in fire with colonial land clearance, the history of fire thereafter varies greatly. It is likely that fire declined as agriculture increased, especially in the era of intensive farming in the late eighteenth and early nineteenth centuries when resources were precious and grazing would have left little fuel in most fields.

Thereafter a record of active burning emerges from the *Vineyard Gazette*, largely centered on the Great Plain and a time when farming was declining, land was abandoned, and brushy areas became widespread (see fire timeline). Large fires are confined to the one-hundred-year period after 1847 when more than twenty fires exceeded 1,000 acres in area, and nine covered more than 5,000 acres. During this period, both seasonal and longer-term trends are apparent. More than 90 percent of fires occurred in spring (late March to June), although two 10,000-acre burns occurred in July. No substantial fires were reported before March, and only four burned after October. There is a long-term trend of declining fire size, with none exceeding one hundred acres since 1965. This drop in the past half-century can be attributed to a reduction and greater care in outdoor burning and the substantial change in the Vineyard landscape and vegetation, and fire control. The open landscape of the nineteenth and early twentieth centuries, filled with open fields, extensive brushlands, and young forests, was considerably more flammable than the heavily forested landscape today.

The large fires were predominantly accidental burns due to various causes: "setting fire to a small plot of blackberry bushes," "the carelessness of a man in smoking a pipe," "some bushes, which were being burned off," "a burning match carelessly thrown into the dry bushes," or "the knocking out of a pipe." Fires were also set maliciously. On May 20, 1870, the *Gazette* ran a note: "Attention is called to the Reward offered . . . for information to detect the persons who have several times, lately, sought to amuse themselves by the dangerous business of setting brush and woods on fire. Last Saturday a great fire raged on the Great Plain, and but for the persistent efforts of several men, a portion of our town would be in ashes. Now, a word of warning to those incendiary spirits:—Cease this sort of work, immediately, unless you desire to be placed where you can be found every day and night, without trouble. . . . There are several bad boys in this town, who will do well to heed these remarks, before they terminate there [*sic*] criminal career in a stone house." Arson was similarly attributed to fires on April 3, 1885, and May 7, 1886. The latter fire burned over three thousand acres of brushland and oak forest.

The Island's geomorphological, vegetational, and meteorological setting controls the broad-scale distribution and pattern of fire (see fire map). With prevailing spring and summer winds from the southwest and west and strong onshore breezes near the coast, fires tend to move inland and to the north and eastern sides of the Island. Physiography reinforces and adds complexity to this pattern. Fires are concentrated on the dry and flat Great Plain. The western moraine is dissected by hills, hollows, ponds, wetlands, and streams that provide local fire breaks and pockets of moist and less flammable vegetation than the scrub oak, oaks, and pines on the Plain. These characteristics, the protection provided by Vineyard Sound to the west, and prevailing winds make the moraine the region least prone to fire.

leave it on the ground. Both approaches share two qualities: neither restores the historical and desired vegetation, and each requires a regular infusion of resources while yielding no consumable product. Clearly, these were not the approaches employed for centuries by Vineyarders struggling to simply live off the land.

Farming originally produced these openlands, and remains the key to revitalizing them. Our meadows, heaths, and shrublands were overwhelmingly produced by deforestation and grazing. Ironically, environmentally destructive processes produced some of our most cherished habitat and scenery. In order to maintain and expand these

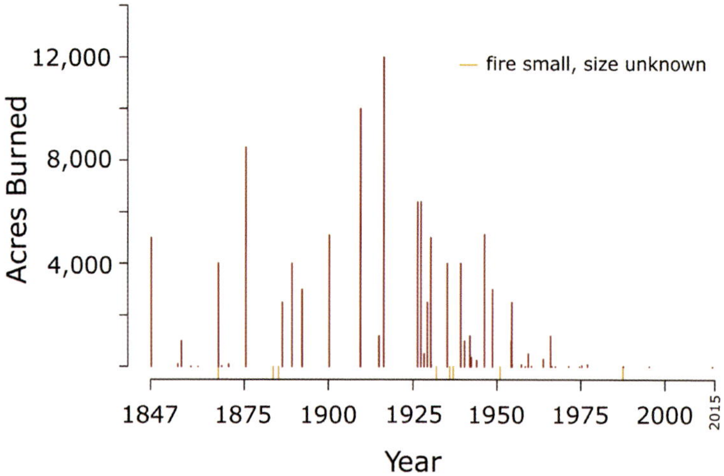

A timeline of fires compiled from the *Vineyard Gazette* and other sources. Burning increased with farmland abandonment as brushy flammable areas developed and fires spread through carelessness and occasional arson. The development of much less flammable forests, a reduction in open burning, monitoring of brush burning by fire departments, and improved firefighting have limited recent wildfires.

legacies of colonial history, we need to encourage parallel practices today, over at least a representative portion of the landscape. Most important, we need to bring on the grazers—sheep, cattle, and goats—along with hogs, chickens, and other poultry. Search the old photos and we generally find two actors in the landscapes—grazers and farmers. Both work incessantly to keep the land open. And both do it as a way of life that yields consumable products.

Today's growing interest in local food provides an opportunity to advance the goals of both conservation and local farmers by expanding agriculture. Conservation groups and farmers should partner for their own good and that of society. This partnership could provide farmers long-term access to much-needed land. Meanwhile, conservationists and landowners could maintain desirable habitats at reduced cost and environmental impact. Societal benefits include food, diversified local employment and commerce, and the attractive character of Island life and land.

The Vineyard provides many opportunities for such partnerships

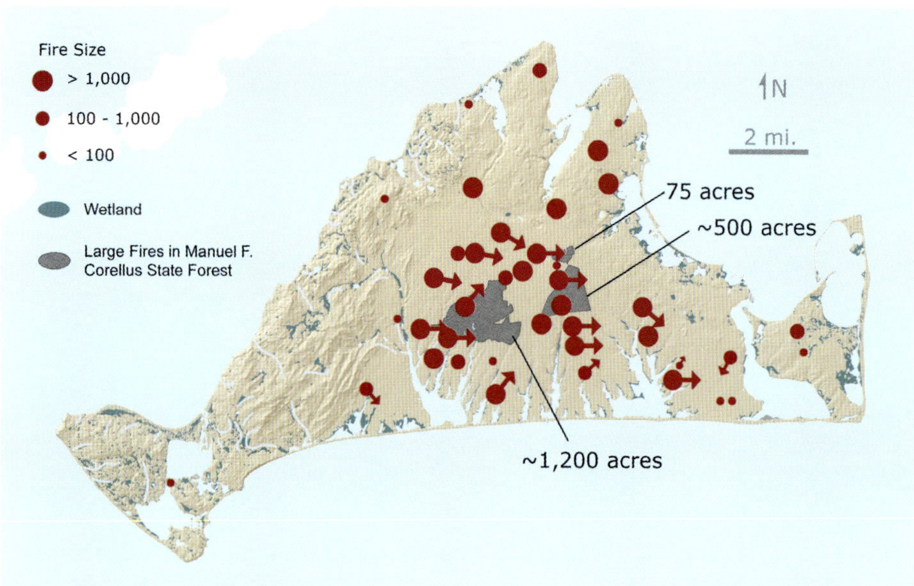

Sizable fires (1850–1950) have been concentrated on the Great Plain, where they were unconstrained by topography or substantial areas of older oak forests and often blown in an easterly direction by prevailing winds. Arrows indicate the direction of movement.

Scorched pines and regrowing vegetation following a prescribed fire by The Nature Conservancy on Katama, June 2009. The silo from Katama Farm and The FARM Institute looms in the background.

Menemsha Hills Reservation along Gosnold Road in Chilmark, where The Trustees of Reservations have cut trees and brush-hogged and chipped scrub oak and other woody vegetation over an expansive area to maintain an open savanna and low shrubland, March 2013. The open field and large home in the background are private.

and could serve as a model for effective farmland conservation management. Interesting examples already exist, though little of it is thought of as conservation stewardship. Slip Away Farm on Chappaquiddick and Tea Lane Farm in Chilmark represent successful efforts by the Land Bank, along with the Martha's Vineyard Preservation Trust and the Town of Chilmark, to partner with farmers. The Land Bank and Sheriff's Meadow Foundation have employed goats to clear fields and brush at Fulling Mill Brook, Waskosims Rock, Cedar Tree Neck, and other sites. The Trustees of Reservations, the largest owner of farmland in Massachusetts, leases land to Grey Barn Farm and has merged with The FARM Institute to run Katama Farm. Seven Gates has leased its hayfields and pastures to local farmers for years. Sheriff's Meadow Foundation is exploring ways to graze land at Quansoo Farm, Wades Field, and Nat's Farm. Some two thousand acres of conservation and private land managed in other ways could join these and triple the existing base of farmland. These include Quansoo Farm, Long Point, Wasque, Woods Preserve, Waskosims Rock, Tisbury Meadow, and expansive private grasslands across the south shore and on Chappaquiddick.

Admittedly, there are challenges. To be successful, this conservation model needs to generate marketable products and profits for the farmer while producing desirable conservation habitats. The farmer seeks productivity and a profit, while ecologists know that in coastal landscapes, low-productivity sites support among the greatest diversity of uncommon species. Farmers may want to harvest hay when the grass peaks in nutritional value in early summer but weeks before grassland birds fledge. Conservation groups accustomed to a predictable management schedule may be frustrated by the farmer's need to juggle many lands and crops, the dictates of weather, and complicated lives. Both groups strive for cost effectiveness and logistical simplicity.

One model pursued off-Island may provide a workable arrangement. It calls for the landowner to bear the cost of the basic infrastructure—the land, robust fencing of the parcel's perimeter, and access to water—while the farmer is responsible for the ownership, movement, and management of the animals and any

Cows share the space with turkeys on Harvard Farm in Petersham, Massachusetts. This former golf course (1922–2012) is being returned to its colonial roots as a small New England farm (ca. 1750–1921) by Harvard University as part of a long-term study and demonstration of farming as conservation management, August 2014.

specialized infrastructure, such as internal fencing to partition the land or manage a rotational grazing regime. The land is made available to the farmer for little or no charge, but with the reasonable expectation that the landowner may save money through reduced costs for personnel, machinery, fuel, and other expenses. Since access is inexpensive, the farmer is willing to accept the land's lower productivity (operating with limited or predominantly organic fertilizers) and to accommodate other conservation objectives, such as the late cutting of hay to allow grassland birds to fledge. One added wrinkle: grazing animals often attract public audiences, certainly more than burned or mowed fields. While this requires oversight, it also yields major opportunities for outreach centered on conservation and agriculture.

A concern for the conservationists and the Massachusetts Natural Heritage and Endangered Species Program is the fate of uncommon species. Changing management regimes to approaches that are not a standard part of the conservationist's tool kit may worry conservation managers. Change can be introduced slowly and with close monitoring. In most cases, openland species should be compatible with farming regimes that are similar to ones that allowed these species to occupy the sites in the first place.

Overall, the conservation benefits of farming can be great. In contrast to the homogenizing action of a mower cutting all plants indiscriminately to a single height, animals diversify the landscape. Each species of grazer has its selective preferences for food and distinctive style of grazing. All leave their own pattern of grazed intensity, tracks, and manure. Varying the type and intensity of grazing, the duration of fallow periods, and other specific practices will produce a kaleidoscopic of habitats that would have been familiar to nineteenth-century farmers and captured on Henry

Goat herd used by the Martha's Vineyard Land Bank to maintain and expand historic pastures on Fulling Mill Brook and other properties, June 2014.

Whiting's map. Farming is historically consistent, is beneficial to society, and provides many lessons about nature, history, culture, and sustainability.

Woodlands

With diverse ownership and environments, Vineyard forests provide many opportunities for active management: production of timber, firewood, mulch, and other products; maintenance of distinctive species or vegetation characteristics; removal of non-native or less desirable species; and aesthetics. In many cases, these objectives are mutually reinforcing. Mulch can be a by-product of the removal of failing conifer plantations or trees encroaching on scrub oak bottoms; the selective harvesting of mature trees to open woodlands and savannas can yield timber; and coppicing or clear-cutting creates successional conditions that support declining bird and mammal species while producing regular supplies of firewood. Firewood, chips, and mulch also result from routine landscaping and maintenance of trails and roadsides.

Many private lands, portions of the State Forest, town forests and water supply land, Land Bank, and other conservation properties lend themselves to active management. These activities can be conducted by staff or through agreements with local firms and residents. Large plantations remain on the State Forest that should be cut to restore the native growth of huckleberry, blueberries, scrub oak, oak, and pine. A more targeted activity involves the removal of non-native spruce and white pine encroaching into bottoms, fields, forest understories, and fire lanes.

Many conservation landowners have undertaken natural resource management. For years, the State Forest ran a cut-a-cord program in which local residents freely cut firewood in areas designated for thinning. On portions of Cedar Tree Neck, Sheriff's Meadow Foundation has undertaken small-scale harvesting to produce lumber, bridge timbers, planking, and posts. The Trustees of Reservations recently created a vast expanse of shrubland and savanna on the Menemsha Hills Reservation, yielding large

Mowing, brush-hogging, and fire have been used by The Trustees of Reservations in an attempt to keep the large expanse at Wasque open. Aggressive sprouting by oaks, scrub oaks, pitch pine, and huckleberry is leading to a gradual increase in woody species over grass and herbs.

quantities of timber, cordwood, chips, and mulch for local operators. The Land Bank produces chips for trail improvement as a by-product of property maintenance. Rapidly growing forests, disappearing openlands, and stewardship of conservation lands present opportunities to produce a small but growing fraction of the Island's firewood, landscaping, and specialty timber needs locally.

Wildlands

Wildlands occupy the other end of the spectrum of conservation management from farmlands and woodlands. These typically sizable reserves are managed in a mostly hands-off manner to promote natural conditions that were embraced passionately by Henry David Thoreau, John Muir, Bob Marshall, and Aldo Leopold. Wildlands are landscapes shaped by natural processes with management limited to access for appreciation—no extraction of resources, no steering the trajectory of vegetation development, and no effort to clean up after windstorms or insect damage. This approach rests comfortably

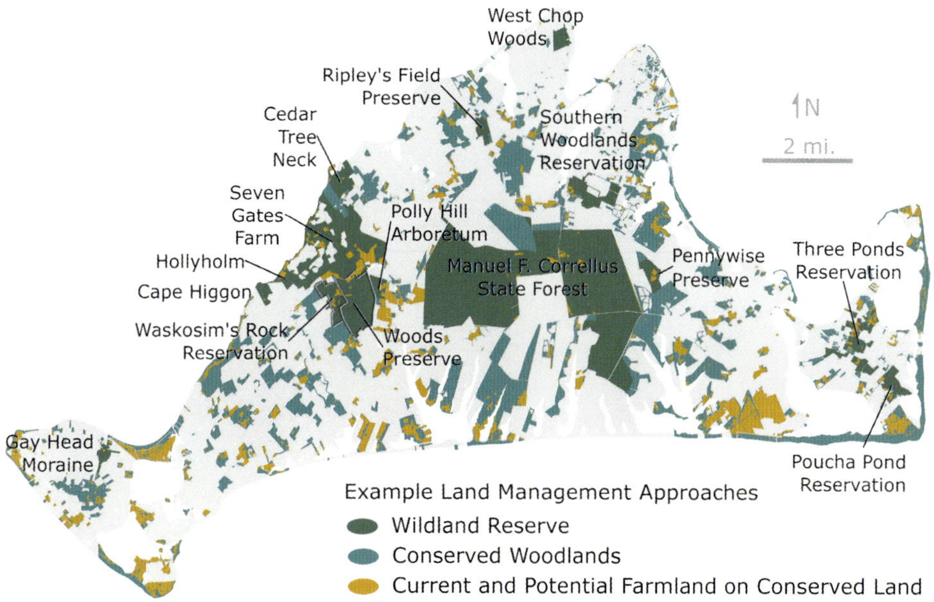

One option for Martha's Vineyard in keeping with the *Wildlands and Woodlands* and *A New England Food Vision*. Given the Island's diverse land and ownership patterns, many feasible options exist for complementary stewardship embracing active woodland management, farming, and wildland reserves. In this schematic, large wildland reserves stretch from the State Forest to much of the Woods Preserve, Waskosim's Rock Reservation, Seven Gates Farm, Cedar Tree Neck, Holly Holm, and Cape Higgon. Dispersed reserves elsewhere on the Island—Ripley's Field Preserve, West Chop Woods, Southern Woodlands Reservation, Gay Head Moraine, and parts of the Three Ponds Reservation—could support fascinating habitat and provide access to wildlands across the Island. Much of the rest of the conserved woodlands and presumably many currently unprotected woodlands may be actively managed for diverse objectives, including aesthetics, habitat, firewood, and timber. Farming could expand to many open conservation areas.

in the knowledge that despite whatever changes arise, nature continues to function and that we benefit immensely by witnessing these dynamics. Indeed, a major objective in securing wildland landscapes is to accommodate diverse conditions from pristine ancient woods to windthrown hillsides and recovering mid-aged stands. In that variation comes a breadth of habitats and wealth of human experience.

Wildlands complement our actively managed landscapes and provide reference to them for comparison and insight. On landscapes with a history of intensive use like the Vineyard, wildlands—and the streams, wetlands, ponds, and woodland habitats they encompass—present a stunning opportunity to witness nature reassert itself over a formerly tamed land. One eventual product of wildland management will be big ancient trees. These will tower overhead in life, stand as skeletons for decades in death, and crash and molder when they fall in woods, streams, ponds, and wetlands. With their diverse forms and the many processes they support, these large trees have great value. They inspire awe, sublime beauty, and many lessons. On the Island they will answer centuries-old questions and address others arising in light of an uncertain future. How "high-timbered" can Island forests become? What does an old-growth pitch pine forest look like, and will such forests yield future generations of pine? Will beech, beetlebung, and hickory gradually assume something akin to their prehistorical abundance? Or, will climate change and the spread of non-native white pine and spruce produce novel woods that the Island has never seen?

Over time these wild spots will yield scenes absent for centuries. Many will be solemn and majestic. But some will be challenging. Tangles of blown-down trees will inhibit easy passage off of maintained trails. The sight of uprooted or dying trees may trigger calls for intervention and salvage. But nature will function fine without these. We can think of wild forests as we do the eroding bluffs at Lucy Vincent or the salt marshes surrounding Poucha Pond and dunes adjacent Squibnocket Pond. All are wild and should be left alone to experience nature as it expresses itself.

As landowners and organizations identify tracts, large and small, that are allowed to become wilder, it would be ideal if they could capture the Island's variation with regard to soils, vegetation, human history, and natural processes. A few informal wild reserves exist, with great potential for more. The oak bluffs at Wasque, much of the Woods Preserve, large woodlands on Seven Gates Farm, and tracts in West Chop Woods and the State Forest have been managed in recent decades in a hands-off way that demonstrates and confirms the many values of this approach. Other sites have been allowed to develop freely as much from benign neglect as thoughtful design. Formalizing the wildland status of some of these sites will ensure that they persist into the future.

One intriguing possibility for a vast interconnected wildland stretches across the western moraine from the edge of the Great Plain to Vineyard Sound, on lands extending from Polly Hill Arboretum, the Woods Preserve, and Waskosim's Rock through parts of Seven Gates Farm and Cedar Tree Neck. Across this sweep of Middle and North ridges lie diverse forests, from dry open oak woods and moist pockets of beetlebung and red maple to pure groves of beech. With its complicated ownership history, this undulating terrain supports many legacies of use, ranging from small areas unlogged for centuries to scattered old woodlots to former pastures that reforested in the past century. The Island's oldest trees are centered in this expanse, as are hundreds of acres of oaks killed in the past decade by caterpillars. The persistence of 350-year-old trees, the abundant saplings beneath dead trees, and the thriving beech alongside crumbling oaks provide lessons in resilience, recovery, and change.

These and adjoining properties also contain focused areas with great potential for active management that would add substantial variation to the landscape. This includes sites currently exposed to grazing and haying, timber harvesting and firewood cutting, and prescribed fire. The potential for complementary and even

The conservation landscape from Polly Hill Arboretum (bottom) to the Woods Preserve (central ridge of dead trees, Fisher Pond, house and barn), Waskosim's Rock Reservation (mixture of openland and forest at center left), and Seven Gates Farm (large field and forest beyond) forms the largest expanse of morainal forest on the Island, October 2011.

coordinated management across this large landscape is compelling. Given that much of the area is not currently accessible to the public, it is especially important that wildlands extend over lands of the Land Bank and organizations that would allow them to be visited and used for education.

Across the Great Plain and centered on the State Forest lies another great opportunity for coordinated wildland and woodland management of an equally vast but distinctly different forest. Here, large areas of oak and pine are publicly accessible for biking, hiking, hunting, education, study, and other uses. Lying on deep sandy soils and comprising the largest block of conserved forest on the Vineyard, the landscape includes young and maturing forests, scrub oak barrens, and scattered plantations. The older

Allen Healy of Mermaid Farm mowing hay at the Polly Hill Arboretum, June 2013.

forests would make intriguing wildlands as their largest trees are beginning to die or blow down and be replaced by new growth that will diversify the forest in age, structure, and composition. Old forests with dense canopies and weak understories are among the least flammable vegetation on the Plain. Allowing forests to mature and locally shade out flammable scrub oak can reinforce efforts to reduce fuels and fire. Other parts of the State Forest warrant active management. Though technically a "reserve" under state guidelines, large portions will likely continue to be managed by the Department of Conservation and Recreation to maintain the bottoms and fire roads, remove non-native plantations, diversify forest structure, control the southern pine beetle, and reduce wildfire risk to adjoining neighborhoods.

Elsewhere, smaller tracts accessible to neighborhoods and outlying communities could provide local reserves with magnificent trees. Examples include the pines in West Chop Woods and Southern Woodlands, mixed pines and oak on Ripley's Field and the Pennywise Preserve, oak woods in the Poucha Pond and Three Ponds reservations, and mesic woods on Gay Head Moraine.

This call for broad-scale land protection coupled with integrated, landscape-scale conservation management is being advanced elsewhere across New England and nationally. A recent study in northern New England indicates that over the past fifteen years, Vermont, New Hampshire, and Maine have conserved land at a rate adequate to reach the *Wildlands and Woodlands* stated goals of protecting 70 percent of New England woodlands by 2100. Moreover, three times in the past century, this region has witnessed major increases in the pace of land protection, suggesting that the ambitious regional goal could be accomplished even earlier. One approach that has increased the efficiency and pace of conservation is the formation of regional conservation partnerships (RCPs)—collaborations among organizations and agencies to advance a common conservation vision in their region. To date, some forty-four RCPs cover more than sixty percent of New England and are making great progress. Nearly 25 percent of Massachusetts is permanently protected from development (second in New England to New Hampshire, at 32 percent and a 50 percent increase in just over ten years), with great advances spurred by town, state, federal, and private funding, and donations of conservation easements by private landowners. The conservation future of the region is under way and in our hands.

This off-Island world would be surprised to learn that the Vineyard has long had its own conservation partnership. Moreover, it would be shocked to recognize that the Vineyard has the potential to develop an increasingly active and thriving farm community, sustain extensive woodlands that contribute meaningful resources to local need, and enjoy the sights, sounds, and solemnity of expansive ancient woodlands with three-hundred-year-old trees. Building on a remarkable history of conservation puts those aspirations within reach. Aggressively increasing the extent of conserved land and approaching the management of the land with an eye toward its history may allow them to be realized.

Perspectives in Island Time

There is a tendency at every important but difficult crossroad to pretend that it's not really there.
— Bill McKibben, *The End of Nature*

2015. It is nearing the end of a delightfully long spring, the first week in June, and despite more than four weeks of effort the white oaks up on the moraine in West Tisbury are still struggling to fully leaf out. Locally, and more broadly across New England, it is a vibrant and expectant time. It is also a spring that seems filled with more uncertainty and even doubt than usual.

We have successfully emerged from a brutally long, frigid, and snow-filled winter. For the first time ever the snowshoes behind the seat of my pickup were put to more use—for three months continually—on the Island than up at our log cabin deep in the Northeast Kingdom of Vermont. With that experience in the rearview mirror and the leaves crackling beneath my feet as I walk Island paths and enjoy the sounds and freshness of spring, the woods can't help but seem vibrant. Expectancy comes with this. We're past that time each April when I stare at those dark, hulking, and seemingly lifeless forms in the woods that we call trees, with complete disbelief that they could ever spring forth with new green growth. They have, and the woods are that dazzlingly electric green once more, that color that lasts just a week or so. The daffodils have come and gone, as have the choruses of pinkletinks. The towhees are loud on the edges of the woods, the first robins are fledging, and the ospreys are back where they belong—out on the Red Gate knoll above Squibnocket Pond, just off the big field at Quansoo Farm, and in an oak leaning out from the Pimpneymouse Farm island in Poucha Pond. They are back in Goodale's Pit as well. Our neighbor Wayne Smith just changed our dinner plans by leaving us with our first bluefish of the year. We took our last walk down on Lucy Vincent Beach before it is locked up to residents for the summer—so little left of that bluff! And, once again, I'm shocked that the Steamship Authority website is coming up blank with ferry openings and that Five Corners is jammed from all directions.

Summer's nigh, the lighthouse is safely moved (Again. Which one? Take your pick, they will all move again in time), and the breach at Norton Point remains closed. While a cold front raises the question whether a late frost one early morning may nail the

Wequobsque cliffs, Chilmark.

scrub oaks out on the Great Plain, the momentum of the season is all forward, building toward the long, active days of summer. Out on North Road, the ridge above Fisher Pond is predominantly green for the first time in years as the new growth of the expanding tree cover shoots upward and oak carcasses fall to the ground. The early cutting of hay is just wrapped in green bales at the Littlefield place near the Agricultural Society. With the removal of the dam boards at the sluiceway on Warren Pond by Prudy Burt, Johnny Hoy, and Chris Kennedy, the herring are shooting up the Tiasquam and past South Road for the first time in more than fifty years. Much is good.

Yet uncertainty is pervasive. For the first time, reports and anecdotal sightings suggest that winter moth damage is solidly established in pockets of the Island and spreading. Defoliation is spied behind the hospital and out on Chappy, a few trees are leafless on our back hill, and the trees lining the stretch of road coming up-Island out of Vineyard Haven display full crowns of skeletonized leaves. A friend responds optimistically that at least the moths are consuming the non-native privets around her place in Oklahoma, toward the head of the Lagoon. State entomologist Bob Childs notes that winter moths eat nearly everything, yet along State Road and in pockets of the Great Plain it is only oaks that seem completely defoliated. For years the chronic process wreaked by this non-native moth has worked the woods in Falmouth and across Plymouth County. Now they seemed ready to settle in on the Vineyard. Will they become part of the place, weakening the oaks each year, and leveraging upward the prominence of hickories, beech, and beetlebung? Around our place, where I'm warming the house with the last of the trees killed seven years ago by the cankerworm, white oaks appear less prone to winter moths, raising hope that this favored species might also benefit. The patter of dropping frass as I rest next to the long, curving woodpile at the woods edge heightens the tension of the moment and uncertainty for the future. Today we have no answers.

More broadly, the official start of the hurricane season was just preempted by the naming of the first storm out in the Pacific and, coincidentally, release of Colorado State Professor William Gray's annual prediction for tropical storm activity on the east coast. Hearing that fewer storms than average is predicted due to a growing El Niño provides a measure of relief until the newscaster reminds us that all it takes is one direct hit to render any average meaningless. Meanwhile, Tom Dunlop at the *Gazette* has been spreading word of a study by Jeff Donnelly's tempestology group at the Woods Hole Oceanographic Institute that reports monster hurricanes during the past few thousand years. Once again, historical insights dramatically alter our thinking about the plausible future. The Great Colonial Hurricane of 1635 no longer reigns as the benchmark for our worst scenario. Much more powerful winds and greater ocean surges are recorded in the sediments of a Falmouth

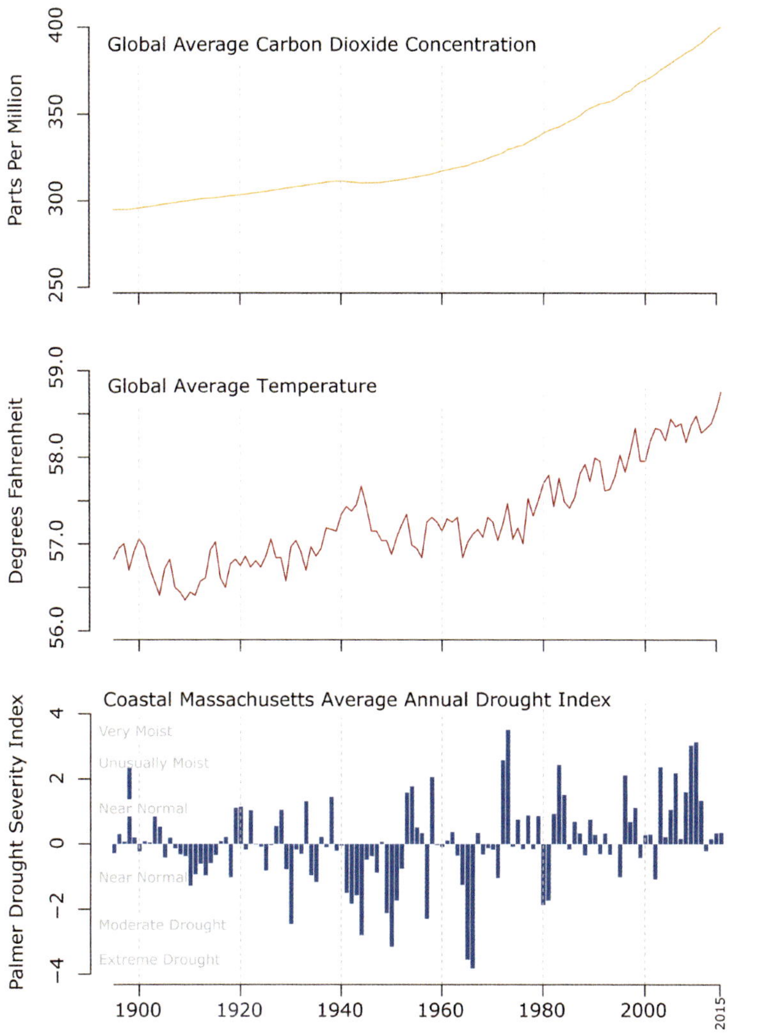

The climate is changing. Since the nineteenth century, the globe and region have experienced changes that influence critical natural and cultural processes. All planning needs to accommodate future changes. Rising levels of carbon dioxide (*top*) are helping to drive a global increase in temperature (*middle*) while also increasing the growth and efficiency of water use by plants. *Bottom:* Although many regions are projected to dry, the fifty-plus year history of increasing moisture and decreasing drought in coastal Massachusetts is expected to continue. Carbon dioxide and global average temperature data from U.S. National Aeronautics and Space Administration; drought data from U.S. National Oceanic and Atmospheric Administration.

Despite the many standing oak carcasses, the lush regrowth of oaks and beeches marks a thriving ecosystem on the Woods Preserve, September 2014.

pond, in the Vineyard's backyard. Yet, just as our efforts to cope with an even more challenging future should be strengthened, another story in the news examines how complacency prevails in coastal communities due to the lengthening period since the last truly strong storm.

Nationally and globally the story is mixed, but generally trending toward the pessimistic. The latest chemical analyses of the nation's precipitation and streams yield surprisingly good news, for they robustly demonstrate the sweeping benefits that can derive from federal regulatory legislation. The Clean Air Act, enacted under Richard Nixon's administration, has worked wonders. Acid rain has been reversed (though not eliminated). The sulfur content of precipitation is plummeting, nitrogen (a polluting fertilizer of our forests and waters) concentrations are declining, rain and snow are increasingly less acidic, and streams, ponds, and stressed

With rising sea levels, salt marshes adjoining Poucha Pond are beginning to break up and salt-tolerant grass species are creeping into adjoining upland fields on Pimpneymouse Farm. An osprey nest sits in the dead top of an oak at the left edge of the wooded island, March 2012.

woods are showing signs of improvements across the northeastern United States and beyond. A colleague from Colorado reports a tenfold reduction in the acidity of rainfall in her Rocky Mountain measurements.

Additional optimism accompanied a report that the cost of solar energy has dropped close to that of fossil fuel and the nation's infrastructure for solar and wind energy is growing at double to triple that of coal. The Science and Policy Exchange at the Harvard Forest released a study documenting that the reduction in coal-based power generation proposed by the Environmental Protection Agency would save four thousand U.S. lives annually. The co-benefits of cleaner energy include increased health and longevity, reduced mercury poisoning of waters, a substantial drop in greenhouse gas emissions, and a slowing of climate change. On the land conservation front, good news also emerged. Over the past decade New England has experienced the fastest pace of land protection ever.

Yet as many positive signs were emerging, the head of the Senate Committee on Science, James Inhofe, brought a snowball into Congress to underscore his resistance to any action acknowledging a human hand in climate change. With Alaska and the Pacific Northwest snowless and absurdly warm, California heading into a megadrought, and a new global temperature record, the arrogance, ignorance, and entrenched power of the forces against environmental action based on science and history remain shocking. Meanwhile, many climate experts have concluded that the goal of limiting global warming to two degrees is already out of reach. The inertia in the climate system and legacy of our history of land use, deforestation, and fossil fuel combustion will push the system—that is, the globe—further regardless of our best efforts. We are reminded that early, aggressive action is always the best approach in addressing any environmental concern. That includes the ongoing destruction of nature in our own backyard.

When I initially outlined this volume, I presumptuously titled this last chapter "Insights." I conceived it as drawing conclusions together from each section of the book into a series of lessons about the Island. Over time at this work my attitude has changed. Though I've learned much and gleaned many personal insights—from the generalizable to the unique—my sense, increasingly, is that of skimming the Island's surface. My own explorations, intense and prolonged as they have been, are just beginning to plumb the depths of land and people. There are many ancient ways to follow, archives to decipher, and individuals left to encounter in history, on a back porch, or when rambling across a new landscape. My plan for insights gave way to a series of perspectives—a few key conclusions accompanied by new leads to pursue beginning tomorrow.

Overwashed portion of East Beach on Chappaquiddick, just north of Wasque, March 2013.

Most important, and surprising given popular thinking these days, is the conclusion that the bulk of Island history was dominated by natural processes. Set aside is the notion of a landscape carefully managed by native people for thousands of years. In its place is

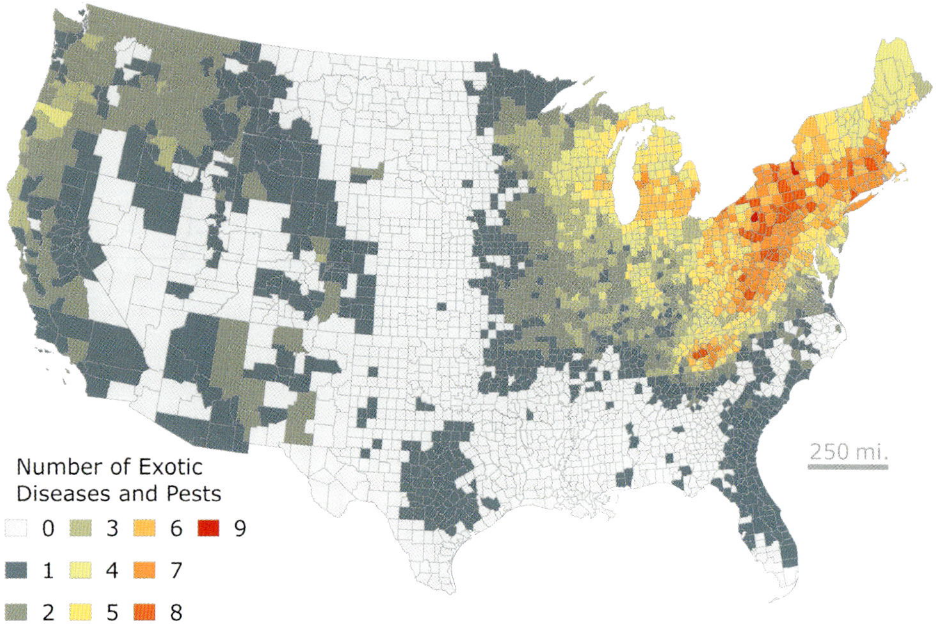

The Northeast has the dubious distinction of being the national epicenter for the arrival and spread of damaging forest diseases and insects. Data from Krist (2014). U.S. Forest Service, National Insect and Disease Forest Risk Assessment.

a quite wild Vineyard that persisted until recent centuries. This perspective lends new awareness and appreciation for native ability to accommodate those occasionally abrupt and stressful changes brought by nature. It also brings greater awe and dismay at the rate and magnitude of changes wrought by humans in recent times. An utterly distinct era arrived on these shores four hundred years ago, a cultural and ecological transition even greater than previously imagined. This was no simple shift in technology or the imposition of will over nature. Rather, cultural relationships to the land and water were transformed. Overnight, human activity shifted from background noise to an increasingly dominant force.

Along with this new perspective comes a change in the approach needed to interpret the land through time. The closer we get to the present, the less the land is shaped by natural processes that have operated for eons to those that people have wrought in recent decades. The emphasis on the inertia of ecological forces and legacies of past events remains, but the focus increasingly shifts to cultural signatures on the land. Hopefully, recognition of this fact will not simply improve our understanding of the land today and our ability to forecast where it may go tomorrow. It should also give us great pause in examining our attitudes and behavior toward the earth. We are now the major actors on this stage. Our decisions, and our every move, affect not just the present and our current environment. Every action sets off a cascade of subsequent reactions that will carry on deep into the future. We are at major crossroads regarding directions for the Island, the region, and the globe. Where we put our energy will determine that future.

On the Island, this work also highlights how much more we can draw from this approach of melding bits of history and science to understand this place and guide sound decisions. The volume has drawn a focus on a few key actors, events, and places—the first insights by Pete Ogden into the landscape's dynamics; the ability to read our landscape from Henry Whiting's map of his in the nineteenth century; the widespread survival of forest and unique

Oak tree still partially rooted in its original location at Wasque after the soil was eroded away by the current, December 2013.

position of the Great Plain in Island history; the role of agriculture in shaping the land and yielding resources, conservation values, and quality of place; and the extraordinary benefits and challenges that come from planning and acting as an Island. But, each avenue of exploration highlights how much more natural and cultural history of the Island remains to be uncovered. There are countless leads to explore, papers to decipher, and knowledge to tap in joy-filled quests to understand more. Lessons emerging from these efforts will contribute greatly to our wise management of the place.

Examples abound. Just think of the changing and thriving landscapes left to explore back and forward through time—the coastal variation of Squibnocket, Red Gate Farm, and Moshup Trail; the vastness of Katama; the Peaked Hill region; and the woodlands, peatlands, and fields of Chappy. These are but signature examples of the many magnificent landscapes with completely different histories whose stories remain to be known, or shared. And then there are the people: Simon Athearn, Daniel Fisher, Whiting and Shaler, Jeremiah Pease and Levitt Thaxter, many Butlers, Nortons, Luces, Cottles,

Plate-like bark on a two-foot-diameter pitch pine in the ancient West Chop Woods.

and others. As engaging puzzles, these topics offer challenges along with the rewards of insight into uncharted territory for scholars at any level. But as these stories of land and people are deciphered and woven into a fabric of Island history they will yield much more—connections to the land past and present, motivation for preservation and conservation, and knowledge to guide decisions and actions.

Such detective work advances through a combination of serendipitous leads, newfound goldmines, and determined scouring of records; all of this could benefit from the good will and effort of Island folk near and far and Island institutions. One intriguing notion is that of the citizen historian: individuals, whether on or distant to the Island these days, scouring their closets, drawers, attics, basements, and barns for Island history. Journals, account books, letters, maps, photographs, and artifacts abound, waiting to be recovered and discovered, interpreted and preserved. Modest efforts in these directions taken in researching this volume uncovered every category of materials lying untapped and often as likely to be discarded as archived. Heroic efforts by Tom Dunlop and the *Gazette* to uncover and preserve early home movies and make these broadly available reveal the value of such efforts. Linsey Lee's sustained work with oral histories underscores the fact that important avenues to the past also lie outside traditional archival materials but require dedicated pursuit and documentation. Individuals can explore, document, or archive. They can also forge their own perceptive histories, as Jim Norton has for Vineyard Haven, Edo Potter has for Pimpneymouse Farm, Tess Bramhall has for Seven Gates, and folks from Chappy have for their frequently separated island.

At our major repositories, opportunities abound to aid this quest to inform the future through explorations of the past. The *Gazette* has a searchable digital archive back to the 1970s, its entire collection back to 1846 sits on (clunky) microfiche readers at Island libraries, and its invaluable archive of file drawers is readily accessible in the second floor on Summer Street. But, the paper's

Sun setting on the Menemsha Harbor breakwater.

goal of a comprehensive digital archive will open immense historical windows that will serve scholars, students, local officials, citizens, and more. To immerse deep in time and accounts of raging wildfires, plover gunning, Norton Point openings, western and global travels, agricultural collapse, and local boosterism will add immeasurable depth to our understanding of the past.

Meanwhile, at the Museum, digital access and curation is creeping slowly into sight. The promise that this holds is gigantic, as shown by the Museum's counterpart on Nantucket or the marvelous Leventhal map center at the Boston Public Library. Maps, photos, art work, transcripts, digital reproductions, and oral histories online convey an aspirational future that would be relished in homes on- and off-Island, schools, and offices. As volunteer efforts by Dick Burt and Jim Richardson with the Museum's native and colonial collections demonstrate, basic curation yields many insights. The effective presentation of Museum collections to all will spur support, the discovery and donation of more collections, and broader connections with Island history.

Finally, and most important, there is the curation and conservation of cultural resources and the land itself. When, in his delightful *Elegy* on Martha's Vineyard, Everett Allen returns to Cromwell Lane, his childhood neighborhood a block from Five Corners, the operative word capturing his experience is "gone": "Its onetime ruts and meanderings, cinders, grasses, and overhanging boughs are no more." Emma Cromwell's roses were replaced by a parking lot; the arbor of Concord grapes was "gone." "The Cromwell house, a venerable, Victorian mansion, has been bulldozed in deference to more practical matters."

It is not practical or desirable to retain everything. Cromwell Lane is an extreme example, after all, for it is a half-step from ground zero in the persistent battle over ferry landings, the Stop and Shop expansion, and the future of Vineyard Haven. But there is a way forward that can conserve the bulk of the remaining land on the Island along with most important structures and features. There is a path that combines the retention of nature and history, yields local food and products, sustains but lessens the impact of the economic forces that threaten to destroy the very qualities of the Island that underlie it, and gives us all access to, and appreciation for, nature, cultural resources, and an understanding of human history.

We can benefit from preserving and understanding old buildings, ancient ways, cellar holes, and ditch and ridge fencelines. We need open fields, grazing animals, and people working and living from the land. We gain a sense of our place on earth from those contorted trees at Quansoo, the ancient stools in ancient woods, and the multi-stemmed trees that tell tales of wood chopping and the annual search for anything that might warm a family in winter. And, we gain new perspective on the Island and ourselves as the trees begin to tower and age, the stonewalls and rusting cars in the woods become lichen-covered and archaic, and as the boards are pulled from the dams, the herring crowd their way inland, and the salters once again return to the sea.

Bibliography

Abrams, M. D. 2002. "The Postglacial History of Oak Forests in Eastern North America." In *Oak Forest Ecosystems: Ecology and Management for Wildlife,* edited by W. J. McShea and W. M. Healy, 34–45. Baltimore, Md.: The Johns Hopkins University Press.

Abrams, M. D. 2007. "Tales from the blackgum, a consummate subordinate tree." *BioScience* 57: 347–359.

Adams, M. B. 1992. "The State Forest: Wasteland in the woodland, new agenda for hope." *Martha's Vineyard Magazine,* Fall/Holiday Issue, 13–19.

Allen, C. D. 2010. "A global overview of drought and heat-induced tree mortality reveals emerging climate change risks for forests." *Forest Ecology and Management* 259:660–684.

Allen, E. S. 1976. *A Wind to Shake the World: The Story of the 1938 Hurricane.* New York: Little, Brown.

Allen, E. S. 1982. *Martha's Vineyard: An Elegy.* Boston, Mass.: Little Brown & Co.

Alley, J., R. Carroll, N. Freydberg, S. Frisch, A. Hale, H. Hancock, H. B. Hough, D. Hull, L. Look, D. Swift, and W. Tyra. 1973. *A Report to the People of Martha's Vineyard on the Development of the Island.* Edgartown, Mass.

Annett, B., G. Gerlach, T. L. King, and A. R. Whiteley. 2012. "Conservation genetics of remnant coastal brook trout populations at the southern limit of their distribution: Population structure and effects of stocking." *Transactions of the American Fisheries Society* 141:1399–1410.

Appleyard, D., K. Lynch, and J. R. Myer. 1964. *A View from the Road.* Cambridge, Mass.: MIT Press.

Askins, R. A. 1987. "Relationship between the regional abundance of forest and the composition of forest bird communities." *Biological Conservation* 39:129–152.

Askins, R. A. 1993. "Population trends in grassland, shrubland, and forest birds in eastern North America." *Current Ornithology* 11:1–34.

Askins, R. A. 2001. "Sustaining biological diversity in early successional communities: The challenge of managing unpopular habitats." *Wildlife Society Bulletin* 29:407–412.

Askins, R. A., F. Chavez-Ramirez, B. C. Dale, C. A. Haas, J. R. Herkert, F. L. Knopf, and P. D. Vickery. 2007. "Conservation of grassland birds in North America: Understanding ecological processes in different regions." *Ornithological Monographs* 2007:1–46.

Athearn, E. B. 1996. *Chronicle of an Islander: Martha's Vineyard History and Adventure.* Centerville, Mass.: Edwin B. Athearn.

Baker, V. R., R. C. Kochel, J. E. Laity, and A. D. Howard. 1990. "Spring sapping and valley network development." *Geological Society of America Special Paper* 252:235–365.

Balco, G. 2011. "Contributions and unrealized potential contributions of cosmogenic-nuclide exposure dating to glacier chronology, 1990–2010." *Quaternary Science Reviews* 30:3–27.

Balmer, O., and A. Erhardt. 2000. "Consequences of succession on extensively grazed grasslands for central European butterfly communities: Rethinking conservation practices." *Conservation Biology* 14:746–757.

Banks, C. E. 1911. *The History of Martha's Vineyard, Dukes County, Massachusetts.* Boston: George H. Dean.

Barrs, B. 1975. "Jurisdictional Allocations in Land Use Planning: The Battle for Control on Martha's Vineyard." Master's thesis, Massachusetts Institute of Technology.

Bellantoni, N. F. 1987. "Faunal Resource Availability and Prehistoric Cultural Selection on Block Island." Ph.D. thesis, University of Connecticut.

Bendremer, J. C. M. 1993. "Late Woodland Settlement and Subsistence in Eastern Connecticut." Ph.D. thesis, University of Connecticut.

Benninghoff, W. S. 1948. "The Late Quaternary Vegetation of No Mans Land, Off Martha's Vineyard, Massachusetts." Ph.D. thesis, Harvard University.

Berlik, M. M., D. B. Kittredge, and D. R. Foster. 2002. "The illusion of preservation: A global environmental argument for the local production of natural resources." *Journal of Biogeography* 29:1557–1568.

Bernstein, D. J. 1993. *Prehistoric Subsistence on the Southern New England Coast: The Record from Narragansett Bay.* New York: Academic Press.

Bierregaard, R. O., B. A. David, L. Gibson, R. S. Kennedy, A. F. Poole, M. S. Scheibel, and

J. Victoria. 2014. "Post-DDT recovery of osprey (*Pandion haliaetus*) populations in southern New England and Long Island, New York, 1970–2013." *Journal of Raptor Research* 48: 361–374.

Birks, H. H., J. B. Birks, P. E. Kaland, and D. Moe. 1988. *The Cultural Landscape: Past, Present and Future.* New York: Cambridge University Press.

Boose, E. R., D. R. Foster, and M. Fluet. 1994. "Hurricane impacts to tropical and temperate forest landscapes." *Ecological Monographs* 64: 369–400.

Bradford, W. 1952. *Of Plymouth Plantation, 1620–1647.* Edited by S. E. Morison. New York: Knopf.

Bragdon, K. J. 1996. *Native People of Southern New England, 1500–1650.* Norman: University of Oklahoma Press.

Bramhall, E. 1988. *Seven Gates Farm: The First Hundred Years, 1887–1987.* Vineyard Haven, Mass.: Seven Gates Farms.

Breiby, W. 2004. "The Mature Woodlands of Martha's Vineyard, Naushon, Nantucket and Tuckernuck Islands, Massachusetts." Master's thesis, University of Massachusetts.

Brereton, J., and L. S. Livingston. 1602. *A Briefe and True Relation of the Discoverie of the North part of Virginia in 1602.* London: Bishop.

Brewster, W. 1885. "The heath hen of Massachusetts." *The Auk* 2:80–84.

Brewster, W. 1890. "The heath hen. Notes on the heath hen of Massachusetts." *Forest and Stream* 35:188.

Bromley, S. W. 1935. "The original forest types of southern New England." *Ecological Monographs* 5:61–89.

Busby, P. E., G. Motzkin, and D. R. Foster. 2008. "Multiple and interacting disturbances lead to *Fagus grandifolia* dominance in coastal New England." *Journal of the Torrey Botanical Society* 135:346–359.

Busby, P. E., G. Motzkin, and B. Hall. 2009. "Distribution and dynamics of American beech in coastal southern New England." *Northeastern Naturalist* 16:159–176.

Butler, W. 1967. "Martha's Vineyard in 1792. A diary, continued." *Dukes County Intelligencer* 8:96–101.

Byers, D. S., and F. Johnson. 1940. *Two Sites on Martha's Vineyard.* Papers of the Robert S. Peabody Foundation for Archaeology. Andover, Mass.: Phillips Academy.

Caldwell, W. A. 2015. *Reflections on Martha's Vineyard.* Edited by T. Dunlop. Edgartown, Mass.: Vineyard Stories.

Capace, J. 2001. *Land-Use History of Cape Poge and Wasque.* Vineyard Haven, Mass.: The Trustees of Reservations, Island Regional Office. Available at http://www.thetrustees.org/assets/documents/places-to-visit/Land-Use.pdf.

Ceci, L. 1977. "The Effect of European Contact and Trade on the Settlement Pattern of Indians in Coastal New York, 1524–1665." Ph.D. thesis, City University of New York.

Ceci, L. 1990. "Squanto and the Pilgrims: On Planting Corn 'in the Manner of the Indians'." In *The Invented Indian,* edited by J. A. Clifton, 71–89. New Brunswick, N.J.: Transaction.

Champlain, Samuel de. 1878. *Voyages of Samuel de Champlain,* Vol. 2. Translated from the French by Charles Pomeroy Otis. With Historical Illustrations, and a Memoir by the Rev. Edmund F. Slafter. Boston: Prince Society.

Chilton, E. S. 2002. "'Towns They Have None': Diverse Subsistence and Settlement Strategies in Native New England." In *Northeast Subsistence: Settlement Change, A.D. 700–1300,* edited by J. P. Hart and C. Reith, 289–300. Albany: New York State Museum.

Chilton, E. S. 2008. "So Little Maize, So Much Time: Understanding Maize Adoption in New England." In *Current Northeast Paleoethnobotany II,* edited by J. P. Hart, 53–58. Albany: New York State Museum.

Chilton, E. S. 2010. "The Origin and Spread of Maize (*Zea Mays*) in New England." In *Nantucket and Other Native Places: The Legacy of Elizabeth Alden Little,* edited by E. S. Chilton and M. R. Rainey, 159–180. Albany: State University of New York Press.

Chilton, E. S., and D. L. Doucette. 2002. "Archaeological Investigations at the Lucy Vincent Beach Site (19-DK-148): Preliminary Results and Interpretations." In *A Lasting Impression: Coastal, Lithic, and Ceramic Research in New England Archaeology,* edited by J. E. Kerber. Westport, Conn.: Praeger.

Chilton, E. S., and D. L. Doucette. 2002. "The archaeology of coastal New England: The view from Martha's Vineyard." *Northeast Anthropology* 64:55–66.

Clarke, G. L., and W. A. Patterson III. 2007. "The distribution of disturbance-dependent rare plants in a coastal Massachusetts sandplain: implications for conservation and management." *Biological Conservation* 136:4–16.

Cogbill, C. V., J. Burk, and G. Motzkin. 2002. "The forests of presettlement New England, USA: Spatial and compositional patterns based on town proprietor surveys." *Journal of Biogeography* 29:1279–1304.

Cokinos, C. 2000. *Hope Is the Thing with Feathers: A Personal Chronicle of Vanished Birds.* New York: Penguin.

Crevecoeur, J. H. 2007 (1782). *Letters from an American Farmer.* Carlisle, Mass.: Applewood Books.

Cronon, W. 1983. *Changes in the Land: Indians, Colonists, and the Ecology of New England.* New York: Hill and Wang.

Daggett, J. T. 1965. *It Began with a Whale: Memories of Cedar Tree Neck, Martha's Vineyard.* Boston: Fleming & Son.

Day, G. M. 1953. "The Indian as an ecological factor in the northeastern forest." *Ecology* 34: 329–346.

Dean, V. K. 1939. "The Economic Geography of Martha's Vineyard." Ph.D. thesis, Clark University.

DeGraaf, R. M., and R. I. Miller. 1996. *Conservation of Faunal Diversity in Forested Landscapes.* Amsterdam: Springer.

Devens, S. A. 1838. *Sketches of Martha's Vineyard and Other Reminiscences of Travel at Home, etc.* Boston: James Munroe & Co.

Donahue, B., J. Burke, M. D. Anderson, A. Beal, T. Kelly, M. Lapping, H. Ramer, R. Libby, and L. Berlin. 2015. *A New England Food Vision: Healthy Food for All, Sustainable Farming and Fishing, Thriving Communities.* Durham: University of New Hampshire.

Donnelly, J. P. 2006. "A revised late Holocene sea-level record for northern Massachusetts, USA." *Journal of Coastal Research* 22:1051–1061.

Donnelly, J. P., A. D. Hawkes, P. Lane, D. MacDonald, B. N. Shuman, M. R. Toomey, P. J. van Hengstrum, and J. D. Woodruff. 2015. "Climate forcing of unprecedented intense-hurricane activity in the last 2000 years." *Earth's Future* 3:49–65.

Dresser, T. 2011. *Wampanoag Tribe of Martha's Vineyard: Colonization to Recognition.* Charleston: The History Press.

Dunford, F., and G. O'Brien. 1997. *Secrets in the Sand: The Archaeology of Cape Cod.* Hyannis, Mass.: Parnassus Imprints.

Dunlop, T. 2012. *The Chappy Ferry Book: Back and Forth Between Two Worlds—527 Feet Apart.* Edgartown, Mass.: Vineyard Stories.

Dunlop, T. 2013. "Farming on Chappaquiddick and a History of the Chappy Schoolhouse." In *A Report for Slip Away Farm of Chappaquiddick.* Edgartown, Mass.: Martha's Vineyard Preservation Trust, Martha's Vineyard Land Bank Commission.

Dunwiddie, P. W. 1989. "Forest and heath: The shaping of vegetation on Nantucket Island." *Journal of Forest History* 3:126–133.

Dunwiddie, P. W. 1992. *Changing Landscapes: A Pictorial Field Guide to a Century of Change on Nantucket.* Nantucket, Mass.: Nantucket Conservation Foundation, Nantucket Historical Association, and Massachusetts Audubon Society.

Dunwiddie, P. W. 1994. *Martha's Vineyard Landscapes: The Nature of Change.* West Tisbury, Mass.: Vineyard Conservation Society.

Duranleau, D. L. 2009. "Flexible Sedentism: The Subsistence and Settlement of Coastal New England and New York." Ph.D. thesis, Harvard University.

Earle, J. M. 1861. *Report to the Governor and Council Concerning the Indians of the Commonwealth. Under the Act of April 1859.* Boston: William White Printer.

Eberhardt, R., D. R. Foster, G. Motzkin, and B. Hall. 2003. "Conservation of changing landscapes: Vegetation and land-use history of Cape Cod National Seashore." *Ecological Applications* 13:68–84.

Emanuel, K., R. Sundararajan, and G. Williams. 2008. "Hurricanes and global warming." *American Meteorological Society* 89:347–367.

Faison, E. K., D. R. Foster, W. W. Oswald, B. C. S. Hansen, and E. D. Doughty. 2006. "Early-Holocene openlands in southern New England." *Ecology* 87:2537–2547.

Feldhamer, G. A. 2002. "Acorns and White-Tailed Deer: Interrelationships." In *Oak Forest Ecosystems: Ecology and Management for Wildlife,* edited by W. J. McShea and W. M. Healy, 215–223. Baltimore: Johns Hopkins University Press.

Fitzgerald, D. M., and S. Van Heteren. 1999. "Classification of paraglacial barrier systems: Coastal New England, USA." *Sedimentology* 46: 1083–1108.

Flender, J. O. 2005. "Chilmark's Paint Mill, an ancient manufactory." *Dukes County Intelligencer* 47:3–17.

Fletcher, P. C., and R. J. Roffmoli. 1986. *Soil survey of Dukes County, Massachusetts.* Holden, Mass.: United States Department of Agriculture, Soil Conservation Service.

Forbush, E. H. 1918. "The heath hen of Martha's Vineyard." *American Museum Journal* 18: 275–285.

Forbush, E. H. 1927. *Birds of Massachusetts and Other New England States, Part II: Land Birds from Bobwhites to Grackles.* Edited by

Massachusetts Department of Agriculture. Norwood, Mass.: Berwick and Smith Company.

Foster, D. R. 1988. "Species and stand response to catastrophic wind in central New England, USA." *Journal of Ecology* 76:135–151.

Foster, D. R. 2002. "Thoreau's country: A historical-ecological perspective on conservation in the New England landscape." *Journal of Biogeography* 29:1537–1555.

Foster, D. R., and J. D. Aber. 2004. *Forests in Time: The Environmental Consequences of 1,000 Years of Change in New England.* New Haven: Yale University Press.

Foster, D. R., J. D. Aber, J. M. Melillo, R. D. Bowden, and F. A. Bazzaz. 1997. "Forest response to disturbance and anthropogenic stress: Rethinking the 1938 Hurricane and the impact of physical disturbance vs chemical and climate stress on forest ecosystems." *BioScience* 47:437–445.

Foster, D. R. (ed.), A. D'Amato, B. Baiser, A. Ellison, D. Orwig, W. Oswald, A. Barker Plotkin, J. Thompson. 2014. *Hemlock: A Forest Giant on the Edge.* New Haven: Yale University Press.

Foster, D. R., B. M. Donahue, D. B. Kittredge, K. F. Lambert, M. L. Hunter, B. R. Hall, L. C. Irland, R. J. Lilieholm, D. A. Orwig, A. W. D'Amato, E. A. Colburn, J. R. Thompson, J. N. Levitt, A. M. Ellison, W. S. Keeton, J. D. Aber, C. V. Cogbill, C. T. Driscoll, T. J. Fahey, and C. M. Hart. 2010. *Wildlands and Woodlands: A Vision for the New England Landscape.* Cambridge, Mass.: Harvard University Press.

Foster, D. R., B. Hall, S. Barry, S. Clayden, and T. Parshall. 2002. "Cultural, environmental and historical controls of vegetation patterns and the modern conservation setting on the island of Martha's Vineyard, USA." *Journal of Biogeography* 29:1381–1400.

Foster, D. R., D. B. Kittredge, B. Donahue, G. Motzkin, D. A. Orwig, A. M. Ellison, B. Hall, E. A. Colburn, and A. D'Amato. 2005. Wildlands and Woodlands: A Vision for the Forests of Massachusetts. Petersham, Mass.: Harvard Forest. Available at http://www.wildlandsandwoodlands.org/sites/default/files/Wildlands%20and%20Woodlands%20New%20England.pdf.

Foster, D. R., and G. Motzkin. 1999. *Historical Influences on the Landscape of Martha's Vineyard: Perspectives on the Management of the Manuel F. Correllus State Forest.* Harvard Forest Paper no. 23. Petersham, Mass.: Harvard Forest. Available at http://harvardforest.fas.harvard.edu/sites/harvardforest.fas.harvard.edu/files/publications/pdfs/vineyard.pdf.

Foster, D. R., and G. Motzkin. 2003. "Interpreting and conserving the openland habitats of coastal New England: Insights from landscape history." *Forest Ecology and Management* 185: 127–150.

Foster, D. R., G. Motzkin, D. Bernardos, and J. Cardoza. 2002. "Wildlife dynamics in the changing New England landscape." *Journal of Biogeography* 29:1337–1357.

Foster, D. R., and D. A. Orwig. 2006. "Preemptive and salvage harvesting of New England forests: When doing nothing is a viable alternative." *Conservation Biology* 20:959–970.

Foster, D. R., W. W. Oswald, E. K. Faison, E. D. Doughty, and B. C. S. Hansen. 2006. "A climatic driver for abrupt mid-Holocene vegetation dynamics and the hemlock decline in New England." *Ecology* 87:2959–2966.

Foster, D. R., F. Swanson, J. D. Aber, I. Burk, N. Brokaw, D. Tilman, and A. Knapp. 2002. "The importance of land-use legacies to ecology and conservation." *BioScience* 53: 77–88.

Freeman, F. 1869. *The History of Cape Cod.* Boston: Piper.

Freeman, J. 1807. "Description of Dukes County." *Dukes County Intelligencer* 12:4.

Gamble, A. (ed.); H. F. Walling (Illustrator). 2009. *The 1858 Map of Cape Cod, Martha's Vineyard and Nantucket.* Dennis, Mass.: On Cape Publications.

Goldstein, P. Z. 1997. "Lepidopteran Assemblages and the Management of Sand Plain Communities on Martha's Vineyard, Massachusetts." In *Grasslands of Northeastern North America: Ecology and Conservation of Native and Agricultural Landscapes,* edited by P. D. Vickery and P. W. Dunwiddie, 217–236. Lincoln: Massachusetts Audubon Society.

Gookin, D. 1970. *Historical Collections of the Indians in New England.* Edited by J. H. Fiske. Worcester, Mass.: Towtaid.

Grinnell, G. B. 1910. *American Game Bird Shooting.* New York: Forest and Stream Publishing Company.

Griscom, L., and G. B. Emerson. 1959. *Birds of Martha's Vineyard.* Martha's Vineyard, Mass.: Privately Printed.

Gross, A. O. 1928. "The heath hen." *Memoirs of the Boston Society of Natural History* 6.

Guernsey, S. J. 1916. "Notes on explorations of Martha's Vineyard." *American Anthropologist* 18:81–97.

Halligan, J. J. 2000. "Maushop's Legacy: Cultural Continuity on Martha's Vineyard, MA, 7000

Years Ago to the Present." Master's thesis, Harvard University.

Halperin, M. R. 2013. "Introduction to *The Smiths of Pohoganut*." In *From the Diaries of Hannah Smith and Rebecca Smith, 1813–1814,* edited by Marian Ragan Halperin. Martha's Vineyard, Mass.: Martha's Vineyard Museum.

Harff, C. H. 2004. *A Dream Fulfilled—Farm Neck Golf Club.* Occoquan, Wisc.: Creating Results Publications.

Herbster, H., and S. G. Cherau. 2000. *Archaeological Reconnaissance Survey Town of Edgartown, Martha's Vineyard, MA.* Pawtucket, R.I.: Public Archaeology Laboratory.

Herbster, H., and S. G. Cherau. 2002. *Archaeological Reconnaissance Survey Town of Aquinnah, Massachusetts.* Pawtucket, R.I.: Public Archaeology Laboratory.

Hine, C. G. 1907. *The History of Cedar Neck (Hines Point, Vineyard Haven).* New York: Hine Bros.

Hine, C. G. 1908. *The Story of Martha's Vineyard.* New York: Hine Bros.

Hitchcock, E. 1832. "Report on the geology of Massachusetts; examined under the direction of the government of that state during the years 1830 and 1831." *American Journal of Science and Arts (1820–1897)* 22:1–70.

Hitchcock, E. 1841. "Final Report on the Geology of Massachusetts." In *Massachusetts Geological Survey.* Amherst, Mass.: J. S. & C. Adams.

Hitchcock, E. 1841. *First Anniversary Address before the Association of American Geologists, at their Second Annual Meeting in Philadelphia, April 5, 1841.* New Haven, Conn.

Hollick, C. A. 1893. "Observations on the geology and botany of Martha's Vineyard. Contributions from the Geological Department of Columbia College." *Transactions of the New York Academy of Science* 13:8–22.

Hornaday, W. T. 1913. *Our Vanishing Wildlife: Its Extermination and Preservation.* New York: Scribners.

Hough, H. B. 1936. *Martha's Vineyard Summer Resort, 1835–1933.* Rutland, Vt.: Tuttle.

Hough, H. B. 1940. *Country Editor.* Martha's Vineyard, Mass.: Doubleday, Doran & Company.

Hough, H. B. 1950. *Once More the Thunderer.* New York: Ives Washburn.

Hough, H. B. 1982. *Martha's Vineyard Summer Resort, After 100 years.* Rutland, Vt.: Academy Books.

Hough, H. B. 1985. *Far Out the Coils: A Personal View of Life and Culture on Martha's Vineyard.* Edited by K. Tweed. Vineyard Haven, Mass.: Tashmoo Press.

Huntington, E. G. 1959. "An Archaeological Study from Martha's Vineyard." In *The Dukes County Intelligencer 3.* Edgartown, Mass.: Dukes County Historical Society.

Hurley, S. 1998. *The Wild Natives of Cape Cod.* Buzzard's Bay: Massachusetts Division of Fisheries and Wildlife.

Jarvinen, B. R. 2006. *Storm Tides in Twelve Tropical Cyclones (Including Four Intense New England Hurricanes).* Miami: National Hurricane Center. Available at http://www.aoml.noaa.gov/hrd/Landsea/12Tides.pdf.

Jelesnianski, C. P., J. Chen, and W. A. Shaffer. 1992. *SLOSH: Sea, Lake and Overland Surges from Hurricanes.* NOAA Technical Report NWS 48. Silver Springs, Md.: NOAA.

Jones, A. L., and P. D. Vickery. 1997. "Distribution and Population Status of Grassland Birds of Massachusetts." In *Grasslands of Northeastern North America: Ecology and Conservation of Native and Agricultural Landscapes,* edited by P. D. Vickery and P. W. Dunwiddie. Lincoln: Massachusetts Audubon Society.

Jones, P. A. 2007. *Oak Bluffs: The Cottage City Years on Martha's Vineyard.* Charleston: Arcadia Publishing.

Josselyn, J. 1883. "Two voyages to New England [1675]." *Massachusetts Historical Society Collections* 33:211–354.

Keith, A. R. 1969. "The mammals of Martha's Vineyard." *Dukes County Intelligencer* 11:47–98.

Keniston, A. 1966. "The last years of the heath hen." *Dukes County Intelligencer* 7:282–287.

Knutson, T. R., J. L. McBride, J. Chan, K. Emanuel, G. Holland, C. Landsea, I. Held, J. P. Kossin, A. K. Srivastava, and M. Sugi. 2010. "Tropical cyclones and climate change." *Nature Geoscience* 3:157–163.

Krech, I. S. 1999. *The Ecological Indian: Myth and History.* New York: W. W. Norton.

Krist, F. J., J. R. Ellenwood, M. E. Woods. 2014. *2013–2027 National Insect and Disease Forest Risk Assessment.* Vol. FHTET 14-01. Fort Collins, Colo.: USDA Forest Service Forest Health Technology Enterprise Team.

Laderman, A. D., M. Brody, and E. Pendleton. 1989. *The Ecology of Atlantic White Cedar Wetlands: A Community Profile.* Woods Hole, Mass.: Woods Hole Oceanographic Institution.

Largy, T. B., P. G. Burns, E. S. Chilton, and D. Doucette. 2002. "Lucy Vincent Beach: Another look at the prehistoric exploitation of piscine resources off the coast of

Massachusetts." *Northeast Anthropology* 64: 67–73.

Lengyel, J. 1991. "Marching towards underdevelopment." *Vineyard Gazette,* December 31.

Leopold, A. 1939. "The farmer as a conservationist." *American Forests and Forest Life* 45:294–299.

Liebhold, A. M., D. G. McCullough, L. M. Blackburn, S. J. Frankel, B. Von Holle, and J. E. Aukema. 2013. "A highly aggregated geographical distribution of forest pest invasions in the USA." *Diversity and Distributions* 19:1208–1216.

Lindenmayer, D. B., D. R. Foster, J. F. Franklin, M. L. Hunter, R. F. Noss, F. A. Schmiegelow, and D. Perry. 2004. "Salvage harvesting policies after natural disturbance." *Science* 303:1303.

Little, E. A., and J. C. Andrews. 2010. "Drift Whales at Nantucket: The Kindness of Moshup." In *Nantucket and Other Native Places: The Legacy of Elizabeth Alden Little,* edited by E. S. Chilton and M. L. Rainey, 63–102. Albany: State University of New York Press.

Little, E. A., and J. E. Kerber. 1988. "Where are the woodland villages on Cape Cod and the Islands?" *Bulletin of the Massachusetts Archaeological Society* 49:72–82.

Livingstone, D. N. 1980. "Nature and man in America: Nathaniel Southgate Shaler and the conservation of natural resources." *Transactions of the Institute of British Geographers* 5:369–382.

Long, S. 2016. *Thirty-Eight: The Hurricane That Transformed New England.* New Haven, Conn.: Yale University Press.

Loss, S. R., T. Will, and P. P. Marra. 2013. "The impact of free-ranging domestic cats on wildlife of the United States." *Nature Communications* 4:1–7.

Ludlum, D. M. 2001. *Early American Hurricanes, 1492–1870.* Boston: American Meteorological Society.

Lynch, K. 1973. *Looking at the Vineyard: A Visual Study for a Changing Island.* West Tisbury, Mass.: Vineyard Open Land Foundation.

Mann, C. C. 2006. *1491: New Revelations of the Americas Before Columbus.* New York: Vintage Books.

Marinelli, L. 2002. *Never Say Die: The Private Life and Political Thunder of a Dying Breed.* Oak Bluffs, Mass.: Linda Marinelli.

Marsicek, J. P., B. Shuman, S. Brewer, D. R. Foster, and W. W. Oswald. 2011. "Quantifying climatic changes associated with the mid-Holocene *Tsuga* decline at ca. 5.5 ka in the northeastern United States." *Quaternary Science Reviews* 80: 129–142.

Martha's Vineyard Airport Master Plan: Final Scope of Work. 2012. Edgartown, Mass. Prepared by Jacobs Company. Available at http://www.mvyairport.com/masterplan/pdf/MVY%20AMP%20Final%205.31.12%20.pdf.

Martha's Vineyard Donor's Collaborative. 2010. *Understanding the Vineyard.* West Tisbury, Mass. Available at http://www.mvdonors.org/documents/UTVforthewebfinal.pdf.

Massachusetts Department of Conservation and Recreation. 2015. *Resource Management Plan: Nickerson Management Complex* [including Manuel F. Correllus State Forest]. Available at http://www.mass.gov/eea/docs/dcr/news/public-meetings/materials/rmp/rmp-draft-nick-summary.pdf.

McBride, K. A. 1990. "The Historical Archaeology of the Mashantucket Pequots, 1637–1900." In *The Pequots of Southern New England,* edited by L. M. Hauptman and J. D. Wherry, 96–116. Norman: University of Oklahoma Press.

McDowell, N. G., and C. D. Allen. 2015. "Darcy's Law predicts widespread forest mortality under climate warming." *Nature Climate Change* 5:669–672.

McEwan, R. W., J. M. Dyer, and N. Pederson. 2010. "Multiple interacting ecosystem drivers: Toward an encompassing hypothesis of oak forest dynamics across eastern North America." *Ecography* 34:244–256.

McMahon, C. C. 2004. "To have and have not—Nantucket, Martha's Vineyard, and the Public Trust Doctrine: Remembering the land that time forgot." *Boston College Environmental Affairs Law Review* 431:431–463.

McShea, W. J., and W. M. Healy. 2002. "Oaks and Acorns as a Foundation for Ecosystem Management." In *Oak Forest Ecosystems: Ecology and Management for Wildlife,* edited by W. J. McShea and W. M. Healy, 1–9. Baltimore: Johns Hopkins University Press.

Méras, P. 2006. *Henry Beetle Hough and the Vineyard Gazette.* Bennington, Vt.: Images from the Past.

Meyer, S. R., C. S. Cronan, R. J. Lilieholm, M. L. Johnson, and D. R. Foster. 2014. "Land conservation in northern New England. Historic trends and alternative conservation futures." *Biological Conservation* 174:152–160.

Millette, T. L. 2013. *Martha's Vineyard Deer Survey.* South Hadley, Mass.: Mount Holyoke

College. Available at http://mvboh.org/images/deersurveysummary.pdf.

Minsinger, W. E. 1992. "Damage in Massachusetts." In *Hurricane Bob: August 16–August 20 1991, A Brief History,* edited by W. E. Minsinger. Milton, Mass.: Blue Hills Meteorological Observatory.

Mitchell, H. 1869. "Concerning Vineyard Haven: Its Character as a Port of Refuge and Its Present Condition." Report of the Superintendent of the Coast Survey, showing the Progress of the Survey during the Year 1869. Appendix No. 15, 239–253. Reports concerning Martha's Vineyard and Nantucket by H. L. Whiting and H. Mitchell. Available at http://docs.lib.noaa.gov/rescue/cgs/001_pdf/CSC-0003.PDF.

Mitchell, H., and G. A. Fairfield. 1854. "Nantucket and Martha's Vineyard Sound Tides." Report of the Superintendent of the Coast Survey, showing the Progress of the Survey during the Year 1854. Appendix No. 29, 35–37. Available at http://docs.lib.noaa.gov/rescue/cgs/001_pdf/CSC-0018.PDF.

Motzkin, G., S. C. Ciccarello, and D. R. Foster. 2002. "Frost pockets on a level sand plain: Does variation in microclimate help maintain persistent vegetation patterns?" *Journal of the Torrey Botanical Society* 129:154–163.

Motzkin, G., R. Eberhardt, B. Hall, D. R. Foster, J. Harrod, and D. MacDonald. 2002. "Vegetation variation across Cape Cod, Massachusetts: Environmental and historical determinants." *Journal of Biogeography* 29: 1439–1454.

Motzkin, G., and D. R. Foster. 2002. "Grasslands, healthlands and shrublands in coastal New England: Historical interpretations and approaches to conservation." *Journal of Biogeography* 29:1569–1590.

Mouw, A. R. 2002. "The Modern and Historic Fire Regimes of Central Martha's Vineyard, Massachusetts." Master's thesis, University of Massachusetts.

Mulholland, M. T., T. Binzen, and C. Donta. 1999. *Community-Wide Archaeological Reconnaissance Survey of Oak Bluffs, West Tisbury, and Chilmark, Massachusetts.* Boston: Massachusetts Historical Commission.

Munoz, S. E., K. Gajewski, and M. C. Peros. 2010. "Synchronous environmental cultural change in the prehistory of the northeastern United States." *Proceedings of the National Academy of Sciences* 107:22008–22013.

Neill, C., B. Von Holle, K. Kleese, K. D. Ivy, A. R. Collins, C. Treat, and M. Dean. 2007. "Historical influences on the vegetation and soils of the Martha's Vineyard, Massachusetts coastal sandplain: Implications for conservation and restoration." *Biological Conservation* 136:17–32.

Newby, P., B. Shuman, J. P. Donnelly, and D. MacDonald. 2011. "Repeated century-scale droughts over the past 13,000 yr near the Hudson River watershed, USA." *Quaternary Research* 75:523–530.

New England Timber Salvage Association. 1943. *Report of the U.S. Forest Service Program resulting from the New England Hurricane of September 21, 1938.* Boston: Northeastern Timber Salvage Administration.

Nuttall, T. 1840. *Manual of the Ornithology of the United States and of Canada.* Boston: Hilliard, Gray and Company.

Oakley, B. A. 2012. "Late Quaternary Depositional Environments, Timing and Recent Deposition: Narragansett Bay, Rhode Island and Massachusetts." Ph.D. thesis, University of Rhode Island.

Ogden, J. G., III. 1958. "Wisconsin Vegetation and Climate of Martha's Vineyard, Massachusetts." Ph.D. thesis, Harvard University.

Ogden, J. G., III. 1959. "A late-glacial pollen sequence from Martha's Vineyard, Massachusetts." *American Journal of Science* 257:366–381.

Ogden, J. G., III. 1961. "Forest history of Martha's Vineyard, Massachusetts. 1. Modern and pre-colonial forests." *American Midland Naturalist* 66:417–430.

Ogden, J. G., III. 1963. "The Squibnocket Cliff Peat: Radiocarbon dates and pollen stratigraphy." *American Journal of Science* 261:344–353.

Ogden, J. G., III. 1974. "Shoreline changes along the southeastern coast of Martha's Vineyard, Massachusetts for the past 200 years." *Quaternary Research* 4:496–508.

Oldale, R. N. 1980. "Pleistocene Stratigraphy of Nantucket, Martha's Vineyard, and the Elizabeth Islands, and Cape Cod, Massachusetts." In *Late Wisconsinan Glaciation of New England,* edited by G. J. Larson and B. D. Stone, 1–34. Dubuque, IA: Kendal/Hunt.

Oldale, R. N., and R. A. Barlow. 1986. "Geologic map of Cape Cod and the Islands, Massachusetts." Miscellaneous Investigations Series, I-1763. Reston, Va.: U.S. Geological Survey.

Oldale, R. N., and C. J. O'Hara. 1984. "Glaciotectonic origin of the Massachusetts coastal end moraines and a fluctuating late

Wisconsinan ice margin." *Geological Society of America Bulletin* 95:61–74.

Oswald, W. W., E. K. Faison, D. R. Foster, E. D. Doughty, B. Hall, and B. C. S. Hansen. 2007. "Post-glacial changes in spatial patterns of vegetation across southern New England." *Journal of Biogeography* 34:900–913.

Parshall, T., and D. R. Foster. 2002. "Fire on the New England landscape: Regional and temporal variation, cultural and environmental controls." *Journal of Biogeography* 29:1305–1317.

Parshall, T., D. R. Foster, E. K. Faison, D. MacDonald, and B. C. S. Hansen. 2003. "Long-term history of vegetation and fire in pitch pine-oak forests on Cape Cod, Massachusetts." *Ecology* 84:736–738.

Patterson, W. A., III and A. E. Backman. 1988. "Fire and Disease History of Forests." In *Vegetation History,* edited by B. Huntley and T. Webb III, 603–632. Dordrecht, Netherlands: Kluwer Academic.

Patterson, W. A., III and K. E. Sassaman. 1988. "Indian Fires in the Prehistory of New England." In *Holocene Human Ecology in Northeastern North American,* edited by G. P. Nicholas. New York: Plenum.

Pease, R. L. 1871. *Report of the Commissioner Appointed to Complete the Examination and Determination of all Questions of Title to Land and of all Boundary Lines Between the Individual Owners, at Gay Head, on the Island of Martha's Vineyard.* Boston: Wright & Potter.

Perley, S. 1891. *Historic Storms of New England.* Salem, Mass.: Salem Press.

Peters, E. L. 1977. "Land Tenure and Subsistence on Martha's Vineyard: An Introduction to a Study of the Pond People." B.A. thesis, Colorado College.

Pike, C., D. Robison, and L. Abrahamson. 2001. *Black Oak Decline on New York's Long Island, 1990–1996.* Newtown Square, Pa.: USDA Forest Service.

Potter, E. 2010. *The Last Farm on Chappaquiddick.* Edgartown, Mass.: Vineyard Stories.

Pyne, S. J. 1982. *Fire in America: A Cultural History of Wildland and Rural Fire.* Princeton, N.J.: Princeton University Press.

Rabinowitz, C. B. 1974. "Martha's Vineyard: The development of a legislative strategy for preservation." *Boston College Environmental Affairs Law Review* 3:2.

Rackham, O. 1986. *The History of the Countryside: The Classic History of Britain's Landscape, Flora and Fauna.* London: J. M. Dent & Sons.

Railton, A. R. 2012. *The History of Martha's Vineyard.* Beverly, Mass.: Commonwealth Editions.

Raleigh, L. 2000. *Land-Use History of Long Point Wildlife Refuge.* Vineyard Haven, Mass.: The Trustees of Reservations. Available at http://www.thetrustees.org/assets/documents/places-to-visit/Land-Use-Long-Point.pdf.

Raleigh, L., J. Capace, and A. Berry. 2003. *Sand Barrens Habitat Management: A Toolbox for Managers.* Vineyard Haven, Mass.: The Trustees of Reservations. Available at http://www.thetrustees.org/assets/documents/what-we-care-about/Toolbox.pdf.

Rawinski, T. J. 2014. *White-Tailed Deer in Northeastern Forests: Understanding and Assessing Impacts.* Newtown Square, Pa.: U.S. Department of Agriculture, Forest Service.

Richardson, J. B. 1974. "Bibliography of the archaeology, anthropology, botany, geology and zoology of Martha's Vineyard." *Dukes County Intelligencer* 15:130–134.

Richardson, J. B. 1985. "Prehistoric man on Martha's Vineyard." *Oceanus* 28:35–42.

Ritchie, W. A. 1969. *The Archaeology of Martha's Vineyard: A Framework for the Prehistory of Southern New England.* Garden City, N.Y.: Natural History Press.

Rivers, W. H. 1997. "Coming Full Circle: Restoring Sand Plain Grassland Communities in the State Forest on Martha's Vineyard, Massachusetts." In *Grasslands of Northeastern North America: Ecology and Conservation of Native and Agricultural Landscapes,* edited by P. D. Vickery and P. W. Dunwiddie, 79–84. Lincoln: Massachusetts Audubon Society.

Rowlands, W. 1939. "A Study of Forest Damage Caused by the 1938 Hurricane in the Vicinity of Petersham, Massachusetts." Master's thesis, Harvard University.

Russell, E. W. B. 1983. "Indian-set fires in the forests of the northeastern United States." *Ecology* 64:78–88.

Schneider, P. 2001. *The Enduring Shore: A History of Cape Cod, Martha's Vineyard, and Nantucket.* New York: Holt, Henry.

Segel, J. D., and R. A. Pierce. 2003. *The Wampanoag Genealogical History of Martha's Vineyard, Massachusetts: Island History, People and Places from Sustained Contact Through the Early Federal Period.* Madison: University of Wisconsin.

Shaler, N. S. 1874. "Martha's Vineyard." *Atlantic Monthly* 34:732–740.

Shaler, N. S. 1888. Report on the Geology of Martha's Vineyard. In *Seventh Annual Report of the Director, 1885–86,* Department of the Interior, U.S. Geological Survey.

Shaler, N. S. 1909. *The Autobiography of Nathaniel Southgate Shaler.* Cambridge, Mass.: Houghton Mifflin.

Shuman, B., P. Newby, and J. P. Donnelly. 2009. "Abrupt climate change as an important agent of ecological change in the northeast US throughout the past 15,000 years." *Quaternary Science Reviews* 28:1570–1572.

Silverman, D. J. 2005. *Faith and Boundaries: Colonists, Christianity and Community Among Wampanoag Indians of Martha's Vineyard, 1600–1871.* Cambridge: Cambridge University Press.

Simon, A. W. 1973. *No Island Is an Island: The Ordeal of Martha's Vineyard.* Garden City, N.Y.: Doubleday.

Smith, D. M. 1946. "Storm damage in New England forests." Thesis. Yale University, New Haven, Connecticut.

Snow, D. R. 1980. *The Archaeology of New England.* New York: Academic Press.

Society for the Protection of New England Antiquities. 1859. *A Visit to Martha's Vineyard.* Boston.

Stacy, B. 2013. "Storm of the century: Remembering the hurricane of 1938." *Dukes County Intelligencer* 555:59–62.

Sterba, J. 2012. *Nature Wars: The Incredible Story of How Wildlife Comebacks Turned Backyards into Battlegrounds.* New York: Crown.

Stevens, A. 1996. "The Paleoecology of Coastal Sand Plain Grasslands on Martha's Vineyard, Massachusetts." Ph.D. thesis, University of Massachusetts.

Stoddard, C. 1980. *A Centennial History of Cottage City.* Oak Bluffs, Mass.: Oak Bluffs Historical Commission.

Swift, W. S., and J. W. Cleveland. 1903. *Records of the Town of Tisbury, Massachusetts. Beginning June 29, 1669, and Ending May 16, 1864.* Boston, Mass.: Wright and Potter.

Taylor, J., and M. London. 2010. *Agricultural Self-Sufficiency on Martha's Vineyard.* Prepared by the Martha's Vineyard Commission. Available at https://mvcommission.org/sites/default/files/docs/agriculturalselfsufficiencyonmarthasvineyard.pdf.

Trefethen, J. B. 1953. "The Massachusetts Land and Its Wildlife: A History of the Resident and Migratory Game Birds and Mammals of Massachusetts." Master's thesis, University of Massachusetts.

Tuck, J. A., and R. Grenier. 1981. "A 16th-century Basque whaling station in Labrador." *Scientific American* 245:180–190.

Tuck, J. A., and R. Grenier. 1985. "Discovery in Labrador: A 16th-century Basque whaling port and its sunken fleet." *National Geographic* 169: 40–71.

Uchupi, E., and R. N. Oldale. 1994. "Spring sapping origin of the enigmatic relict valleys of Cape Cod and Martha's Vineyard and Nantucket Islands, Massachusetts." *Geomorphology* 9:83–95.

U.S. Department of the Interior. 1970. *Islands of America.* Washington, D.C.

Van Tassel, J. 1974. "The Economic Transition of Martha's Vineyard." Master's thesis, Tufts University.

Vickery, P. D., B. Zuckerberg, A. L. Jones, W. G. Shriver, and A. P. Weik. 2005. "Influence of fire and other anthropogenic practices on grassland and shrubland birds in New England." *Studies in Avian Biology* 30:139–146.

Vineyard Open Land Foundation. 1973. *Looking at the Vineyard.* West Tisbury, Mass.

Vispo, C. 2015. *The Nature of the Place: A History of Living with the Land in Columbia County, NY.* New York: Adonis Press.

Wagner, D. L., M. W. Nelson, and D. F. Schweitzer. 2003. "Shrubland Lepidoptera of southern New England and southeastern New York: Ecology, conservation and management." *Forest Ecology and Management* 185:95–112.

Waisanen, P. J., and N. B. Bliss. 2002. "Changes in population and agricultural land in conterminous United States counties, 1790 to 1997." *Global Biogeochemical Cycles* 16:1–19.

Wampanoag Way—An Aquinnah Cultural Trail. 1999. In *Map and brochure produced by Wampanoag Tribe of Gay Head (Aquinnah).* Aquinnah, Mass.

Weiss, E. 1975. "Robert Morris Copeland's plans for Oak Bluffs." *Journal of the Society of Architectural Historians* 34:60–66.

Westveld, M. 1956. "Natural forest vegetation zones of New England." *Journal of Forestry* 54:332–338.

White, M. J. 2002. *Eldridge Tide and Pilot Book, 2003.* Medfield, Mass.: Robert E. White Instruments.

Whiting, H. L. 1869. "Edgartown Harbor" Report of the Superintendent of the Coast Survey, showing the Progress of the Survey during the Year 1869. Appendix 15, 236–239, in Reports concerning Martha's Vineyard and Nantucket by H. L. Whiting and H. Mitchell. Available from http://docs.lib.noaa.gov/rescue/cgs/002_pdf/CSC-0085.PDF.

Whiting, H. L. 1886. "Report of Changes in the Shore Line and Beaches of Martha's Vineyard as Derived from Comparisons of Recent with Former Surveys." Appendix 9, 263–266, in Report of the Superintendent of the Coast Survey, showing the Progress of the Work during the Fiscal Year Ending with June, 1886. Available from http://docs.lib.noaa.gov/rescue/cgs/002_pdf/CSC-0085.PDF.

Whitney, G. G. 1994. *From Coastal Wilderness to Fruited Plain: A History of Environmental Change in Temperate North America from 1500 to the Present.* New York: Cambridge University Press.

Winthrop, J. 1649. *The History of New England from 1630 to 1649.* Boston: Little, Brown and Co.

Willoughby, G. 2008. The diaries of George Haskell Willoughby. Unpublished archives in the Martha's Vineyard Museum, Edgartown, Mass.

Woodworth, J. B., and E. Wigglesworth. 1934. *Geography and Geology of the Region Including Cape Cod, the Elizabeth Islands, Nantucket, Martha's Vineyard, No Mans Land and Block Island.* Vol. LII: *Memoirs of Museum of Comparative Zoology at Harvard College.* Cambridge, Mass.

Index

Images are indicated by *italic* page numbers.

access, 3, 6, 203–5, 209–10, 249
acorns, 92–93, 245–46, 249
affordable housing, 202, 206, 263, 275
Agassiz, Louis, 41, 44–45, 170
Agricultural Preservation Restrictions, 211
Agricultural Self-Sufficiency on Martha's Vineyard (Taylor and London), 291–92
Agricultural Society, 130, *156*, 176–77, 235, 244
agriculture: and conservation, ix–x, 210, 211, *288*, 289–99, *300;* daily activities, 156, 157, 163–64; decline, ix, 147, 174–78, *179*, 216, 252; glacial history and, 20; growing degree days, *25;* hurricanes and, 120; Native American, x, 21, 81–85, 96–97, 99–102; nature's recovery from, 12–13, 147–48, *286*, 293 (*see also* forests: reforestation); in New England generally, *30;* nineteenth-century, x, 147–48, 156–61, 165, 166, 170, 171, 175–78, *179;* and pollen records, *76*, 215–16; resurgence, ix–x; and wetlands, 156, 160, 161, 215; and wildlife populations, 239–42. *See also* farmland; fences; fields; hay; pasture land
airport, 49, 222, *245*, 252, 261–62, 264
Airport Bog, 63, 67
Allen, Everett, 311
Allen Farm, Chilmark, *43*, 158, *202*, 211, *278*, *283*, 292
Alley, James, 269
Angel Gabriel (ship), 115–16
Aquinnah, *4*, 26; conservation, *35*, 211; development, *190;* erosion and deposition, *134;* Kennedy estate, 204; vegetation, 180, *227*, 234; Wampanoag lands, *94*. *See also specific locations*
archaeology, x, 5, *82*, *84*, 85, 87–99, 101–4
Archaeology of Martha's Vineyard (Ritchie), 95–96
Askins, Bob, 252
Athearn, Edwin, 193
Athearn, James, 206, *289*
Athearn, Simon, 162, 309
Atlantic Monthly, 177–78
Audubon Society (Massachusetts), 80, 199–200, 241. *See also* Felix Neck Wildlife Sanctuary

Baldwin, Liz, 258
Banks, Charles, 59, 63, 246
Barlow, Roger, *47*
barrier beaches, *208*, *210*, *258;* and coastal pond waters, 36, 97, 128; movement and mechanisms of, 54, 128, 134–39; Norton Point breach's effects, 127–33, *137*, 140–45, *205*, *206*, 286; opening/closing breaches artificially, 36, 138, 143, 160; vegetation, 78. *See also* coastal ponds
bathtub, 284
bayberry, 33, *72*, 232
beaches, 3, 56–58, 136, 203–4, 209, 265–67, *286*. *See also* barrier beaches; coastal erosion; *and specific beaches*
bears, 82, 83, 96, *239*
beavers, 82, 83, 96, *239*, 240, 242
bedrock, 17–18
beech trees: age and size, *225*, *226;* beech woods, 68, *209*, 232; climate change and, 66; harvesting, *163;* in modern woodlands, 224–25, 231–32; in New England, 31, *32;* numbers, 69, *72*, 74–77, 98, 161, 217, 218; nuts, 92–93; oak mortality and, 64–68, 70, 286, 301; and reforestation, 218, 219; winds and, *118*, 121, *124*. *See also* forests
beetlebung (black gum) trees, *35*, *68*, *222;* age and size, 224, *225*, *226;* decline in, 217, 218; in modern woodlands, 225, 231–32; name, 12; numbers, *72*, 74–76, 161; oak mortality and, 67; in secondary forests, 219. *See also* forests
Benninghoff, William (Bill), *63*, 64
Bert, Dick, 95
Bethaven Farm, *289*, *292*
Bierregaard, Rob, 256, 257
Bierstadt, Albert, *81*, 82
Bigelow, Robert, 192–93
Billingsgate Island, 56
Binzen, Tim, 97
birds: domestic fowl, 158, 189, 215, 291, 292, 296; grasslands birds, 240–42, 250–52; habitat loss, 79, 241–42, 252; hunting, *15*, 166, 187, 189, *191*, 240–41, 246, 250–51; migrating birds, 29, 251, 257; non-native, 242, 284; population over time, *239*, 240–41; predators of, 242, 257–58. *See also specific groups and species*
black gum trees. *See* beetlebung trees
black oak, 70, 161, *226*, 229. *See also* oak trees
Black Pond, 45, *63*, 69, *72*, 73, 75, *76*
Blaney's Pond (Naushon), *63*, 64, *69*
Block Island, 18–19, 24, *27*, *43*, 74, 98, 161
blueberry, 67, 216, 217, 219, 232, 237, 246, 299

bogs. *See* cranberry bogs; wetlands
Boldwater development, 189, 201
Boose, Emery, 112, *113,* 116, 117, 122
Boston Globe, 272, 273
bottoms, 57, *229,* 230–31
Bradford, William, 87, 109, 112–15, 117, 118
Bramhall, Tess, 310
Brereton, John, 75
Brewster, William, 245–46
brickyards, 7, 33, *167,* 170, 222
Brooke, Edward, 270
brook trout (salter), 242, 253–54
Burt, Dick, 95, 311
Burt, Prudy, 304
Butler, William, 157
Byers, Douglas, 95

Cabot, Sebastian, 102
Cagney, James, 262
camp meetings, 166, 175, 181
cankerworm, 66, 68. *See also* oak trees: mortality
Cape Cod, *7;* access and land use, 24–25; climate, *27;* European exploration, 21; geology and geomorphology, 17–19, *20,* 21, 41, *42,* 43–44, *48,* 51, 54; hurricanes, 108, 112, 118, 122; National Seashore, 36, 263, 265; Native American sites, 101; paleoecological studies, *63,* 64, 69, 73; pamets, 57; sea level and, 54–56; vegetation, 21, 27, 31
Cape Poge, *19, 86,* 147; archaeological sites, 97; formation, 54; lighthouse, *167,* 168; remnant chimneys, *285;* vegetation, 76, 180; watershed, *276*
Cape Poge Bay, *37, 53*
Cape Poge Wildlife Refuge, 199, 210
Caroline Tuthill Preserve, *197,* 206, 224, 275
Carpenter, Erastes, 181, 183, 185
Carroll, Bob, 269
catbrier, 33, 67, 217, 219, 232
caterpillars, 66, 68, 301. *See also* moths; oak trees: mortality
cats, 242, 243, 257–58, 259
cattle: bedding and manure, 160; declining numbers, 12, *179;* first imported, 158; hay for, 148; impact on land/vegetation, 215; and modern land management, 258, *287,* 296, *298;* pastures and fences, 7, 160–61; for slaughter, 290, 292. *See also* livestock; pasture land
cedar, 27, 76, 234
Cedar Neck, 181, *184*
Cedar Tree Neck, 56, *63,* 64, 68, 75, 203
Cedar Tree Neck Sanctuary, 199, 203, *205, 300,* 301
Champlain, Samuel, *87,* 101, *103*
Changes in the Land (Cronon), 83
Chappaquiddick, *4,* 17, *86, 153,* 307; agriculture, 157, *293;* archaeological sites, 97; beaches, 56–58, 204, 209; conservation, *35;* development, 189, *190;* geology, 45–47; in the nineteenth century, 147, 149–54, 157, 163, 164, *167;* paleoecological studies, 63–64, 75; un-development, 285; vegetation, 180, *216,* 219, *227;* Wampanoags on, 26; weather, *116;* woodlands, 222, 234. *See also* Cape Poge; Poucha Pond; Wasque
Chappaquiddick Open Space Fund, 211
Chase, Tom, 285
Cherau, Suzanne, 97
cherry trees, 218, 219, 232
chestnut trees, 31, *32,* 92–93, *163,* 245
chickens, 158, 160, 291, 292, 296
Childs, Bob, 304
Chilmark, *4,* 25, 26, *153;* affordable housing, 202, 206; agriculture, 20, 161; conservation, *35,* 209, 211, 297; development, *190, 270;* forests, 20; in the nineteenth century, 147, 149, *153,* 154–55, 159, 161, 163, *164,* 166, *167,* 168; vegetation, 149, *227. See also specific locations and landmarks*
Chilmark Pond, 3, *4, 63,* 64, 128, *134, 202,* 204
Chilton, Elizabeth, *84,* 96–97, 98, 99–100
chimneys, *12, 34, 285*
Christiantown, 26, *94, 97,* 154, 207
Chronicles of an Islander (Athearn), 193
Clean Air Act, 305–6
clean energy, *257,* 279
cliffs, 56. *See also specific locations*
climate, *25,* 27–31, 50–51, 66, 70–75, 180. *See also* climate change; weather
climate change, 307; global statistics, *304;* historical and paleoecological record and, 61–64, 69–78; increase in storms, 140, 304; Little Ice Age, 100; postglacial change, 70–78, 98; predictability, 286; sea level rise, 5–6, 38, 139–40, 286, *306;* vegetation shifts, 64, 66, 69–75, *76, 77. See also* environmental change
coastal erosion, ix, 5–6, 21; artificial barriers, *134, 140, 142, 143, 144,* 145, 286; East Beach, *307;* entire island, 133–34; future of, 38, 139–42; during hurricanes, 67, 108, *113;* Lucy Vincent Beach, 3, *5;* mechanisms, 134–39; Menemsha Harbor, *113;* Ogden and, 67; prehistoric, 53–54; south shore, 128, *133, 138,* 140; Wasque and Norton Point, 127–33, *137,* 140–45, *206,* 286, *309*
coastal plain, 49, 53–55
coastal ponds, 36, *37, 50, 51,* 56, 78, 94–95, 97–98, 138–39. *See also* Norton Point breach
Coastal Survey, 128, 130, 149–52. *See also* Gilbert, Samuel Augustus; Martha's Vineyard: 1850 map; Whiting, Henry L.
coastline: evolution of, 133–34; native occupation, 93; sea level and, 53–54, 62; storm surges, 108, *113,* 114–15, 117, 118, *125. See also* barrier beaches; beaches; coastal erosion

Coffin, Grace Cousens, 170
Coffin, Thomas M., 165
colonial era. *See* European settlement
conservation, ix–x, 175, 195–211; and access, *199*, 203–5, 209, 210 (*see also* access); and agriculture, ix–x, 210, 211, 288–99, *300;* benefits, 311; challenges of, 198, 281–82; development vs., 36–37, 195–97, 199–209; early efforts, 177, 191, 241; easements on private lands, 204, 274; failures, 286; federal government and, 263, 265–73, 305; fire as tool, 79–80; future of, 281–302; goals, 37–38, 284; greenways, 275; historical perspective, 284, 285–89; history of, xi, 25, 33, 36–37, 197–209; and indigenous land management, 82–83; Island Trust approach, 200, 206, 263, 268–73; Island-wide planning and, ix, xi, 36–37, 200, 206, 260–80; land management approaches, 289–300; maps of protected areas, *4, 35, 197, 199, 269, 274, 280, 294;* modern landscape, 209–11; in New England generally, 33–36; pace of, 282; priority habitat designation, 31; pyramid, *282;* surface water and ownership and, 48–49; varied types, protection levels, and motivations, 198; and wildlife, 239, 241, *245*, 246, 249, 284, 298–300 (*see also specific species*); woodlands and wildlands, *223*, 299–300. *See also* clean energy; Land Bank; Martha's Vineyard Commission; planning; *and specific organizations and protected areas*
Conservation Partnership of Martha's Vineyard, 198. *See also specific member organizations*
consumption, 289–91
continental shelf, 18, 49, 53–56
Copeland, Robert Morris, 181, *182*, 183
corn (maize), 83, 88, 96, 99–101
Correllus State Forest: access, *199*, 203; history, 198–99, *245;* hurricane damage, 121, 125; incursion attempts, 209; land management approaches, 284, *300*, 301, 302; plantations, 218, 284, 299; reforestation, 219; woodland type, 79, 226, *227*, 228, *229*
Cottage City, 166, 183, *184*, 185, 187, *190*. *See also* Oak Bluffs
Cottle, Charles, 165
coyotes, 27, *239*, 242, 258–59, 287
cranberry bogs, 33, 45, 218
Crocker Pond, 155, 170, 171
Cromwell Lane, 311
Cronon, Bill, 83
cultural resource management (CRM), 97
Cunningham dig site, *90*
cynipid wasp, 70

Daggett, Thomas, 92
dams, 33, 176, 218–19, 254, 304
David, Gus Ben, *256*
Davis, Everett Allen, 187
Day, Gordon, 79, 83
Day, Lindsay, *73*
DCPEDC. *See* Dukes County Planning and Economic Development Commission
DDT, 256
de Crevecoeur, Hector St. John, *155*, 161
Deep Bottom Pond development, *192*
Deep Pond (Falmouth), *63*, 64, 69
deer, 27, 68, 90–93, 96, 238–40, 242, 247–50, 258–59
Deevey, Edward, 67
deforestation: after 1850, 192–93, *202, 220, 221;* benefits, 171; graphs, *72, 77, 239;* before 1850, 12, 32, *33*, 59, 156, 171, 215–16, 221–22; regrowth following, 12–13 (*see also* forests: reforestation). *See also* forests; woodcutting
DeGraff, Dick, 284
De Raiseres, Isaac, 100
development, ix, 33, 175; 1800–1900, 181–87, *190;* 1900–1950, 187–93; 1950–present, *190, 192*, 193–94, 200–206, 261, 264–65, *268*, 270, 273–75; coastal erosion and, 140, *141* (*see also* Norton Point breach); conservation vs., 36–37, 195–97, 199–209; and deforestation, 192–93, *202, 221*, 222; Opportunity Areas for redevelopment, *276*. *See also* conservation; planning; un-development
Devens, S. A., 156
Division of Fisheries and Wildlife, 254
Donahue, Brian, 291
Donnelly, Jeff, 304
Donta, Chris, 97
Doucette, Dianna, *82, 84, 96*, 96–97
Downing, Maria Ann, 157
Dr. Daniel Fisher house, *154*, 170
Dr. Fisher's Road, *167*, 168, 170
drill holes, in boulders, 10, 171
drought, 66, 68, 162, *304*
Duarte Cranberry Bog, *63*, 64, 67
Duarte Pond, 63, *68, 69*, 73, 75, 77, 204, 222, 254, 285
ducks, *15*, 96, 189, *191*, 241, 284
Dukes County Planning and Economic Development Commission (DCPEDC), 201, 262, 263, 267, 268, 270, 273
dunes, *6*, 136, *138*
Dunlop, Tom, 127–28, 133, 145, 304
Dunwiddie, Peter, 64
Duranleau, Deena, 93, 97–98, *100*, 102

East Beach, *86, 200, 307*
East Chop, *4*, 14, 149, *153*, 154, 161, *167, 190*, 222, 228
Eastville, 154, 181
Ecological Indian, The, 85

ecological studies, *8;* history and, 7, 59–64; tree age and size, 222–25. *See also* paleoecological studies
Edey, Maitland, 274
Edgartown: affordable housing, 202; beach and pond access, 204; Boldwater development, 189, 201; and conservation, *35,* 211; division of, 185; fence pattern, 159; golf course and yacht club, 189; groins, *143;* hurricane damage, 143; lighthouse, *167;* map, *4;* and the MVC, 263; in the nineteenth century, 147, 151, *153,* 154, 161, 163, *164,* 166, *167, 169,* 178–79, 181, *185, 186;* origins, setting, and character, 21, 25, 26; pagoda tree, 234; predicted surge flooding, *125;* solar array, *278;* vegetation, 161, *227. See also* Cottage City; Edgartown Harbor; Oak Bluffs
Edgartown Great Pond, 97, 128, *133,* 204, *208,* 209. *See also* Herring Creek Farm
Edgartown Harbor, *24,* 108, 138, 139, 154, 178–79, *185. See also* Katama Bay
Edgartown–Vineyard Haven Road, *150,* 200–201, 228
Edgartown–West Tisbury Road, *150,* 170, 218, 228, 231
Elizabeth Islands: agriculture, 161; birds, 244, 252; climate, *27;* conservation, 36; deer hunting, *249;* European exploration, 21; federal government and, 263, 265; geology and geomorphology, 17–19, *20,* 21, *43,* 54; hurricane damage, 122; included in Chilmark, 154; sea level and, 54–56, 74
energy, clean, *278,* 307
environmental change, 66–68, 88, 98. *See also* climate change; paleoecological studies
environmental studies. *See* ecological studies; paleoecological studies
erosion, 136, 216. *See also* coastal erosion
European settlement: deforestation, 12, *77,* 215–16 (*see also* deforestation); exploration and first settlements, 21–22, 75, 81; and land use and ownership, 156; land variation and, 25–26; native peoples influenced/impacted, 85–87, 102–4; ruins and artifacts, 33; and wildlife populations, 239–40
EXPOS model, 122

farm animals. *See* livestock
farmland, *179,* 292–99. *See also* agriculture; fields; pasture land; *and specific farms*
Farm Neck Golf Course, *191,* 273–74
Federal Emergency Management Agency (FEMA), 143
federal government, conservation role of, 263, 265–73, 305. *See also* Island Trust Bill
Felix Neck Wildlife Sanctuary, 80, *197,* 199–200, 273, *274*
fences, *11, 29,* 148, 159, *173,* 219. *See also* stonewalls
ferries, 139, 185, *194, 261,* 303
fertilizers, 160, 263, 279, 298
fields, 12, 158–60, 175, 178, 180, *196,* 216–17, 289–99. *See also* agriculture; development; farmland; pasture land
fire: land management through, 79–80, 82–85, 215, *290,* 293, *297;* nineteenth-century fires, 166, 183, 295; prehistoric, *72,* 73, *74, 76;* trees vs. scrub oak, 295, 302; wildfires, 295, *296*
fish and shellfish, 93, 95–97, *100,* 242, 252–54, *279. See also* fishing; shell middens
Fisher, Daniel (Dr.), *14, 154,* 166, 168–70, *171,* 176, 185, 251, 309
Fisher, Jeremy, 230
Fisher Pond, 170, 171, *301*
fish hawks. *See* ospreys
fishing, 142, 165, 170, 191, *206,* 242. *See also* fish and shellfish
Flynn, George Daniel, 189
Flynn family holdings, 189, 201, 204
food production (modern), 289–92, *298. See also* agriculture
Forbes, John Murray, 170
Forbush, E. H., 251
forests (woodlands): ancient vs. secondary, 76, *104,* 216–18, *220,* 221–32; beech woods, 68, *209,* 232; coastal erosion and, 127, *128, 309;* conservation management, 289, 299–302 (*see also* conservation); diseases and pests, 66, 68, 70, 301, 304, *308;* diversity, 16; and fire, 295, 302 (*see also* fire); glacial geology and, 20, 31; grazing in, *160;* harvesting, 32, 161–65, 218, 222–24, 299 (*see also* woodcutting); houses in, *201;* as legacies, 8; modern woodlands, 220–34; need for, 37; in New England generally, *30,* 31, *32;* nineteenth-century, 147–49, *153, 158, 159,* 161–65, *167, 188;* oak forests, 64–68, 225–28, 245, *292;* Ogden's studies, 67; plantations, 218, 234–37, 284; prehistoric, 69, 72–76, 85, 98, 104 (*see also* paleoecological studies); reforestation, ix, x, 12–13, 32–33, 79, 147–48, 175–80, 216–20, *239,* 252, 293; 1782 Vineyard map, *155;* storm damage and recovery, 76, 112, *113,* 114, 117–18, 120–25; at time of European settlement, 75–78; types, 225–29, *227;* and wildlife populations, *239;* and wind, 31. *See also* deforestation; trees; *and specific tree types*
fossils, 59–60. *See also* paleoecological studies
Foster, Ava and David, *11*
Frances Newhall Woods Nature Preserve. *See* Woods Preserve
Franklin Street (Forest Avenue), 187, *188, 192*

Freeman, James (Rev.), 79, 159, 160, 162, 163, 195, 238
freshwater, 36, *51*. *See also* ponds; streams; water
Frisch, Shirley, 269
frost, 230–31, *232*
fuel wood, 13, 163–64, 222–24. *See also* woodcutting
Fulling Mill Brook, 165, 168–70, 254, *298*
furs, 103, 240

Ganz, Robert and Anne, *166*
Gay Head, 26, *108;* beach, 56; cliffs, *19,* 26, 56, *134;* conservation, 200, *300,* 302; lighthouse, *19, 167;* in the nineteenth century, 149–54, 159, 161, 163–64, *167, 185, 190*. *See also* Aquinnah; Wampanoag people; *and specific locations*
Gay Head Community Baptist Church, *89*
geese, 96, 189, 239, *241,* 259
geology and geomorphology: Cape and Vineyard geomorphology compared, *48;* and climate, 27–31; glacial geology, 5, 17–21, 41–52, 56–58, *233;* and human settlement, 21–26; hydric soils, *233;* isostasy, 49–50, 52; legacy concept and, 7; ocean processes and, 52–56; Shaler and, 174; topography (elevation, slope), *47;* and vegetation, 19–20, 31, 231–32
Gifford, K. Dun, 268
Gilbert, Samuel Augustus, 151, 250–51
glaciers, 41–42. *See also* geology and geomorphology; Ice Age; ice sheets
Goat Rocks, *58*
goats, 158, 215, 296, *298*. *See also* livestock
golf courses, 189, *191,* 192–93, *197,* 209, 222, 273–74
Goodale's gravel pit, 7, *8,* 10, 255, *257,* 303
Gosnold, Bartholomew, ix, 21, 23, 75, *81,* 82, 87, 91, *103*
grasses, *72, 76,* 215–16, 234
grasslands, 33, 79–80, 171, 216, *290, 291,* 293. *See also* birds; pasture land
Great Plain, the, *4,* 162; agriculture, 20; birds, 251 (*see also* heath hen); bottoms and hollows, 57, *229,* 230–31; conservation lands, 198–99, *223;* fires, 295, *296;* forest planting, 235–37; glacial origins, *47, 48;* mid-nineteenth century, 147, *153,* 161–62; native land use, 97; paleoecological studies, 63, 77; roads, *167,* 170; surface water lacking, 48–49, 57, 77, 162; vegetation, 20, *219* (*see also* scrub oak); wind exposure, 122; woodlands, *155,* 161–62, 178, *180, 219,* 221, 228, 230–31, *233*. *See also specific locations*
Great Woods, 154, 161
Grey Barn Farm, 210, 297
Gross, Alfred, 162, 244, 246
groundwater, 36, 57, 70, *71,* 73, 275. *See also* ponds; water
Guernsey, Samuel, 93–95
Gulf Stream, 28, 50, 70, 73, 139
Guthrie, Woody, 203

Hall, Brian, *224*
Hancock, Herbie, 269
harbors, 154–55. *See also specific harbors*
Harlock Pond, 45, 63, *69,* 75
Harris, Lewis, 201
Hart, John, 88
Hart, William, 187
Harthaven, 187
Harvard Farm, *298*
Harvard Forest, 63, 98, *100,* 120–22, *126,* 307. *See also* paleoecological studies; *Wildlands and Woodlands*
Harvard University Research Bureau, 267
Hassler, Ferdinand Rudolph, 150–52
Hawthorne, Nathaniel, 166, 170
hay, 148, *158,* 158, 160, 161, *196,* 216, *302,* 304
Healy, Allen, *302*
heath hen, 49, 85, 96, 198, 238, *239,* 242–46
Heath Hen Reservation, 162, 198–99, *242*
heathlands, 79, 171, 180, 216, 226–28, 232–34, *291,* 293
hemlock trees, 31, *32,* 69, 73, 98
Herbster, Holly, 97
herring, 242, 252–53, 304
Herring Creek Farm, *197,* 204–5, *286*
Hickory Cove, 204, 285
hickory trees: decline, 217, 218; in modern woodlands, 224–25, 232; in New England generally, 31, *32;* numbers, *72,* 74–77, 98, 161; nuts, 92–93; oak mortality and, 67, 70; role of, 31; winds and, 121. *See also* forests
Higginson, Frances, 90
Highland House, 183
Hine, C. G., 162
Hitchcock, Edward A., 41, 44–45, 60
hogs, 7, 148, 158, 160, 215, 292, 296. *See also* livestock
hollows and bottoms, 57, 230–31
holly, 232, *236*
Holly Holm, 232, *236, 300*
Holmes Hole, 159, 166, *184,* 185–87. *See also* Vineyard Haven
Holmes Hole Harbor, 154, 178–79
Hornaday, William, 241, 248, 258
horses, 148, 158, 170, *192,* 215, 241
hotels, 183
Hough, Betty, 191
Hough, Henry Beetle, 174–75; books, 192, 199; and conservation, 143, 199, 269, 274; and the

Hough, Henry Beetle (continued)
Gazette, ix, 174, 191, 269; on the highway, 228; and planning, 267, 269; on the resort boom, 193; on trespassing, 203. *See also* Sheriff's Meadow Foundation; *Vineyard Gazette*
housing, 202–6, 263, 268, 270, *271,* 275. *See also* development
Hoy, Johnny, 304
huckleberry, 67, 76, *104,* 162, 216, 219, 224–26, 228, 232, 237, 299
Hudsonia ericoides (false heather), *217, 220*
hunting: birds, 166, 187, 189, *191,* 240–41, 246, 250–51; clubs, 187–89, 198; deer, 240, 247–50; fur trade, 240; by Native Americans, 83, 90, *100,* 240, 247; wildlife populations threatened, 241, 242
Hurley, Steve, 254
HURRECON model, 116, 117
hurricanes, 28, 107–26; Bob (1991), *106,* 107–8, 112, *120,* 121, 125, 126, 237; Edna (1953), 67; forest damage, 76, 112, *113,* 114, 117–18, 120–25; Fujita scale, *110;* future storms, 38, 124–25, 140, 286, 304–5; Great Colonial (1635), xi, 38, 109, 112–18, *120,* 126, 304; historical reconstruction, 112, 116–17, 304–5; human activity and, 123–25; hurricane disturbance regime, 118–23, 125–26; Irene (2011 tropical storm), 108, *111, 120, 122;* 1938 hurricane, 107–9, 112, *113,* 117, *120, 121,* 124–25, 126, 143; 1944 hurricane, *124,* 125, 126; Sandy (2012), 3, *5,* 29, 108; shoreline and, 3, *5,* 37; storm surges, 108, *113,* 114–15, *115,* 117, 118, *125;* wind direction, *110,* 111–12, 114, 117, 122, *123*

Ice Age, 5, 17–18, 41–47, 50–54, 73. *See also* geology and geomorphology
ice sheets, 41–47, 50–51, 73. *See also* Ice Age
Illusion of Preservation, The (booklet), 289–90
industry, nineteenth-century, 147, 165–70
inertia, 10–14, 219
Inhofe, James, 307
insects, 66, 68, 70, 231, 239, 242, 301, 304. *See also* oak trees: mortality
Iroquois people, 21, 83, 85, 99–101. *See also* Native Americans
Island Action Committee, 269, 270
Island Grown Initiative, 209, 291
Island Properties (Strock company), 261, 263, 273, *274*
Islands of America (Dept. of the Interior), 265–67
Island Trust Bill, 200, 206, 268–73, 275–77
isostasy, 49–50, 52

James (ship), 115–16
Jefferson, Thomas, x, 150
Jeffer's Pond, *63,* 64, 75
Jernegan, Nathaniel (Capt.), 191
Jernegan Pond, 63, *70,* 75, 77
Johnson, Frederick, 95
Johnson, Luanne, 258
Jorgensen, Eric, *11*
Josselyn, John, 239–40
Justinian, 203

Katama, *155,* 161, 232–34, 252, *286, 294*
Katama Air Field, 80, *197,* 209, 252, *286*
Katama Bay, 97, 138–39, 142–43, 170, *183, 185, 216, 276, 279. See also* Edgartown Harbor; Norton Point breach
Katama Farm, *197,* 202, 209, 210, 251, *286,* 297
Kaye, Clifford, 133, *134*
Keith, Allan, 284
Keith Farm, *196,* 202, 224
Kennedy, Chris, 304
Kennedy, John F., 265
Kennedy, Ted, xi, 200, 206, 263, 265–72, 275–77
killdeer, 241
Krech, Shepard, 85
Kupersmith, Corey, *12,* 200

Lagerman, Jay, *194*
Lagoon Heights, 181, 183, *184,* 189
Lagoon Pond, *63,* 97, 170, 185, *188, 190,* 222, 254, *276,* 279
Lake Tashmoo, *63,* 97, 170
Lambert's Cove, *35,* 45, 165
Land Bank, ix, 198, 207, 263, 274; and access, 204; conservation lands, *35,* 200, 207, 210–22; and farming, *293,* 297, *298;* funding and acres conserved, *207;* goals, 38; impact of, 205; Island Open Space plan, 263, 268; wood products, 299. *See also specific tracts*
land loss, ix, 54–56, 133–34, 139–40. *See also* coastal erosion
land management: approaches varied with time/context, 142–43; conservation goals and, 284; farming as conservation tool, 289–99, *300;* maintaining open land, 13, 289–99; native practices, 78, 80–85, 88; wildlands, 299–302. *See also* agriculture; conservation
land ownership: and access, 203–5; colonial settlement and, 156; conservation and, *199,* 209–11, 269, 297–98 (*see also specific tracts and organizations*); fences and, 159 (*see also* fences; stonewalls); Great Plain and, 49, 161; by new residents, 187; Wampanoag, 154, 159, 164; water accessibility and, 49
landscape: acceptable norms changing, 206; change inevitable, 285–86; colonial, 12, 13; diversity, 17, 18, 149, 171; grazing animals and, 296–98 (*see also* livestock); human activity and hurricane impact on, 123; modern conserva-

tion landscape, 209–11 (*see also* conservation); modern farm landscapes, 292 (*see also* farmland); modern landscape's creation, 174–94, 285; Native Americans and, 78, 80–82, 307–8; postglacial changes, 71, 74; pre-colonial, 12; threatened by tourism and development, 261–67, 270–73 (*see also* planning). *See also* agriculture; development; geology and geomorphology; landscape, nineteenth-century; *and specific features and locations*
landscape, nineteenth-century, 13, 147–73, 181–87; agriculture and, 148, 156–61 (*see also* agriculture); described by contemporaries, 147; development, 181–87 (*see also* development); diversity, 149, 171; 1850 Vineyard map, x, 149–52, *153, 159, 190, 294;* forest decline and recovery, 147–48 (*see also* deforestation; forests: reforestation); industry and, 165–70; remnants (legacies), *12, 14,* 148, 168, 170–71, 178, 224, *285* (*see also specific types*); Shaler on, 174, 177–78; towns, villages, and harbors, 154–56. *See also* agriculture; landscape; *and specific features and locations*
land use: access and, 24–25, *199,* 209, 210 (*see also* access); colonial era, 156; conservation as, 195–97; modern landscape shaped by, 174–94, 285; Native American, 78, 80–85, 88, 97, 104, 307–8; in the nineteenth century, *158;* summer tourism and, 189–91. *See also* agriculture; conservation; development; hunting; pasture land; tourism; woodcutting
Largy, Tonya, 96
Last Farm on Chappaquiddick (Edo Potter), 189
legacy concept, 7–9, 14. *See also* landscape, nineteenth-century: remnants
Lengyel, James, 205, 211. *See also* Land Bank
Leon, Gon, 258
Leopold, Aldo, 248–49, 292, 299
Lewis, Elisha, 244
lighthouses, *19, 167,* 168, 170, *185,* 303
Lily Pond (Edgartown), 63, 75, 77
Lily Pond (Red Gate Farm), *62, 63*
Little, Betty, 91–92, 104
Littlefield Greenways. *See* Margaret Littlefield Greenways
livestock, x, 27, 148, 158, 160–61, 215, 292, 296. *See also* hay; pasture land; *and specific animals*
Livingston corer, 61
local production, 289–92
London, Mark, 260, 279–80, 291
Long Cove, *50*
Longfellow, Henry Wadsworth, 82
Long Island, 17–19, *20,* 21, 24–25, *27,* 31, 43, 108, 117, 255
Long Point, *54,* 80, 228, 252, *290,* 297
Long Point Reservation, 189, 203, 210
longshore drift, 134–35
Lucy Vincent Beach, *2,* 3–6, 8, *52,* 303; archaeological excavation, x, 5, *82, 84,* 96–97
Ludlum, David, 116
lumber, 165, 299. *See also* woodcutting
Lutz, Harold, 67

Manual of Ornithology of the United States (Nuttall), 244–45
maple trees, 31, *32,* 64–67, 77, 121, *163. See also* red maple
Margaret Littlefield Greenways, *197,* 209, 226, 228, *232,* 275
maritime activity, 165, 166, 168, 169, *172,* 174, 178–79, 189. *See also* harbors; ships; *and specific harbors*
Marsh, George Perkins, 36
Martha's Vineyard (the Island): 1850 map, x, 149–52, *153, 159, 190, 294;* historical research sources, x, 310–11; name, ix, 21; population, 166, 178, 193; residents' diversity, 17; 1782 map, *155,* 161; 2016 map, 4; uniqueness, xi, 17, 18; from Woods Hole, *16. See also specific topics, such as* agriculture *or* landscape
Martha's Vineyard Commission (MVC), 200, 273; Agricultural Self-Sufficiency study, 291–92; and coastal vulnerability, 140, 141; Districts of Critical Planning Concern, 140, 263, 273, 275; goals, 38; Island Open Space plan, 263, 268; Island Plan, 200, 206, 260, 263, *266,* 274–80, 282 (*see also* planning); and the Waterview development, 263, 273–74. *See also* Athearn, James
Martha's Vineyard Land Bank. *See* Land Bank
Martha's Vineyard Museum, x, 311
Martha's Vineyard National Bank, 169
Martha's Vineyard Preservation Trust, *293,* 297
Martin, Nicholas, 170
Massachusetts, *7;* agriculture's decline, 174; birds, 251–52; coastal drought index, *304;* coastal erosion, 139; colonization, 81, 85–87; and conservation, *35,* 209, 241–42, 252, 272–74, 281 (*see also* Correllus State Forest); deer management, 248–49; forests, *30,* 31, *32,* 33, 147–48, *163;* hurricanes, 107–8, 112–18 (*see also* hurricanes); mid-nineteenth-century population, 165; and shoreline access, 204; wildlife populations, *239. See also* New England
Massachusetts Coastal Erosion Commission, 141
Mather, Increase, 116
Mather, Richard, 115–16
Mattakeset Lodge, *183,* 183, *185*
Mayflower, 85–87. *See also* Pilgrims
Mayhew, Thomas (elder), 21
Mayhew, Thomas (the Younger), 21–22
Mayhew, Thomas, Jr., 95

Mayhew family, 21–22
McBride, Kevin, 97, 101
McCleary, Robert, 151, 250–51
McKibben, Bill, 175, 303
McPhee, John, 248
meadows. *See* farmland; fields; grasslands; openlands; pasture land
Meetinghouse Road, *20*, 202
Menemsha Beach, *262*
Menemsha Harbor, *113*, 117, *190*, *310*
Menemsha Hills reservation, 8, 56, 200, 210, *217*, 218, *222*, 228, *297*, 299
Mermaid Farm, 158, 292, *302*
Merry's Hill view, 147
Metcalf and Eddy Comprehensive Plan, 260, 262–68, *269*, *271*, 273–76
mice, 68, 250
Middle Line neighborhood, 202, 206
Middle Ridge, 66, 286. *See also* oak trees: mortality
Middle Road, 168
Mill Brook, *56*, 76, 97, 170, 171, 176, 253, 254
Millette, Thomas, *247*, 247
mills, 33, *34*, *167*, 168–70, 176, 178
Milton, Thomas (Capt.), 234
Mink Meadows Golf Course, 189, 192–93, 222
Mitchell, Henry, 136–39, 141, 143, 154
Mohu estate, *12*
moose, 82, *239*, 240
moraines, 18–21, *40*, 41, *43*, 45–47, *48*, 49, 51, 56
Morning Glory Farm, 158, *197*, 211, *278*, 292
Morton, Thomas, 85, 87
Moshup's Pond, *62*, *63*, 64
Moshup Trail, 180, 209, 219, *220*, 232–34
moths, 66, 231, 239, 304. *See also* caterpillars
Motzkin, Glenn, 284
Mourt's Revelation, 89
Mulholland, Mitch, 97
muskrat, 96, 238, 240, 247
MVC. *See* Martha's Vineyard Commission
MVLB. *See* Land Bank

Nantucket: agriculture, 161, *179*; birds, 251; climate, *27*, *28*; conservation, 36–37, 270, 274; development, 267, 275; fire (1846), 166; forest plantations, 234; fuel wood imported, 163; geology and geomorphology, 17–19, *20*, 21, 41, *42*, 43, 54; hurricanes, 108, 117; Native American diet, 98; paleoecological studies, 60, 64; population, 166; sea level and, 54–56; separation from mainland, 74; stranded whales, 92; vegetation, 21
Nantucket Sound, 56, *135*
Nashaquitsa Cliffs, 128
Nashaquitsa Pond, *121*. *See also* Quitsa Pond native settlement
National Hurricane Center, 118
National Oceanic and Atmospheric Administration (NOAA), 115, 118
Native Americans: agriculture, 21, 81–85, 96–97, 99–102; archaeological record, 5, 85, 93–99 (*see also* archaeology); diet, x, 21, 79, 88, 90–93, 95–102; early European contacts, 21, 81, 85–87; and environmental change, 88; European influences on, 85–87, 102–4; historical-archaeological discrepancy, 85, 101–4; hunting, 83, 90, *100*, 240, 247; and hurricanes, 114, 115; and land use, *72*, 78, 80–85, 88, 104, 307–8; pre-European populations, 75, 78, 85, 98–99; settlement practices, *87*, 88–90, 101. *See also* Iroquois people; Wampanoag people
Nat's Farm, 202, 297
nature, resiliency of, 10–13, 286, 299–301. *See also* forests: reforestation
Nature Conservancy, The, 80, 130, 197, 198, 204–5, 207, 210, 263, *297*. *See also* Woods Preserve
nature study. *See* ecological studies; paleoecological studies
Naushon (island): coyotes, sheep, and cattle, 258–59, *287*; geology and geomorphology, 19, 21; paleoecological studies, 60, 63–64; pasture land, 23–24, *287*; post-hurricane timber salvage, 125; vegetation, 27, 68, 69, 73, 224, 232. *See also* Elizabeth Islands
New Bedford Mercury, 169
New England, *7*; agriculture, *30*, 175–77, *179*; archaeological record (*see* archaeology); climate, 27–29, 50; conservation in, 33–36, 302 (see also *Wildlands and Woodlands*); development, *30*; European settlement, 21–23 (*see also* European settlement); forests, *30*, 31, *32*, *33*, 147–48, 175, *308*; geology and geomorphology, 17–19, *20*, 43–45, 49–50 (*see also* geology and geomorphology); heath hen in, 244; hurricanes, 107–8, 112–18, 124–25 (*see also* hurricanes); industrialization and development, 148, 165–66; population, *176*; postglacial sea level rise, *55*; prehistoric vegetation, *77* (*see also* paleoecological studies); topography, *18*; wave height and tidal range, *135*; whale-strandings, 91–92; wildlife depredation, 239–40. *See also* Massachusetts; Native Americans
New England Farmer, The, 177
New England Food Vision, A, *288*, 289, 291
Newhall, Virginia Whiting, 130
nitrogen, 27, 36, 37, 259, 275, 305
NOAA. *See* National Oceanic and Atmospheric Administration

No Island Is an Island (Simon), 17, 200, *264*, 270–72
Noman's Land, 5, 59, *60*, 63, 64, 74, 75, 154
nor'easters, 28, 76, *116*, 122, *143*
North Neck, *53*, *116*
Norton, Ansall, 157
Norton, Jim, 310
Norton family, 158, 309
Norton Point breach, 67, 127–33, *137*, 140–45, *205*, *206*, 286, *309*
nuts, 92–93. *See also* acorns
Nuttall, Thomas, 244–45, 255

Oak Bluffs, 17; conservation, *35*, 209; golf course, 189; map, *4;* and the MVC, 263; and nineteenth-century tourism, 170; origins, setting, and character, 25, 26, 181, *182*, 183, 185; tourism, 279; vegetation, *227;* and the Waterview development, 262, 263, 273; woodlands, *12*, *164*, 200, *201*, 222. *See also* Cottage City; Lagoon Heights
Oakes, John, 201
oak trees, *11*, *20*, *196;* acorns, 92–93, 245–46, 249; coastal erosion and, 127, *128*, *309;* dominance, 31, 224; and fire, 73; in modern woodlands, 224–28, 231–32; mortality (die-off), *9*, 64–68, 70, 286, 301, 304; in New England generally, 31, *32;* nineteenth-century woodlands, 149, 161, 162, *163;* oak forests, 64–67, 225–28, 245, *292;* prehistoric population, 69, 72–73, 75–77, 98; and reforestation, 217, 219–20; salt damage, *121;* winds and, *30*, 114, 121. *See also* forests
Obama, Barack, 194
ocean (sea): and climate, 27–31, 50, 70, 73, 74; and Island vegetation, 26–27; local currents, *135*, 139; ocean processes and Island configuration, 21, 52–56; salt's effect on vegetation, *121*, 122–23; sea floor, 53; sea level rise (present/future), 5–6, 38, 139–40, 286, *306;* sea levels, (pre)historic, 49, 52–56, 73, 74; storm surges, 108, *113*, 114–15, 117, 118, *125;* tides, *135*, 139; wave height and tidal range, *135;* and weather, 29–31, 109–10 (*see also* hurricanes). *See also* coastal erosion; coastline
Ogden, Pete (James Gordon Ogden III), 63–64, 67, 128, 133, 145, 163, 180, 216, 308
oil and candle factory, 166, 169
Oklahoma, 181, 183, 189
Oldale, Robert, *47*
Old Mill Pond, 171, 254
Old Town. *See* Edgartown
Old Whaling Church, *154*, *185*
O'Neill, Brendan, 205–6
openlands, 13, 79–80, 218, 242, *291*, 293–98. *See also* farmland; fields; grasslands; heathlands; pasture land; shrublands
"Origin and Spread of Maize, The" (Chilton), 96–97
Ornithology of Massachusetts (Peabody), 241
Orwig, Dave, *11*
Osborne, Abraham, 169
ospreys (fish hawks), 242, 255–57, 303
Oswald, Wyatt, 69, *73*
otter, 96, 238, *240*, 240
outwash and outwash plains, 18–19, *43*, 47–49, *50*, *51*, *53*. *See also* Great Plain, the
oxen, *156*, 158

paint mill, 33, *166*, *167*, 170
paleoecological studies, *8*, 60–64, 67, 69–78, 104, 215–16, 239. *See also* climate change: post-glacial change
pamets, 57
passenger pigeon, *239*, 241, 242
pasture land, *22*, *196*, *283;* access and, 23–25; conservation and management, *287*, *289*, *290*, 293–99; human/animal impacts on, 170–71, 215–16; nineteenth-century, 149, *158*, 160–61, 178; reforestation and, 32–33, 175, 179–80, 216, 218. *See also* farmland; hay
Peabody, William, 241
Peabody Museum, Harvard, 93–94
Peaked Hill region, *11*, 56, 161, 224
Pease, Jeremiah, 166, 168, 181, 309
peat, 59, 164
Pennywise Preserve, *197*, *300*, 302
Perley, Sidney, 107, 116
pigeons (wild), 240
pigs. *See* hogs
Pilgrims, 81, 85–87, 89, 103, 109, 244. *See also* European settlement; hurricanes: Great Colonial; *and specific individuals*
Pimpneymouse Farm, 143, 189, *200*, 303, 310
Pimpneymouse Pond, *63*, 64, *73*, 75
pine beetle, 70
pine trees: increase in, 217; in New England generally, 31, *32;* nineteenth-century woodlands, 149, 161, *163;* planted, 234–37; prehistoric population, 72–73, 75–76; wind damage, 121. *See also* pitch pines; white pine
piping plover, 239, 257
pitch pines, *12*, *234;* age and size, 76, 224, *225*, *226*, 229, *309;* coastal erosion and, 127; cultivation, 234–35, *236;* in modern woodlands, 224, 228–29; in New England generally, 31; numbers, 69, *72*, 73, 75–77, 161; storms and, 121, *122*. *See also* pine trees
Pitman, Theodore, 93–94

planning, ix, xi, 36–37, 260–80; federal role debated, 263, 265–73; Island Open Space plan, 263, 268; Metcalf and Eddy Comprehensive Plan, 260, 262–68, *271*, 273, 274, 275–76; MVC's Island Plan, 200, 206, 260, 263, 274–80, 282; timeline (history), 263. *See also* Land Bank; Martha's Vineyard Commission
plover, 151, 251
Pohogonot tract, 79, 149, *155*, 189, 219, 226, 275
poison ivy, 67, 217, 219, 232, 237
pollen records, 67, 69, 72–78, 215–16. *See also* paleoecological studies
pollution: acid rain, 27, 305; Clean Air Act, 305–6; DDT, 256; long-term impacts, 10; natural absorption, 37; in New England, 27–28; water contamination, 36, 275, *276*, 305 (*see also* nitrogen)
Polly Hill Arboretum, 66, 68, *118*, 207, *300*, 301, *302*. *See also* oak trees: mortality
ponds: agricultural use, 155; artificial, *189*, 284; damming of, 171 (*see also* dams); depth, *71*; function, 37; morainal ponds, 48–49, *54*; origins, 45, 73–74; sand and, 128; sediments, 61; Shaler on, 178; study sites, 63; water quality, 36, 38, 259 (*see also* nitrogen). *See also* coastal ponds; paleoecological studies; water; *and specific ponds*
population (Vineyard), 166, 178, 193
post oak, 70, 161, *219*. *See also* oak trees
Potter, Edo, 143, 189, 310
Potter, Hatsy, 189
Poucha Pond, *15*, 159, 170, 189, *200*, *276*, *300*, 302, *306*
Powell, John Wesley, 44
power lines, *273*
precipitation, *25*. *See also* climate; hurricanes
prehistoric occupation, 21
Priester's Pond (Athearn Pond), 67, 155, 170, 171
private property and access, *203*, 203–5, 209. *See also* access
Prospect House, 183, *184*
Pynchon, John, 240

Quanames, *294*
Quansoo, *53*, 149, 219, 224, 252, *294*
Quansoo Farm, 8, *29*, *30*, 80, 203, 218, 297, 303
Quitsa Pond native settlement, 94–95

rabbits, 238
ragweed, *72*, *76*, *77*, 215
railroads, *167*, 170, 183, 185
Railton, Arthur, 192, 193
recreation, 193–94, *198*, *206*, 211. *See also* access; golf courses; hunting; yacht clubs
Red Gate Farm, *6*, *47*, *62*, 63, *136*, 303
red maple, 31, 64–67, *72*, 74–76, 217, 225, 231, 232. *See also* forests; maple trees
Redstone, Ed, 200
regional conservation partnerships, 302. *See also* conservation; planning; *and specific organizations*
regional institutions, 264, *275*
Reston, James, 270
Reston, Richard, 201, 205
Richardson, James (Jim), 95, 311
Ripley's Field Preserve, 226, *300*, 302
Ritchie, William, *90*, 95–96, 104
roads: ancient ways, *20*, 148, *149*, 171, *192*; highway, 200–201, 261; maps, *4*, *38*, *153*, *169*, *186*, *188*; nineteenth-century, 148, 151, 156, 162, *167*, 168, 185, 187; traffic, *38*, 261, 264–65, 275; woodlands and, 202, 226–28. *See also specific roads and town maps*
Roaring Brook, 170, 254
Roland, Willett, 120–21
Roosevelt, Tweed, 281
root cellar, *14*
Rosbeck, Peter, *192*
Rotch, Caleb, *155*

salmon, 245
Saltonstall, Leverett, 265
saltworks, 33, *167*, 170
Samoset, 103
sand. *See* barrier beaches; beaches
sand sapping, 57
Sargent, Francis, 263, 270–73
sassafras trees, 64–67, 70, 158, 219, *226*, 231–32. *See also* forests
saw pits, 165
scarlet oak, 70, 161, *226*. *See also* oak trees
Schifter, Rick and Jennifer, 127, 131–32, 144–45
Schifter house and compound, 126–27, 131–32, 141–42, *144*, *145*. *See also* Norton Point breach
Scrubby Neck Farm, 187–89
scrub oak, 33, 228; frost-killed, 230–31, *232*; resilience, 217–18; restoring, 299; scrub oak woodland, 161, 171, 228, *229*, 237, 246; where found, 77, 149, 161, 162, 225–26, 228, *230*, 231. *See also* shrublands
sea. *See* ocean
Sears, Paul, 67
sediments, 42–43, 60, *70*. *See also* paleoecological studies; ponds
seeds, 93, 218, 234
Sengekontacket Pond: adjacent beach, *142*; archaeological sites near, 97; development on, *190*, *191*, 261, 263, *268*, 273; geese, *241*; meadows, 80, 160; watershed, *276*
Seth's Pond. *See* Uncle Seth's Pond
Seven Gates Farm, *58*, *301*; beech wood, *235*; and conservation, 175, 177, 206–7, *300*, 301; fields

and pastures, 297; history written, 310; and interconnected wildland, 301; legacies of previous land use, 14; oak mortality, 66, 67–68, 218; oldest tree, 224; Shaler and, 174, *175*, 187, 206–7; size, 209; tree species, 232. *See also* Shaler, Nathaniel Southgate
shad, 242, 252–53
Shaler, Nathaniel Southgate, *175*, 309; on barrier beaches, 128; and conservation, 177, 198, 206–7, 292; on development and tourism, 183–85; on the division of Tisbury, 187; and the 1850 map, 130; on Holmes Hole Harbor, 154; on the Island, 17, 28; on pitch pine, 162; on reforestation, 174; and Seven Gates Farm, 174, *175*, 187, 206–7
Shear Pen Pond, *37*, 166
sheep, 178, *283;* coyotes and, 27, 258; declining numbers, 12, 166, *179;* exportation of, 292; first imported, 158; fodder, 148; impact on land/vegetation, x, 215, 296; pastures and fences, 7, 161. *See also* livestock
shellfish. *See* fish and shellfish
shell middens, 94, 95, *95*
Sheriff's Meadow Foundation, 175, 198, 199, 207; conservation lands, 193, 206, 209, 210; and grazing, 297; land management, 80; un-development, 285; wood harvesting, 299
ships (sailing ships): exploration and early settlement, 81, 87; and Island harbors, *28*, 142, 154–55, *186, 188;* numbers of, 166, 168, 178; 1635 hurricane and, 114, 115–16; whaling ships, 169, 170, 178. *See also* European settlement; whaling
shoals, formation of, 135, 136
shoreline. *See* coastline
shrublands, 232–34, *286, 291*, 293–96, *297*, 299. *See also* scrub oak
Shuman, Bryan, 64, 69
Silliman, Benjamin, 44
Silva, George, 143
Simon, Anne, 17, 200, *264*, 270–73
Skiff's Island, 56
skunks, 238, 257
Slip Away Farm, 292, *293*, 297
SLOSH model, 118, 141
SMF. *See* Sheriff's Meadow Foundation
Smith, Daniel, 168
Smith, David M., 116
Smith, Jerome (Dr.), 253–54
Smith, John, 240
solar power, *257, 278*, 307
sorrel, 215, 218
South Beach, 143, 201, 209, 263, *286*
Southern Woodlands (of Oak Bluffs), 222, 275; development and, *12*, 200, 209; forest type, 226, 228, 231; land management approaches, *300*, 302
sparrows, 241, 251, 252
spring, 303–4
spring sapping, 57
spruce trees: as non-native species, 220, 237, 284, 299, 300; planted, 218, 228, 237, 284; in prehistoric woods, 31, 59, 72; timber, 165, 289
Squanto, 81, 82, 87
Squibnocket, *155*, 161, 180, 219, *294*
Squibnocket hill/cliffs, 5, 56, *63*
Squibnocket Pond, 97, *210, 276, 294*
Squibnocket Ridge, 63, *134*, 210, 232–34, *270*
squirrels, 68, 82, 96, 238, 241, 242
Standish, Miles, 87, 109
Stanford, Mrs. Lucy, 187
Stanley, O. G., 185, 205
State Forest. *See* Correllus State Forest
Stevens, Andrea, *63*, 64
Stonewall Beach, 56
stonewalls, 7–8, 14, 33, 159, 171, *173*, 175, 219, *270*
stools (rings of trees), 8–10, *13*
storms. *See* hurricanes; nor'easters; weather; winds
streams, *253;* dams, 33, 155, 170, 176, 218–19, 254, 304; fish in, 253–54; map, *51;* nineteenth-century water power, 155, *167, 168*. *See also* water
Strock, Alvin and Moses, 261, 263, 273, *274*
Strong, John, 100
Strother, David, 155
Studds, Gary, 270
summer visitors. *See* camp meetings; development; tourism
surveying methods, 150–52, *151*
sustainability, x, 260, 289–92. *See also* conservation: and agriculture; planning
swamps. *See* wetlands
Swan Pond, *129*, 131, *132*
Swift, Dean, 267
Sylvia Beach, *142*

tanneries, 165, *167*
Tashmoo Pond, 234, *276*
Taylor, Jo-Ann, 291
telegraph, *167*, 168, 170
temperature, *27*, 28–29, *71*, 74, *304*. *See also* climate; climate change
Thanksgiving, first, 81–82, 85, 244
Thaxter, Leavitt, 164, 176–77, 309
Thimble Farm, 209, 211
Thoreau, Henry David: concept of primitive America, 82; and conservation, 33–36, 241, 292, 299; environment studied, 14, 59–60, 62; and the forest, 147–48, 234; on landscape history, 27, 281, 285; on stranded whales, 91; on tamed forests, 215; and wildlife, 247, 252

Three Ponds Reservation, *209, 300,* 302
Tiasquam River, 76, 155, 222, *253,* 254, 304
ticks, 248, 250, 259
tides, *135,* 139. *See also* ocean
Tilton, Samuel, 170
Tisbury, *4;* beach and pond access, 204; conservation, *35;* division of, 185–87; and the MVC, 263; in the nineteenth century, *153,* 154, 163, 164, 166, *167;* origins, setting, and character, 25, 26; solar array, *278;* vegetation, *227. See also* Holmes Hole; Vineyard Haven; West Tisbury
Tisbury Great Pond, 97, 204, *276*
Tisbury Meadow, 297
Tisbury Pond Club, 189
Tom's Neck, *86*
topography, *18, 47,* 53–54. *See also* geology and geomorphology
tourism: 1800–1900, 166–68, 170, 181–85; 1900–1950, 187–93; 1950–present, 193–94, 201, 261, *262,* 264; as environmental and cultural threat, 261–67, 270–73 (*see also* planning). *See also* fishing; golf courses; hunting
townships, towns, and villages: and access, 203–4; and conservation, 207–11, 274; diversity and character, 17, 25–26, 154, 260–61, 274; fence-lines, 159; Indian, 154 (*see also* Wampanoag people); in the mid-nineteenth century, 147, 154–56; tensions between, 183. *See also specific localities*
trails, 79, 203, 204, *209,* 211, *277,* 279. *See also* Moshup Trail
transportation, 260, 268, 275. *See also* airport; ferries; railroads; roads; ships
trees: age and size, 218, 222–25, 229, 300, 301, *309;* broad vs. tall, 218; deer and, *250;* drought and, 9 (*see also* drought); gnarled vs. straight, 8; introduced species, 218, 234, 237; in New England generally, *32;* nuts and acorns, 92–93, 245–246, 249; plantations, 234–37; rings of (stools), 8–10, *13,* 218; storm damage, *76,* 112, *113,* 114, 117–18, 120–25, 237; uprooted, 118, 121–22, *124;* wind and, *30,* 31, 122, 180. *See also* forests; *and specific types of trees*
trout, 253–54
Trustees of Reservations, The (TTOR), 197, 199–200, 207, 282; conservation lands, 127, 189, *198,* 199–200, 206, 207, 210, *290,* 297, 299
Tuck, James, 95
turkeys, 96, 238, *239,* 240–42, 244, *298*
turtles, 96, 239, *242*

Uncle Seth's Pond, 45, 63, *69,* 75, 77, 254
un-development, 282–85
upland sandpiper, 251, 252

VCS. *See* Vineyard Conservation Society
vegetation: agriculture's effect on, 171, 215–18; climate change and, 64, 66, 69–70; glacial geology and, 19–20, 31; in the Indian diet, 92–93, *100;* introduced species, 218, 234, 237; legacies of human activity, 218–19; oak mortality and, 67–68, 286; ocean as barrier, 26–27; openlands species, 232–34; post-agriculture recovery, 32–33, 216–20 (*see also* forests: reforestation); postglacial (prehistoric) vegetation, 69, 71–78; rare and endangered species, 31–32; types (map), *227;* wind and, *30,* 31; zonation, *216. See also* agriculture; forests; grasslands; openlands; pasture land; shrublands; trees; *and specific species*
Verrazano, Giovanni da, 83, 89, 102
viburnum, 33, 67, 217, 232
Vincent, Lucinda Poole Mosher, 3. *See also* Lucy Vincent Beach
Vineyard Conservation Society, *58,* 141, 198, 200, 205–6, 207, 209, 270
Vineyard Gazette: advertisements, 165, 168, 170, 236; archives, 310–11; and conservation, 191, 201, 265, 269, 274; on deer, 248; on the 1850 map, x, 152; on fire, 295, *296;* and forest planting, 234–36; on heath hen reintroduction, 244; Hough and, ix, 174, 191, 269 (*see also* Hough, Henry Beetle); and Island name convention, ix; on maritime activity, 168; and the Metcalf & Eddy plan, 265–67; nineteenth-century landscape described, 147; and tourism, 166–68, 183; and un-development, 282; on the Wasque erosion, 127–28
Vineyard Golf Club, *197,* 209
Vineyard Haven, 17, 310; deforestation, 222; development, *188, 190;* harbor, *23, 28;* Oak Bluffs and, 25; planned breakwater (1887), *140;* trail system, 279; yacht club, 189. *See also* Holmes Hole; Tisbury
Vineyard Open Land Foundation (VOLF), 201–2, 207, 270
Vineyard Sound, *55,* 56, *135*
Vineyard Transit Authority, 260, 275
VOLF. *See* Vineyard Open Land Foundation

Wack house, *137, 144*
Wakeman, Mary, 207, 274
Wakeman Conservation Center Trust, 207
Wallace, Tom, 205
Wampanoag people: agriculture, 161; archaeological sites, x, 5, *82, 84,* 93–99, *96;* burials, 5, *97;* and conservation, *35, 199,* 200, 211; diet, 94, 95–97; land management practices, 80–82, 104,

164; Moshup legend, 92; nineteenth-century lands, 154, 159, 161, 164; place names, *94*, 133; post-European settlements, 25–26, 94–95; pre-European settlement practices, 21, 94–95, 101; schoolhouse, *293;* today, 25–26, 88, *94*. *See also* Aquinnah; Christiantown; Gay Head; Native Americans
Warren Pond, 304
Waskosim's Rock Reservation, *22*, 66, 207, 224, 232, 297, *300*, 301. *See also* oak trees: mortality
Wasque: conservation on, 80, 200, 203, 210, *294*, 297, 301; Norton Point breach and coastal erosion, 127–33, *137*, 140–45, *205*, *206*, 286, *309;* otter, *240;* outwash deposits, *50;* proposed development (1913), *141*, 189; recreation, *198;* reforestation, 180
Watcha Club, 187–89
Watcha Pond, *63*, *202*, *276*
water: contamination, 36, 275, *276*, 305 (*see also* nitrogen); freshwater, 36, *51*, 185, 253–54; glacial meltwater and sediments, 41–43; groundwater, 36, 57, 70, *71*, 73, 275; and movement, 26–27; nineteenth-century water power, 155, *167*, *168;* salty and brackish, 36, *51*, 97, 253, *306;* watersheds, *276*. *See also* coastal ponds; fish and shellfish; ocean; ponds; streams; wetlands
Waterview development, 261, 263, 273
waves, *135*. *See also* coastal erosion; ocean
weather: frost, 230–31, *232;* ocean and, 29–31, 109–10, *110;* storms and coastal erosion, 131, 138 (*see also* coastal erosion); and tree growth, 122, 180; winds, 28–31, *110*, *111*, 118, 122, 180. *See also* climate; hurricanes; nor'easters
Webb, Willoughby, 177, 207
Webster, Daniel, 155, 156, 162, 166, 170, 251
weeds (pre-settlement), *72*, *76*, *77*. *See also specific weeds*
Weeks, Charles, 36
Weeks, William, 92
Welch, Bob, *15*, 189. *See also* Pimpneymouse Farm
Wequobsque hills/cliffs, 5, *46*, 56, *141*, 187, *304*
Wesleyan Grove, 154, 181. *See also* camp meetings
Wesley House, 183
West, departure for the, 147, *155*
West Chop, *4*, 17; development (1850–2010), *190*, *192;* 1850 map, *153;* golf course, 189; lighthouse, *167;* pine woods, 14; power lines, *273;* woodlands, 149, 161, 222, 228
West Chop Woods: development avoided, 209, 222; forest type, 76, 193, 203, 224, 228–29, 231; land management approaches, *300*, 301, 302; pitch pines, 76, 224, *235*, *309*
West Tisbury, *4;* affordable housing, 202; agriculture, 20; beach and pond access, 204; cemetery, *155;* and conservation, *35*, 211, 269; development, *190;* fence pattern, 159; forests, 20; mills, 168; in the nineteenth century, 149, *153*, 155, 159, *167*, 168, *190;* origins, setting, and character, 25, 26, 185–87; real estate costs, 204; Shaler on, 178; solar array, *278;* vegetation, *227*. *See also* oak trees: mortality; Tisbury; *and specific locations*
West Tisbury–Edgartown Road. *See* Edgartown–West Tisbury Road
wetlands: agricultural use, 156, 160, 161, 215; climate change and, *306;* ditches, 218; maps, *51*, *96*, *167;* origins, 45, 73; paleoecological record in, 59–60 (*see also* paleoecological studies); peat harvesting, 164; study sites, 63; uprooted trees, 121–22; vegetation, 78. *See also* cranberry bogs; ponds
Weymouth, George, 87
whales, stranded, 91–92
whaling, 166, 169, *172*, 174, 178. *See also* oil and candle factory
whippoorwill, 79, 241, 251
white oak, *13*, 70, 161, 224, *225*, *226*, 229, 304. *See also* oak trees
white pine: cultivation, 234, 237; in New England generally, 31; prehistoric population, 69, 72; removing, 299; salt spray and, 123; storm damage, *113;* winds and, 121. *See also* pine trees
Whiting, Everett, *292*
Whiting, Henry L., 128, *130*, 130, 309; Agricultural Society founded, 130, 176–77; coastal erosion studied, 131–39, 141–43, 145; and the Coastal Survey, 128, 150–52; descendants' gift, 130, 207; and the 1850 map, x, 149, *153*, 161, 162, *190*, 308; hunting by, 151, 250–51
Whiting family, x, 128, 130
wigwams, *87*, 88–89, 96
Wildland, Woodland, and Farmland vision, 36, 289–99
wildlands, 299–302
Wildlands and Woodlands: A Vision for the Forests of Massachusetts, 281, *288*, 289, 302
wildlife, 238–59; conservation efforts, 241–42, *245*, 246; habitat loss, 79, 241, 242, 293 (*see also* agriculture: and conservation); human activity and, 92, 238–42; in native diet, 90–92, 96; nonnative species, 242, 246, 254; population over time, *239;* rare and endangered species, 31–32; revival of extinct species, 243–44. *See also* birds; fish and shellfish; hunting; *and specific species*
Williams, Roger, 103
Willis, Nathaniel Parker, 169

Willoughby, George Haskell, 157, 162, 163–64
wind energy, *278,* 279, *283,* 307
windmills, 33, *167,* 170
winds, 28–31, *110, 111,* 118, 122, 180. *See also* hurricanes
Windy Gates estate, 187, *189*
Winthrop, John, 89–90, 112–15
Winthrop, Thomas (Gov.), 244–45
Wompesket Preserve, West Tisbury, *54*
woodcutting (timber cutting), 32; legacies of, 11, 218; modern harvesting, 289–90, 292, 299, 301; nineteenth-century, 13, 161–65, *163, 164,* 222–24; salvage logging, 124–25; Thoreau on, 147
woodlands. *See* forests
Woods, Edwin Newhall, 130
Woods Preserve, *10, 14, 146;* creation of, 130, 207; farmland, 297; and interconnected wildland, 301; land management approaches, *300;* oak mortality, 9, 64–67, *66,* 218, *301, 305;* size, 207, 209
Woodworth, J. B., *42*
Wright, Herb, 61

yacht clubs, 17, 189